AAPG REPRINT SERIES NO. 10

Facies and the Reconstruction of Environments

Selected Papers Reprinted from

AAPG BULLETIN

Published by The American Association of Petroleum Geologists
Tulsa, Oklahoma, U.S.A., 1974

Contents

 Published, April 1974. Fourth printing, May 1982.
Library of Congress Catalog Card No. 74-76996. ISBN: 0-89181-534-1.

Facies and the Reconstruction of Environments: Preface

The environments of deposition of sedimentary rocks, as manifested in their facies, are primary factors in the formation of source rocks and reservoir rocks. The papers which make up this volume (1) treat the concepts of facies and the reconstruction of environments, (2) show methods used to analyze and display the geologic data needed to investigate such environments, and (3) demonstrate applications to specific areas.

Many additional pertinent papers have been published in the *Bulletin*. My intent has been to choose papers of broad, general application. I have avoided most of those which concern carbonate rocks because of previous and planned Reprint Series volumes devoted to carbonates, and I stopped my selections with Volume 52 (1968) in the belief that newer volumes are readily available to interested readers.

The original suggestion for this volume came from Charles A. Brinkley. The papers were assembled by E. M. Tidwell, of the AAPG Editorial Staff.

JULES BRAUNSTEIN
Shell Oil Company
New Orleans, Louisiana
November 27, 1973

Reprinted from:
BULLETIN OF THE AMERICAN ASSOCIATION OF PETROLEUM GEOLOGISTS
VOL. 26, NO. 11 (NOVEMBER, 1942), PP. 1697–1770, 13 FIGS., 1 PLATE, 8 TABLES

STRATIGRAPHICAL ANALYSIS AND ENVIRONMENTAL RECONSTRUCTION[1]

THOMAS G. PAYNE[2]
Chicago, Illinois

ABSTRACT

This paper presents a plan for comprehensive study and analysis of a sedimentary deposit, including petrographic and paleontologic description of the fundamental attributes, properties, and structures of the sediment and its life assemblage and the interpretive description of genesis of the deposit. The genetic phase embraces evaluation of the environmental factors, ecologic and geographic, and the dynamic processes and resultants involved in sedimentary petrogenesis. Methods, principles and concepts are illustrated throughout the paper by use of the crinoidal phase (Bed 9) of the Grand Tower limestone of Ste. Genevieve County, southeastern Missouri, as a test case.

I. INTRODUCTION

Paleoecology.—With the exception of relatively small areas of bare rock and artificial structures, the subaerial and subaqueous surface of the earth constitutes a life-sediment sphere. In this sphere living organisms and sediments are not only coexistent and interrelated but are influenced to different degrees by the same set of environmental factors. The term *ecology*, in its broadest connotation, may be applied to the environmental relationships of organisms and of the sediments with which they are associated.

Paleoecology, both organic and sedimentary, is a fertile field for geologic research. It is coming to be realized that the life and rock records must be interpreted from the environmental viewpoint, utilizing the ever-expanding background of knowledge of recent sediments and organisms. At present many earth historians have become so bogged down in their own mire of formations, correlation charts, species, controversies, and the like that some of the more fundamental issues have been sidetracked. It is often forgotten that stratigraphic classifications, correlation charts, and faunal lists are not ends in themselves but are rather means to an end, an end that should recreate and picture the organic and physical earth throughout the flow of geologic time.

Stratigraphic analysis.—If stratigraphy is ever to become much more than a "library" science, its exponents must concern themselves less with problems of arbitrary stratigraphic classification and nomenclature and more with funda-

[1] Submitted as a Ph.D. thesis, University of Chicago, Chicago, Illinois, September, 1941. Manuscript received, April 3, 1942.

[2] Present address: Kansas State Geological Survey, Lawrence, Kansas.

mental lithologic description and classification, less with problems involving the correlation of strata hundreds of miles apart and more with the interpretive analysis of the environmental settings represented by the fundamental rock types. Stratigraphic classification, nomenclature, and correlation are important but never will be accomplished satisfactorily until the rocks themselves and the environments which they represent are described and understood—and we are far from such an understanding at the present time.

The literature of stratigraphy is characterized by superficial lithologic descriptions—field descriptions which have served their immediate purpose but have failed to lead the way to a rigorous understanding of the stratigraphic record. The past twenty years have brought forth detailed studies in the field of sedimentary petrology, studies which have been mathematically rigorous in their methods and logical in their approach, but they too have failed in certain respects, namely, that they have been highly specialized and restricted in scope and too time-consuming for widespread stratigraphic application.

Recent petrologic work has paved the way but has not as yet led to and provided a plan for comprehensive analysis and understanding of a sedimentary rock. Rather than considering a sedimentary assemblage in the light of the whole, recent studies have treated only certain sedimentary attributes and properties and have been concerned only with individual, isolated approaches, such as insoluble-residue studies, heavy-mineral analyses, size or shape analyses, porosity determinations, studies of structures, and so forth. Environmental interpretations have been random and inconclusive. The whole story has been told for only few sedimentary deposits, but partial pictures have been presented for many. One notable exception is Krynine's comprehensive and classic treatment of the "Petrology and Genesis of the Third Bradford Sand."[3]

It is now coming to be realized that the future must bring a new form of stratigraphic research, middle-road research which will bridge the gap between the *old stratigraphy* and the *new stratigraphy* that is now unfolding. This middle-road policy will call for complete restudy of the stratigraphic record and for lithologic descriptions and environmental interpretations which recognize and evaluate in a simple semi-quantitative manner the fundamental attributes, properties, and structures of a sediment and the environmental factors and sedimentational processes which control its genesis. With stratigraphic or sedimentary traps becoming of increasing importance in petroleum exploration, the new stratigraphy with its lithologic and environmental research must undergo rapid growth in the near future.

North American Paleozoic rocks are characterized by recurrence in time and space of certain fundamental lithologic and environmental types. It would seem that Werner's medieval concept of "universal formations" might well be revived and altered to a concept of "universal lithologic types." The new stratigraphy

[3] P. Krynine, "Petrology and Genesis of the Third Bradford Sand," *Pennsylvania State College Bull. 29* (1940), pp. 1–134.

must include a breakdown of the stratigraphic record into lithologic and environmental types and their varieties and must place emphasis on *lithologic classification* as well as stratigraphic classification.

The present paper.—This contribution represents an attempt to devise a logical and orderly plan for rapid but rigorous analysis of the sedimentary record. The procedure set forth and illustrated in this paper is tentative in nature, and it is hoped that interested persons will contribute to its revision and improvement.

The writer has been engaged in a study of Middle Devonian lithologic and faunal assemblage *types* and in the interpretive derivation of the environmental setting represented by each. The crinoidal limestone used as an illustration or test case throughout the present paper was chosen at random from a large body of Middle Devonian lithologic material; it represents the crinoidal phase, Croneis' Bed 9,[4] of the lower portion of the Grand Tower limestone exposed near Ozora, Ste. Genevieve County, Missouri. Twelve samples of this limestone were taken at 1-foot (time) intervals in a vertical section exposed on Quarry Hill and in the east bank of Little Saline Creek at the foot of Quarry Hill.

Acknowledgments.—The writer is indebted to Carey Croneis and Francis J. Pettijohn of the University of Chicago faculty in geology and paleontology. Under the tutelage of Dr. Croneis the dissertation and field studies which it necessitated were planned and carried out and the manuscript prepared. Dr. Pettijohn supervised and contributed liberally to the sedimentological phase of the problem. The laboratory and office work was done at and with the facilities of Walker Museum, University of Chicago. This paper represents the essential portion of a Ph.D. dissertation submitted to the faculty of the division of physical sciences, University of Chicago.

II. Sedimentary Petrography (Petrographic Analysis)

A. STUDY METHODS

The form of stratigraphic analysis which this paper propounds involves the use of the manifold study procedure graphically outlined in Figure 1. The procedure brings into play a series of checks and balances by which errors tend to be averaged out. The use and coördination of the various methods in the study of a sedimentary rock allow a convergence of evidence with regard to the attributes, properties, and structures. Any one method alone, such as thin-section examination, may not yield complete and reliable results, but when combined and coördinated with binocular microscopic examination of rock fragments and polished sections, with insoluble-residue studies, simple chemical tests, sieve and pipette analysis, and so forth, a degree of order and certainty emerges, and data are provided for the painting of a comprehensive picture of the sedimentary assemblage.

[4] C. Croneis, "The Devonian of Southeastern Missouri," *Illinois Geol. Survey*, Devonian Symposium Volume (in press).

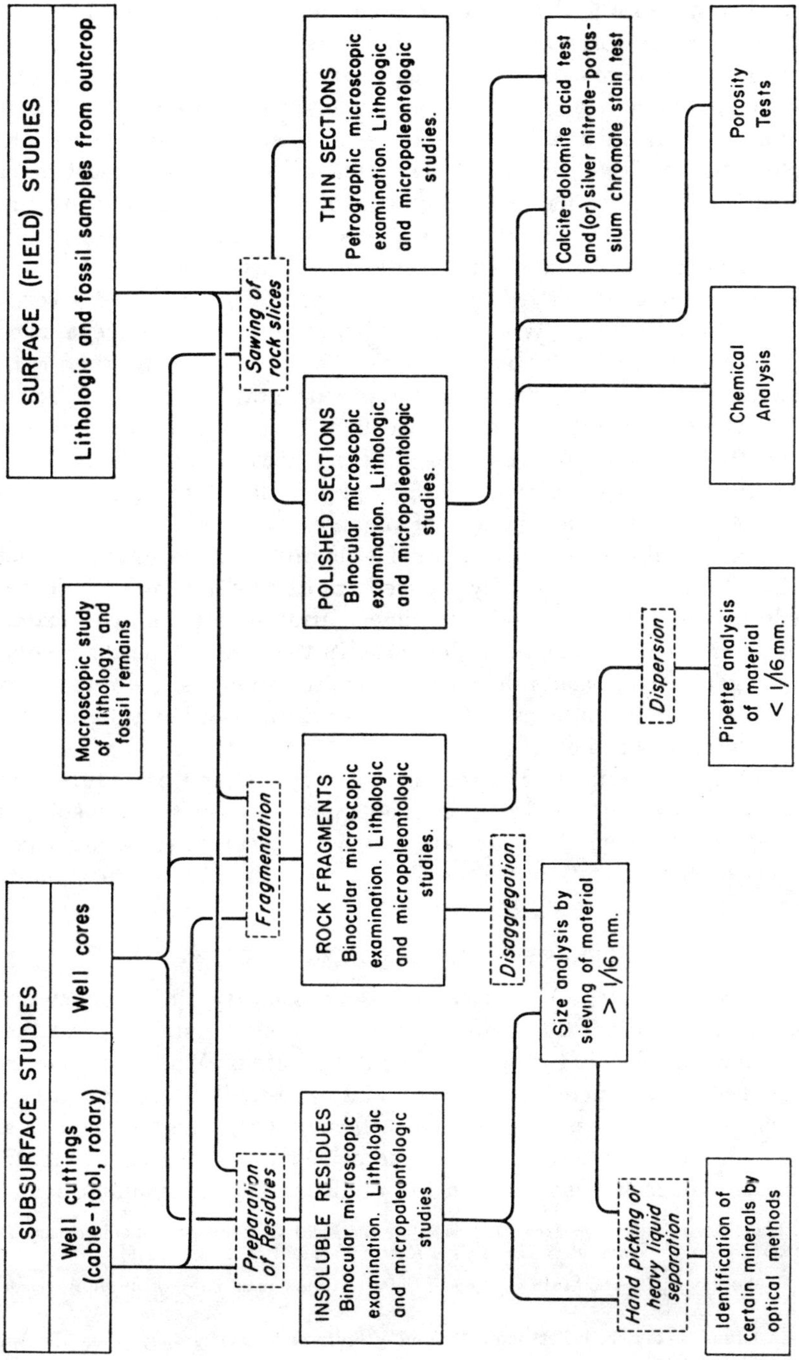

FIG. 1.—Flow outline of laboratory operations.

A handy outline for field study and description has been prepared by Goldman and Hewett.[5] Field observations should include study of the orientation and distribution of fossil remains and determination of the relative abundance of different morphologic types. Lithologic and fossil samples should be collected according to a plan and should be marked so as to show orientation with respect to bedding planes. Study of large-scale sedimentary structures must be carried out in the field, as must also the description of the attributes and properties of the coarse or rudaceous materials of cobble and boulder grades. The finer-grained sediments may be sampled and studied in the laboratory.

Insoluble residues employed by the writer have been prepared according to a procedure formulated by Workman.[6] This procedure includes separation of a residue into two fractions: (1) a sand fraction with grains greater than 1/16 mm. in diameter, and (2) a silt and clay fraction with grains less than 1/16 mm. in diameter. The separation is accomplished by a decantation procedure utilizing the differential settling velocities of the different sizes of components.

Chemical analyses of sediments are helpful in determining calcite-dolomite ratios, in determining the amount of organic (carbonaceous or bituminous) matter present, and in other qualitative and quantitative phases of compositional study. Chemical tests employed by the study procedure here propounded include the hydrochloric acid test and the silver nitrate-potassium chromate stain test[7] for differentiating calcite and dolomite. In the case of limestones composed of an aggregate of calcite and dolomite grains, the stain test permits determination of the calcite-dolomite ratio by a simple method of counting stained and unstained grains in a sample.

The degree of accuracy permitted by the methods and procedure here propounded obviously is not high, but it is adequate to meet the immediate needs of stratigraphic research. The data are in general accurate to within 5 per cent for the major constituents and to within 1 or 2 per cent for the minor components.

B. GENESIS OF SEDIMENTARY COMPONENTS

Genetic classification.—A rigorous genetic classification of sedimentary rock components must provide the answer to three questions. First, *where* was the component formed in relation to the site of final deposition? Was it formed in place or was it formed elsewhere, transported, and deposited? In other words, is the component autochthonous (Au) or allochthonous (Al)? Second, *when* was the component formed in relation to the time of deposition of the sediment? It is obvious that allochthonous components originate before accumulation of the

[5] M. I. Goldman and D. F. Hewett, "Schedule for Field Description of Sedimentary Rocks," *Natl. Research Council Comm. on Sedimentation Rept.* (Washington, D. C., 1928).

[6] L. E. Workman, Illinois State Geological Survey, typewritten copy of insoluble-residue procedure.

[7] W. C. Krumbein and F. J. Pettijohn, *Manual of Sedimentary Petrography*, p. 496. D. Appleton-Century Company, New York (1938).

deposit of which they are a part and are therefore *progenetic*. On the other hand, autochthonous components may originate either approximately contemporaneous with (*syngenetic*) or after (*epigenetic*) the genesis of the deposit. The third and final question pertains to *how* the component was formed, to the source of its material and to the process by which it originated.

The genetic classification of components which this paper introduces is here outlined. Examples are cited of deposits belonging for the most part in the various categories outlined.

I. ALLOCHTHONOUS COMPONENTS (those formed elsewhere and transported to site) (*Al*)
- A. *Progenetic* (*Pro*)
  - 1. Preëxisting rock source (weathering) (*pr*)
    - a. Katamorphic alteration products
      Ex: marine clay from igneous rock source
    - b. Differential solution products
      Ex: quartz sand from quartzose sandy limestone
    - c. Products of mechanical agencies
      Ex: transported arkose from granite terrane
  - 2. Transported chemical precipitate (*cp*)
    Ex: some lithographic limestones
  - 3. Volcanic source (pyroclastic) (*vol*)
    Ex: volcanic ash
  - 4. Transported material of organic origin (*org*)
    Ex: some crinoidal limestones, *Globigerina* ooze, *et cetera*

II. AUTOCHTHONOUS COMPONENTS (those formed in place and untransported) (*Au*)
- A. *Syngenetic* (*Syn*)
  - 1. Preëxisting rock source (weathering) (*pr*)
    - a. Katamorphic alteration products
      Ex: residual clay soil from granite
    - b. Differential solution products
      Ex: residual chert breccia from cherty limestone, tripoli from siliceous limestone
    - c. Products of mechanical agencies
      Ex: intraformational limestone breccia
  - 2. Primary chemical precipitate (*cp*)
    Ex: geyserite, calcareous tufa
  - 3. Primary chemical replacement (*re*)
    Ex: certain dolomites, phosphorite, *et cetera*
  - 4. Organic origin (*org*)
    Ex: coral and algal colonies of reef core
- B. *Epigenetic* (*Epi*)
  - 1. Preëxisting rock source (*pr*)
    - a. Katamorphic alteration products
      Ex: limonite altered from primary detrital magnetite
    - b. Anamorphic alteration products
      - (1) Recrystallization *without* mineral change
        Ex: marble from limestone
      - (2) Recrystallization *with* mineral change
        Ex: wollastonite from quartzose limestone
    - c. Differential solution products
      Ex: clay from phreatic solution of limestone
    - d. Products of mechanical agencies (diastrophic)
      Ex: fault breccias, cataclastic breccias, *et cetera*
  - 2. Secondary chemical precipitate from vadose or phreatic waters (meteoric or juvenile) (*cp*)
    - a. Introduced in voids (*cement*)
      Ex: calcite cement in quartz sandstone
    - b. Secondary growths
      Ex: quartz secondary enlargements in quartz sandstone
  - 3. Secondary chemical replacement (*re*)
    Ex: chert replacing limestone

Sedimentary components of cosmic origin are not considered in the foregoing genetic outline. Some of the red clay and associated small glass beads of deep-sea deposits are believed to be of this origin. This material and meteoritic fragments are, however, of little importance from the stratigraphic viewpoint.

Genetic analysis.—Some sediments are *end-member types* in that they are composed almost entirely of components from but a single genetic category. On the other hand, many sedimentary deposits derive their constituents from several categories and are therefore genetically complex.

The illustrative crinoidal sediment (Pl. I) is not genetically homogeneous inasmuch as it is a composite of several genetically distinct component types, as indicated by the analysis of Table I.

TABLE I

GENETIC ANALYSIS OF CRINOIDAL SEDIMENT

Genetic Class	*Sedimentary Material*	*Per Cent (Average of 12 Samples)*
Transported preëxisting rock material (*Al, Pro, pr*)	Quartz, tourmaline, rutile, magnetite	8–10
Transported material of organic origin (*Al, Pro, org*)	Crinoidal debris—columnals, *et cetera*	75–80
Transported preëxisting rock origin (?) or formed in place by chemical and (or) organic agencies (?)	Glauconite; finely divided and in aggregates	1 (approx.)
Autochthonous organic remains, untransported (*Au, Syn, org*)	Colonial corals, cup corals, brachiopods, pelecypods, *et cetera*	5–10
Epigenetic katamorphic alteration products (*Au, Epi, pr*)	Limonite from alteration of detrital magnetite	Less than 1
Secondary chemical precipitate (*Au, Epi, cp*)	Calcite cement in interstices	5–10

C. ATTRIBUTES OF SEDIMENTARY COMPONENTS

Sedimentary components have eight fundamental attributes which must be evaluated in petrographic analysis. These attributes pertain both to *individual* component grains and to components in the *aggregate*. For example, the attribute of *size* (mean diameter) of an individual quartz grain may be described in millimeters; this grain may be part of a quartz sand or sandstone, in which case the size distribution of associated grains in the sandstone aggregate also must be evaluated. Aggregate size determination must include an evaluation of *average size* and *sorting as to size* and may be presented statistically in terms of a frequency distribution or histogram; this also applies to the aggregate description of the other fundamental attributes.

1. COMPOSITION

Composition is an important attribute in that it determines certain properties both of the component grains and of the rock mass or aggregate. These include optical properties such as color, mechanical properties such as hardness and specific gravity, chemical properties such as solubility in water and reactivity with acid, and so forth.

Composition of sedimentary components may be designated in terms of mineral-species names or rock-type names (in case of rock fragments). Some sedimentary rocks are homogeneous or well sorted as to composition, an example being the Devonian Sylvanian sandstone of Michigan which in general consists of more than 99 per cent quartz grains. As indicated by Table II, the crinoidal sediment which this paper describes is rather homogeneous and shows good sorting as to composition, more than 96 per cent of the aggregate consisting of two mineral species, calcite and quartz.

The aggregate mineral composition of a sediment may be described by the use of mineral adjectives, such as quartzose, calcitic, dolomitic, argillaceous, and micaceous, which represent the dominant mineral species composing the aggregate. If the aggregate is composed largely of rock fragments, rock-type adjectives should be employed, such as *granitic* conglomerate, *basaltic* breccia, and *lithographic limestone* breccia.

TABLE II

COMPOSITION AND STRUCTURE OF COMPONENTS

Composition	*Structural Type of Components*	*Per Cent**
Quartz	Uni-grained mineral particles; crystalline structure	7–8
Tourmaline, rutile, magnetite	(same as foregoing)	Less than 1
Limonite	Multi-grained pellets; no definite internal structure	Less than 1
Glauconite	Multi-grained particles and aggregates; no definite internal structure	1 (approx.)
Bituminous (hydrocarbon) particles		Less than 1
Calcite	Uni-grained particles of organic (crinoidal) origin	75–80
Calcite	Multi-grained organic structural components; remains of brachiopods, corals, *et cetera*	5–10
Calcite	Finely crystalline interstitial cement	5–10

* Average of 12 samples collected at 1-foot vertical (time) intervals in section exposed at Quarry Hill and in east bank of Little Saline Creek.

2. INTERNAL STRUCTURE

A sedimentary component, whether uni-grained or multi-grained, is an individual entity which reacts as a unit during transportation or is formed in place

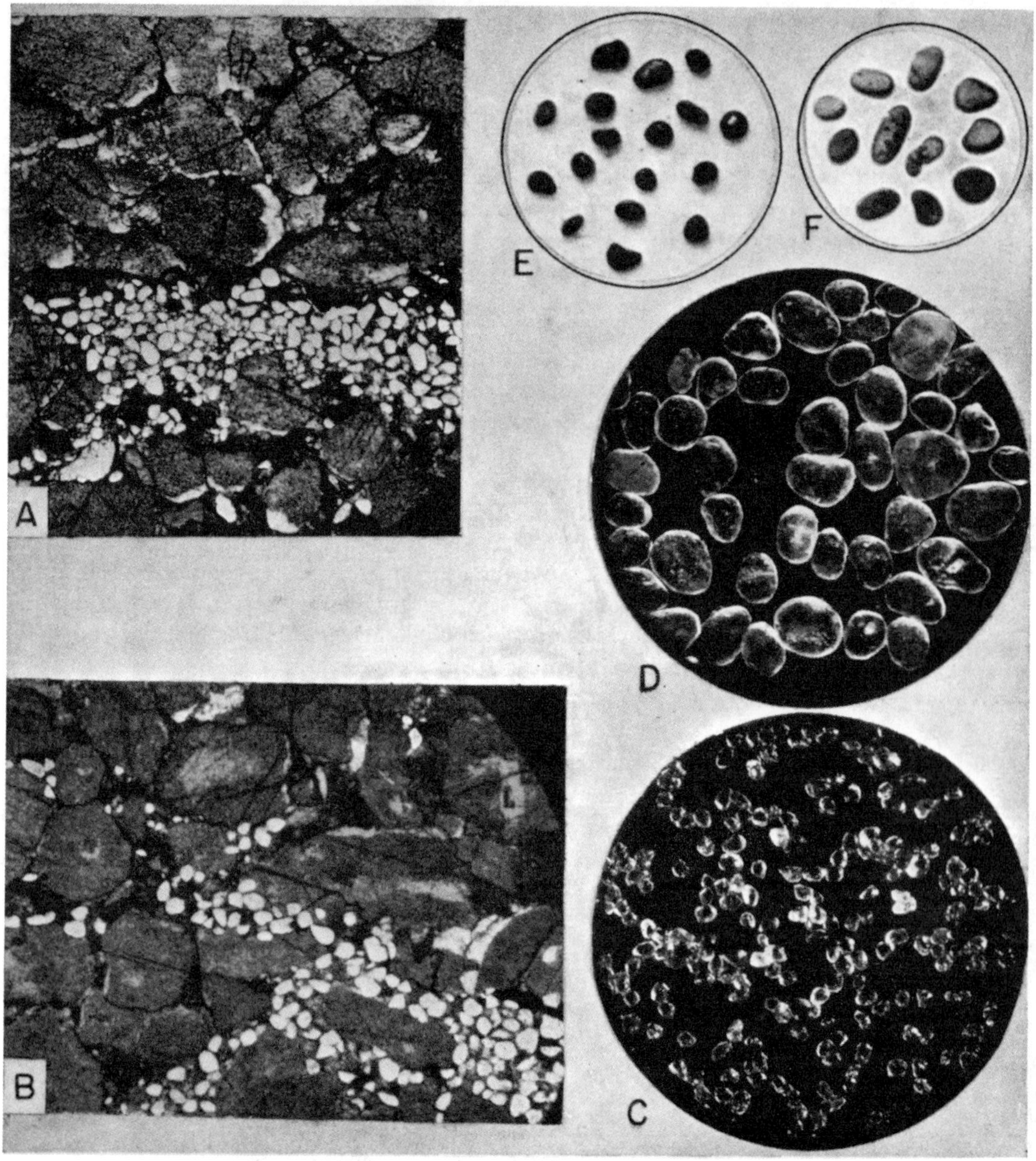

PLATE I

Photomicrographs of thin sections and insoluble-residue material. Grand Tower limestone, crinoidal phase (Bed 9), Quarry Hill, Ozora, Missouri.

A, B—Thin-section views (×16). Crinoidal débris (gray), smaller size quartz grains (white), calcite cement (black).

C, D—Views of insoluble-residue quartz sand (both ×16) showing two distinct size distributions represented.

E—Grains of detrital magnetite and limonite altered from detrital magnetite (×25).

F—Grains of pale brown, detrital tourmaline (×40, transmitted light).

as an entity either during or after genesis of the sediment of which it is a part. The *internal structure* of a component is one of the eight fundamental attributes; the various types of structures are listed and illustrated in the outline classification which follows. The internal structure of the component grains of the illustrative crinoidal sediment is described in the middle column of Table II.

I. Uni-grained Mineral Particles
  A. Crystalline structure
    Ex: quartz grains in sandstone; calcite grains in limestone
  B. Amorphous structure
    Ex: grains of limonite
II. Multi-grained Components of Mechanical Origin
  A. Rock fragments (derived from preëxisting rock)
    Ex: granite cobbles, schist pebbles, limestone granules
  B. Clay galls
  C. Mud balls
III. Multi-grained Components of Volcanic (Magmatic) Origin
  Ex: bombs, lapilli, *et cetera*
IV. Multi-grained Components of Organic Origin
  A. Individual structures or fragments thereof
    Ex: shells and tests of invertebrates; bones of vertebrates
  B. Colonial structures or fragments thereof
    Ex: colonial coral structures, algal structures, *et cetera*
V. Multi-grained Components of Concretionary Origin
  A. Concretionary bodies with *concentric structure*
    1. Oölites (calcite, *et cetera*)
    2. Pisolites (bauxite, *et cetera*)
    3. Concentric concretions
  B. Concretionary bodies with *radiate structure*
    1. Spherulites (pyrite, calcite, *et cetera*)
    2. Radiate concretions
  C. Concretionary bodies with no definite internal structure
    1. Pellets
      Ex: glauconite pellets, siderite pellets, faecal pellets, *et cetera*
    2. Nodules
      Ex: phosphatic nodules, chert nodules
    3. Structureless concretions
      Ex: limestone concretions

The internal structure of sedimentary components is a textural attribute; its mass or aggregate aspect, in the texture of a sediment, may be described by the use of such structural adjectives as oölitic, pisolitic, concretionary, spherulitic, nodular, uni-grained fragmental, and multi-grained fragmental (composed of rock fragments). If the sediment contains a significant portion of multi-grained components of organic origin, the mass textural aspect may be described by the use of such adjectives as foraminiferal, coralliferous, and pelecypodan.

3. SIZE

Size is a textural attribute and is one of the most important of the eight fundamental attributes. If sedimentary particles were perfect spheres, a definition of size would be simple; natural particles, however, are highly variable in shape, and evaluation of grain size must therefore consider grain shape. Grain size may be expressed in terms of mean diameter, which is the arithmetic average of the three principal diameters, length, breadth, and thickness.

Size analysis.—Size analysis of a sedimentary aggregate may be accomplished by the use of one or more of the methods here listed and described. The goal is subdivision of a sample into size classes or grades and calculation of the relative proportion in each class. Several grade scales of size have been proposed; Wentworth's modification of Udden's scale[8] is the one most widely used in this country.

TABLE III

Size Classification of Fragmental and Crystalline-Textured Sediments

Mean Size Range (in Mm., Wentworth's Scale)	*Wentworth Grade Name*		*Fragmental-Textured Sediments; Aggregate Size-Roundness Designation*		*Crystalline-Textured Sediments; Aggregate Size Designation*
Greater than 256	Boulder (*bl*)	Rudites	Boulder gravel, rubble (boulder conglomerate, breccia)		
256–64	Cobble (*cb*)		Cobble gravel, rubble (cobble conglomerate, breccia)		
64– 4	Pebble (*pb*)		Pebble gravel, rubble (pebble conglomerate, breccia)		
4– 2	Granule (*gn*)		Granule gravel, rubble (granule conglomerate, breccia)	Phanerocrystalline	Granular crystalline (*gnc*)
2– 1	Very coarse sand (*vcs*)	Arenites	Very coarse sand, grit (very coarse sandstone, gritstone)		Very coarsely crystalline (*vcc*)
1–1/2	Coarse sand (*cs*)		Coarse sand, grit (coarse sandstone, gritstone)		Coarsely crystalline (*cc*)
1/2 –1/4	Medium sand (*ms*)		Medium sand, grit (medium sandstone, gritstone)		Medium crystalline (*mc*)
1/4 –1/8	Fine sand (*fs*)		Fine sand, grit (fine sandstone, gritstone)		Finely crystalline (*fc*)
1/8 –1/16	Very fine sand (*vfs*)		Very fine sand, grit (very fine sandstone, gritstone)		Very finely crystalline (*vfc*)
1/16–1/256	Silt (*sl*)	Lutites	Silt (siltstone)	Micro-crystalline	Sublithographic (*sli*)
1/256–1 micron	Clay (*cl*)		Clay (claystone)		Lithographic (*li*)
Less than 1 micron	Colloidal particles		Colloidal aggregates		Cryptocrystalline aggregates

Grade scales and their use are described on pages 76–90 of the *Manual of Sedimentary Petrography* by Krumbein and Pettijohn. Wentworth's grade scale and its application to fragmental and crystalline-textured sediments are outlined in Table III. The more important methods of size analysis are as follows.

1. Sieve analysis of disaggregated rock sample or of insoluble residue. Components greater than 1/16 mm. in diameter thus separated into Wentworth grade sizes; sieve separates weighed and recorded.

[8] C. K. Wentworth, "A Scale of Grade and Class Terms for Clastic Sediments," *Jour. Geol.*, Vol. 30 (1922), pp. 377–92.

This method is applicable to insoluble residues only when the residue of sand size or larger constitutes at least 2 grams of the initial 50-gram sample. For residues smaller than this, visual measurement and estimation by a counting and tallying method are satisfactory.

2. Pipette analysis[9] of disaggregated rock sample or of insoluble residue. The form of stratigraphic analysis here propounded merely calls for separation and determination of the relative proportions of the clay and silt fractions. Sieved material less than 1/16 mm. in diameter is dispersed, and a water suspension of about 2% concentration is prepared in a liter cylinder graduate; the suspension is agitated, set at rest, and 2 hours and 3 minutes thereafter is sampled 10 cm. below the surface with a 20 cc. pipette; pipette sample is then dried and weighed and the weight multiplied by a factor (50) which converts it into terms of the total liter sample, i.e., *weight of clay fraction.* This weight is then subtracted from the weight of the total initial sample, and the remainder represents *weight of silt fraction.*
3. Visual measurement of disaggregated material. If dry grains are sprinkled on a slide, they tend to assume positions such that the shortest diameter (thickness) is approximately vertical. Thus the plane of the longest and intermediate diameters tends to be at right angles to the microscope tube; these diameters can be measured with the use of a binocular microscope fitted with a micrometer ocular. Sedimentary components larger than sand size may be measured with a ruler or caliper. The relative proportion of the sample in each of the size classes may be calculated by simple counting and tallying methods.
4. Visual measurement of grains in thin section or polished section or on surface of rock fragment. Grain diameters measured with the use of a micrometer ocular; observed radii corrected in accordance with theory of thin-section size analysis,[10] i.e., multiplied by 1.27. Relative proportion of the section sample in each of the size classes estimated by counting and tallying methods.

Presentation of size data.—A size analysis in the form of a histogram, such as that of Figure 2, should be accompanied by an evaluation of the *average* size and the degree of *sorting* as to size. Average size may be expressed in terms of the arithmetic mean of the diameter distribution, a measure which may be computed[11] readily from the size-frequency distribution or histogram. Sorting may be defined as the degree of concentration of data about an average, a high degree of concentration implying good sorting as to size. There are several statistical measures of sorting, but all are too time-consuming for widespread use in stratigraphical analysis. Sorting is evaluated by the writer as follows.

1. *Good sorting,* if 90 per cent of the sediment is concentrated in 1 or 2 size classes
2. *Fair sorting,* if 90 per cent is distributed through 3 or 4 size classes
3. *Poor sorting,* if 90 per cent is distributed through 5 or more classes

The mass textural aspect of grain size of a sedimentary aggregate may be described by the use of size adjectives, following the classification of Table III. In the case of fragmental-textured aggregates such adjectives should be employed as very coarse sandy, medium sandy, fine sandy, and silty; whereas for crystalline-textured aggregates such equivalent adjectives as very coarsely crystalline, medium crystalline, finely crystalline, and sublithographic may be employed. Where two or more size classes predominate in the aggregate, size may be described in a manner illustrated by the following examples: fine to coarse sandy, sublithographic to finely crystalline, silty medium sand, and coarse sandy pebble gravel.

Size analysis of crinoidal sediment.—The crinoidal phase of the Grand Tower limestone (Pl. I) described in the present paper, although now of crystalline

[9] W. C. Krumbein, "The Mechanical Analysis of Fine-Grained Sediments," *Jour. Sed. Petrology,* Vol. 2 (1932), pp. 140–49.

[10] Krumbein and Pettijohn, *op. cit.,* pp. 129–32.

[11] Krumbein and Pettijohn, *op. cit.,* p. 240.

texture, was deposited in the Devonian sea as a fragmental sediment, a quartzose crinoidal sand, the size distribution of which is portrayed graphically by Figure 2. The total volume of the crinoidal rock in the sampled section is divisible as follows.

	Per Cent, Volume
Transported fragmental material (shown in histogram, Fig. 2)	80–90
Untransported coral, brachiopod, pelecypod, *et cetera* remains	5–10
Finely crystalline calcite cement	5–10
Total	100

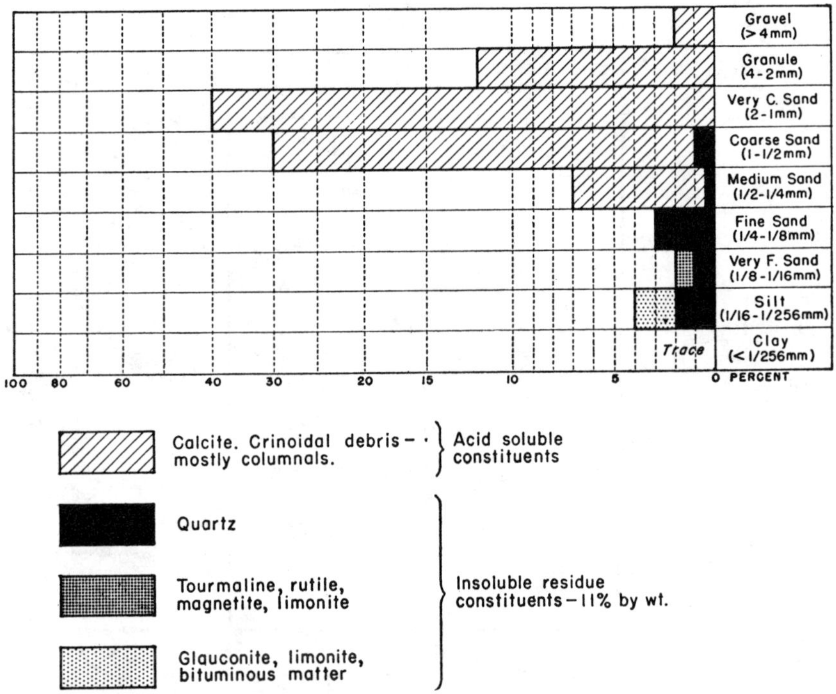

FIG. 2.—Size histogram of transported fragmental components, crinoidal phase of Grand Tower limestone. Average of 12 sets of samples collected at 1-foot vertical (time) intervals from section exposed on Quarry Hill and on east bank of Little Saline Creek.

The size distribution of the crinoidal débris was measured by thin section and polished section methods and by simple measurement of the columnals and other plates standing out in sharp weathered relief on rock surfaces. The grain size of the quartz and other acid-insoluble minerals was measured by sieve analysis of the insoluble residue, which constituted 11 per cent of the rock by weight. The arithmetic mean diameter, as calculated from the size-frequency distribution, is 1.24 mm. This crinoidal sediment shows only a *fair* degree of sorting as to size, 90 per cent of the grains being spread through four size classes (granule through medium sand). The modal class, that showing maximum frequency, is the very coarse sand class (2–1 mm.). At the time of deposition the crinoidal sediment would have been classified as a very coarse to coarse sand (Table III); after deposition and burial it assumed a crystalline texture as a result of cementation and recrystallization, and thus may now be classified as a very coarsely to coarsely crystalline limestone.

4. SHAPE

Grain shape, like size, is a textural attribute. Zingg's shape classification diagram[12] has been modified by the writer so as to include nine shape classes (Fig. 3), to which grains of sand size or larger may be referred.

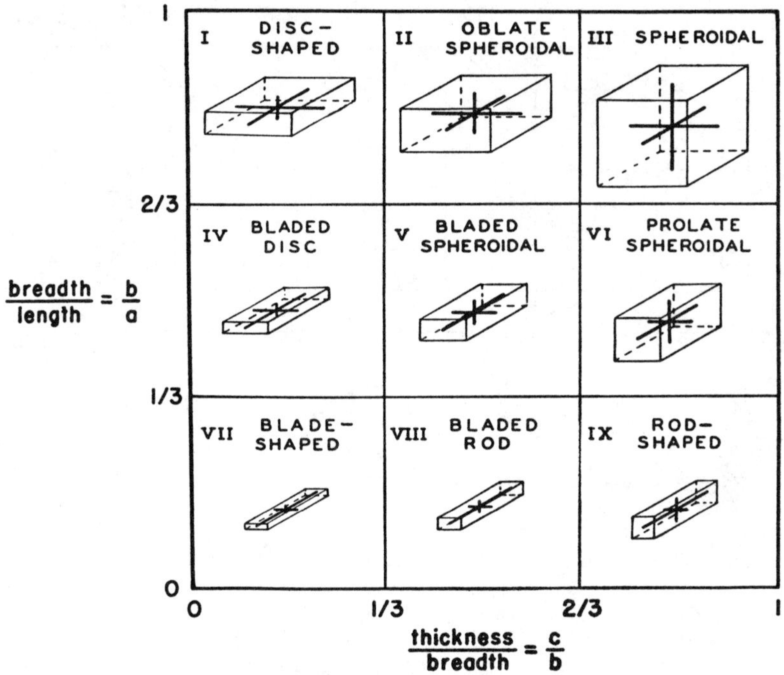

FIG. 3.—Shape-classification diagram modified from original plan of Zingg. *a*, length; *b*, breadth; and *c*, thickness. Boxes (parallelograms) which enclose the three diameters should be ignored; they are included so that shapes represented by diameters will be more apparent.

Shape analysis.—Shapes designated by Zingg's diameter-proportion scheme are determined by estimating the ratios between the three principal diameters (*a*, *b*, *c*). In the case of pebble, cobble, and boulder components, the diameters may be measured with a ruler or caliper; diameters of sand- and granule-size grains may be measured by means of a micrometer ocular and fine-adjustment screw. The long (*a*) and intermediate (*b*) diameters, approximately at right angles to the microscope tube, may be measured with the micrometer ocular. The thick-

[12] Th. Zingg, "Beitrag zur Schotteranalyse," *Schweiz. Min. u. Pet. Mitt.*, Bd. 15 (1935), pp. 39–140.

ness or short diameter (*c*) may be measured with a petrographic microscope by means of the fine-adjustment screw; the instrument is first focussed on the surface on which the grain rests and then on the upper surface of the grain itself, the difference of the micrometer readings representing the measure of thickness.

Zingg's diameter-proportion method of shape classification fails to provide a complete picture of the outline geometry of a grain. In some cases it may be desirable to supplement the shape-class designation with a term describing the geometric form of the grain. For example, in the plane of the *a* and *b* axes or diameters disc-shaped components (class I) may tend to be circular, pentagonal, triangular, square, and so forth. Wentworth[13] has assembled a series of well known geometric form designations which may be used to describe grain outlines. In the same paper he also lists three-dimensional shape terms, such as pyramidal, prismoidal, and wedge-shaped, which may be used in certain cases to describe the shape of grains as solid or three-dimensional bodies.

A shape analysis of a sediment may be presented in the form of a scatter diagram of points plotted on a diameter-ratio grid such as that of Figure 3, each point representing an individual component of the aggregate. In addition, shape analysis should evaluate the *average* shape and the degree of *sorting* as to shape. The shape class or classes having the maximum concentration of points may be designated as the average. If the grains of a given sediment tend to be grouped along a boundary between two classes, the average shape may be expressed by combining the two class numerals. Sorting as to shape may be evaluated in the same manner as sorting as to size. If 90 per cent of the grains is concentrated in 1 or 2 shape classes, the sedimentary aggregate shows *good* sorting; if 90 per cent is distributed through 2 or 3 shape classes, *fair* sorting is indicated; if 90 per cent is distributed through 4 or more shape classes, a condition of *poor* shape sorting obtains.

Shape analysis of crinoidal sediment.—The average shape of the coarse to medium quartz sand (Pl. I, D) in the illustrative crinoidal sediment is approximately along the boundary between the spheroidal and the prolate spheroidal classes (III–VI); the shapes of these grains, however, range into the bladed spheroidal (V) and oblate spheroidal (III) classes. The average shape of the crinoid columnals and other plates (Pl. I, A, B) is somewhere near the juncture of classes I, II, and IV, disc-shaped columnals being dominant. The tourmaline grains (Pl. I, F) are dominantly prolate spheroidal (VI), and the limonite and magnetite grains are for the most part bladed spheroidal (V). The crinoidal sediment thus shows fair to poor sorting as to shape, 90 per cent of the grains being spread through 4 or 5 shape classes; this condition is due to the fact that the grains are of heterogeneous derivation, some representing little transported organic débris and others derived from preëxisting rocks and having undergone several cycles of transport.

[13] C. K. Wentworth, "An Analysis of the Shapes of Glacial Cobbles," *Jour. Sed. Petrology*, Vol. 6 (1936), pp. 85–96.

5. ROUNDNESS

The attribute of grain roundness pertains to the degree of regularity of the surface or outline and to the degree of sharpness (radius of curvature) of edges and corners. Roundness is distinct from shape inasmuch as two grains of the same diameter ratio or shape may show different degrees of roundness.

Roundness analysis.—Several mathematical measures of roundness have been proposed, but none is suitable for use in the rapid study procedure here propounded. In this procedure roundness analysis must be accomplished by simple visual comparison with a set of standards depicting various roundness classes.

A familiar roundness classification method involves placing the components in one or more of four classes, the limits of which are arbitrary; the classes are as follows.

1. *Angular* (*A*)—irregular grain surface dominated by edges and corners all of which are sharp; grain either represents an unworn crystal or a fractured particle
2. *Subangular* (*SA*)—edges and corners still prominent on grain surface but tend to be less sharp and slightly curved or rounded (larger radius of curvature)
3. *Subround* (*SR*)—edges and corners mostly curved; grain outline still somewhat irregular
4. *Round* (*R*)—edges and corners absent; surface outline of grain a smooth curve tending toward an oval or circular shape.

Classification may be accomplished by visual comparison of the grains with the set of illustrations of Figure 2 of the paper by Russell and Taylor[14] on Mississippi River sands.

The other visual roundness classification method involves placing the components of a given sediment in one or more of the nine roundness classes illustrated by Figure 4. This image diagram was presented in a paper by Krumbein,[15] the roundness of the pebbles selected to illustrate each of the nine classes (0.1–0.9) was measured from grain projections utilizing Wadell's formula for roundness of plane figures. Although this image diagram was prepared to make possible visual estimation of pebble roundness, it also is applicable to components in the granule and sand grades of Wentworth's size scale. Roundness classes 0.1–0.2 are roughly equivalent to the *angular* class of the other classification system; 0.3–0.4 to the *subangular* class; 0.5–0.7 to the *subround;* and 0.8–0.9 to the *round.*

A problem is presented by grains the outlines of which show composite roundness, such as those at the lower right of Figure 4. Such grains combine two or more roundness classes and may constitute a significant portion of a sedimentary deposit. They are formed by fracture of grains which previously had undergone rounding. In classifying them it is best to employ a dual number designation, the first denoting the roundness of the major portion of the grain outline and the second the minor portion. For example, the grain marked .3 in the diagram would be classified as .7//.1. This provides a clearer picture of the actual composite roundness condition than does the misleading single number.

[14] R. D. Russell and R. E. Taylor, "Roundness and Shape of Mississippi River Sands," *Jour. Geol.*, Vol. 45 (1937), pp. 225–67.

[15] W. C. Krumbein, "Measurement and Geological Significance of Shape and Roundness of Sedimentary Particles," *Jour. Sed. Petrology*, Vol. 11 (1941), Pl. 1, p. 68.

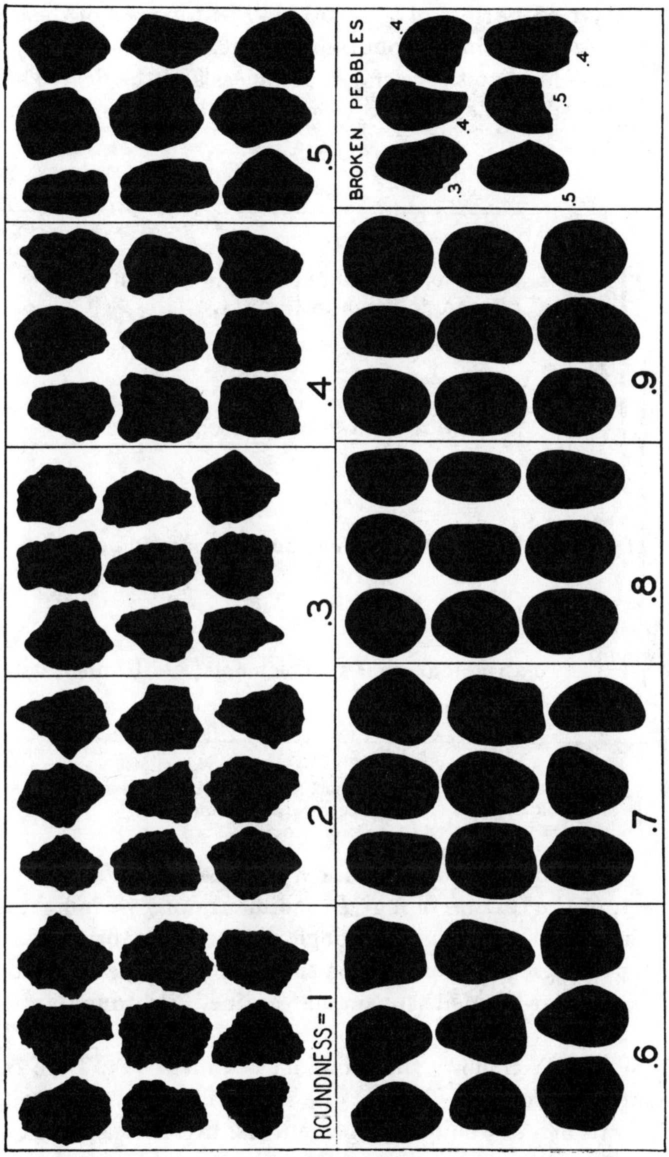

FIG. 4.—Images for visual comparison in classifying grains according to roundness. Diagram prepared by Krumbein for estimating pebble roundness; also applicable to granule and sand sizes.

A roundness analysis, utilizing the classification scheme of Figure 4, may be presented in the form of a histogram or in the form of a diagram such as that of Figure 5. The relative proportion of the total sediment in each of the roundness classes may be estimated by a simple counting and tallying method. The analysis should include a statement as to the average roundness and the degree of sorting, the latter being expressed as good, fair, or poor, following the procedure used in the case of size sorting.

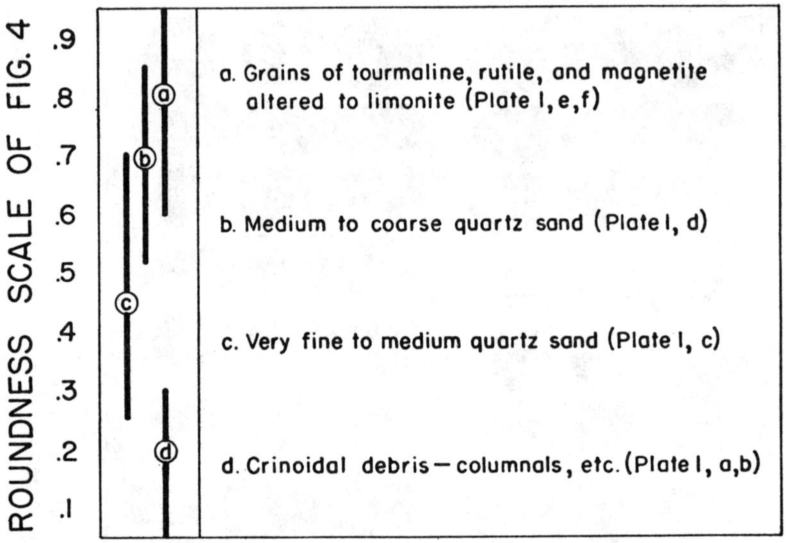

FIG. 5.—Roundness analysis of transported fragmental components of crinoidal phase (Bed 9), Grand Tower limestone.

Roundness of sedimentary components is a textural attribute, and its mass or aggregate aspect, in the texture of a given sediment, may be indicated by the use of such terms as gravel *versus* rubble, conglomerate *versus* breccia, and sandstone *versus* gritstone. Aggregate roundness also may be designated by the use of adjectives such as angular-grained, subangular-grained, subround-grained, and round-grained.

The analysis of Figure 5 shows the crinoidal sediment (Pl. I) to be poorly sorted as to roundness, a feature which is due to the fact that the grains are of heterogeneous derivation. As would be expected, the little transported crinoidal débris is for the most part angular, whereas the quartz and heavy mineral grains, having undergone several cycles of transport, are subround to round. The position of the encircled letters (*a–d*) indicates the average roundness for the different component types, and the extent of the lines suggests the range of variation.

6. SURFACE TEXTURE

The attribute of surface texture, unlike roundness, refers to the finer details of grain surfaces, those features that do not affect the surface outlines of the particles. The surface texture classification here employed is patterned after that of Williams.[16] As indicated by the following outline, the surface of a sedimentary particle has two fundamental textural aspects: degree of surface *relief* and degree of *luster*.

- I. Surface Relief of Particle
  - A. Smooth (*sm*)
  - B. Rough
    1. Frosted (ground-glass surface) (*fr*)
    2. Etched (result of solvent action) (*et*)
    3. Faceted (result of secondary growths) (*fa*)
    4. Striated, scratched or furrowed (*st*)
    5. Percussion or chatter marked (*pm*) (*cm*)
    6. Pitted (*pi*)
    7. Ridged (*ri*)
- II Surface Luster of Particle
  - A. Dull (*D*)
  - B. Polished (glossy) (*P*)

A grain surface may be smooth or rough independent of the luster; it may likewise be dull or polished independent of the relief. Classification of components according to surface texture must therefore be binary, in that it must evaluate both the surface relief and degree of luster; for example, quartz grains may be smooth and polished (*sm*, *P*), and frosted and dull (*fr*, *D*). In some cases different types of relief may be coëxistent on the same grain; that is, the grain may be both striated and pitted or both etched and smooth, and so forth. Furthermore, certain component types within the aggregate may show variation in surface texture.

Inasmuch as geologic agents leave their impress on the surfaces of grains, surface textures are significant in genetic studies. Some textures can be formed in quite different ways; for example, a frosted surface may be formed by rigorous wind (sand-blast) action, by chemical etching, or by incipient secondary enlargement. Surface textures may be inherited, formed during transportation, or formed intrastratally by post-depositional agents.

Surface texture analysis of crinoidal sediment.—The coarse to medium quartz sand grains (Pl. I, D) show a somewhat polished luster and vary in relief from smooth to etched and frosted; their surface texture might be expressed in abbreviated form as follows: *P*, *sm* to *et* and *fr*. The medium to very fine quartz sand grains (Pl. I, C) are *P*, *sm* to *et*. The tourmaline, rutile, magnetite, and limonite grains are *P*, *sm*. The original surface texture of the crinoidal particles has been greatly altered as a result of post-depositional recrystallization, solution, and cementation. The columnals and other plates stand out in sharp relief on weathered rock surfaces and are dull in luster and have a rough surface relief marked by etched, striated, and pitted irregularities.

[16] L. Williams, "Classification and Selected Bibliography of the Surface Textures of Sedimentary Fragments," *Nat. Research Council Comm. on Sedimentation Rept.* (1937), pp. 114–28.

7. RELATIVE POSITION OF COMPONENTS IN SPACE

The position in space of an individual component, although constituting a fundamental sedimentary attribute, is of little significance when considered from the viewpoint of an individual grain standing alone. When considered from the aggregate point of view, however, and applied to the mutual space arrangement of individuals in the aggregate, this attribute is of descriptive importance. It is of greatest importance in the case of deposits, such as the crinoidal sediment described in this paper, the components of which show differentiation as to mineral composition, and (or) size, and (or) shape, and so forth. The space distribution of minor component types, those constituting less than half of the aggregate, may be described in a manner exemplified by the following phrases: in *irregular patches* or *stringers* in the aggregate, in the form of *tabular bodies* (vertical to horizontal bands), *evenly disseminated* throughout the aggregate, and distributed *at random* within the aggregate.

The space arrangement attribute also is significant in the case of sediments composed mostly of one component type, such as a pure quartz sand which shows only little differentiation in composition, size, or shape. Space arrangement under such homogeneous conditions has been studied experimentally by means of aggregates of artificial spheres. When the spheres are arranged in a cubic grid pattern, so that lines connecting their centers are at right angles to each other, the lowest degree of packing and the highest porosity result. When arranged in rhombohedral or other grid patterns, in which lines connecting sphere centers diverge from the right-angle condition, higher packing and lower porosity result because of decrease in the volume of intergrain voids.

Thus far, this attribute of mutual space arrangement has not been studied in detail, is not amenable to simple classification and must therefore be described verbally in each case. It, together with the attributes of size, shape, and fabric, controls the property of packing.

Space arrangement of components in crinoidal sediment.—In the crinoidal limestone, pictured in thin section in Pl. I, A, B, the crinoidal débris constitutes the major component type, and the other component types show an interesting space distribution within the aggregate. The medium to coarse quartz sand occurs as isolated grains distributed at random like the crinoid columnals which they approach in size. In contrast, the silt and fine sand, composed of quartz, tourmaline, rutile, limonite, magnetite, and glauconite, occur throughout the coarser aggregate in the form of irregular patches, stringers, and bands.

This attribute of relative space distribution of components reveals much with regard to the environment and mode of transportation and deposition and is therefore deserving of study. The coarse quartz sand previously mentioned was deposited probably from traction transport in the same manner as the crinoidal particles; thus, the space distribution of these two component types is similar. As described in subsequent pages, the silt and fine sand probably were deposited from suspension transport, during times of quiet water, in the form of thin layers, some of the grains percolating down through the interstices of the larger crinoidal grains.

8. ORIENTATION OR FABRIC

The attribute of *orientation* applies to an *individual* component and is expressed as the angle of dip and strike of the maximum projection plane, this representing the plane defined by the long and intermediate diameters (length and breadth) of the component. The term *fabric* denotes orientation in the aggregate; that is, the fabric of a sedimentary deposit represents a *composite* of the orientations of certain or all of the individual components. Measurement of orientation or fabric is possible only in the case of sediments of sand size or larger (sand through boulder sizes). This attribute obviously is related to grain shape and pertains only to sediments whose grains are not equidimensional; spherical grains have no orientation. *Primary fabric* refers to the orientation of the components that is instituted during deposition; this may be changed during or after burial and compaction.

An interesting example of sedimentary fabric is provided by the basal conglomeratic phase of the Devonian St. Laurent limestone formation at Quarry Hill, near Ozora, Missouri. The pebbles of this conglomerate consist of three preëxisting rock types: (1) sandstone fragments, probably derived from the underlying Beauvais sandstone, (2) sublithographic limestone, possibly eroded from the upper phase of the Grand Tower limestone, and (3) chert pebbles, the character of which suggests derivation from the earlier Devonian Bailey chert. These pebbles are for the most part in mutual contact and are set in a matrix of quartz sand cemented with calcite; they vary in shape through the shape classes II, III, V, and VI (Fig. 3). Large sawed sections of the conglomerate indicate that the pebbles have a preferred orientation throughout the aggregate; that is, their long diameters tend to be similarly oriented in random cross sections and thus show a parallel type of fabric.

The fabric of the afore-described conglomerate is significant in environmental interpretations. The sediment probably was deposited as a beach gravel in which the most common shape class was the prolate spheroidal (VI). The orientation of beach pebbles has been studied by Fraser,[17] who found by statistical studies that roller-shaped pebbles commonly lie with their long diameters parallel with each other and with the shoreline. Fraser attributes this parallel fabric to the tendency for waves to swing pebbles into such a position or to the tendency for the pebbles to roll with their long diameters perpendicular to the direction of current movement.

Fabric classification and analysis.—Description and classification of a sedimentary deposit according to its fabric is a difficult problem, an understanding

[17] H. J. Fraser, "Experimental Study of the Porosity and Permeability of Clastic Sediments," *Jour. Geol.*, Vol. 32 (1935), pp. 910–1010.

of which may be gained from a paper by Krumbein.[18] The classification scheme proposed by the present writer is as follows.

A. Random orientation (non-parallelism) of individual components with respect to both their maximum projection planes and their long diameters
B. Preferred orientation (parallelism) of maximum projection planes of components (long diameters in random orientation)
C. Preferred orientation (parallelism) of long diameters of components (maximum projection planes in random orientation)
D. Preferred orientation (parallelism) of both maximum projection planes and long diameters of components.

Classification of a sedimentary aggregate which shows a preferred orientation (classes B, C, D) calls for two further descriptive considerations: (1) the *angle of inclination* (dip) of the maximum projection planes and (or) long diameters with respect to horizontal or with respect to the bedding planes, and (2) the *compass direction* (strike) of the maximum projection planes. The orientation of components of a deposit may be expected to vary, and this variation, as in the case of the other attributes, calls for evaluation of the *average* angle of dip and direction of strike and the degree of *sorting*. Detailed fabric analyses are beyond the scope of the present paper; methods of analysis and presentation, including orientation histograms and polar coördinate diagrams of dip and strike, are described in the afore-cited paper by Krumbein.

Imbrication is a type of fabric, formed under certain conditions in stream beds and along beaches, which is characterized by a preferred orientation (parallelism) of the maximum projection planes of the components at an angle to the bedding planes. Imbrication is not a sedimentary structure, as is commonly supposed, inasmuch as it involves grain-to-grain relationships rather than mass relationships of grain aggregates. Another type of fabric phenomenon in sedimentary rocks is represented by the parallel orientation of clay flakes in certain clay shales, especially those composed of illite minerals; this fabric renders such rocks cleavable in planes ordinarily parallel with the bedding planes.

Fabric of crinoidal sediment.—The illustrative crinoidal sediment (Pl. I) shows a random orientation of individual components with respect to both their maximum projection planes and their long diameters. Thus the fabric of this rock belongs in category *A* of the classification scheme.

D. PROPERTIES OF SEDIMENTARY AGGREGATES

Sediments and sedimentary rocks have certain mass features, here termed *properties* which call for description in petrographic analysis. Just as the mass characteristics of a society are determined by the attributes of its individuals, so also are the mass properties of a sedimentary aggregate governed by the basic attributes of the particles of which it is made. In other words sedimentary properties and their values are functions of the parameters of the attributes.

[18] W. C. Krumbein, "Preferred Orientation of Pebbles in Sedimentary Deposits," *Jour. Geol.*, Vol. 47 (1939), pp. 673–706.

Eight properties of stratigraphic significance are described here in summary fashion. Such properties as plasticity, hygroscopicity, fusibility, and specific gravity are not particularly important from the stratigraphic point of view and therefore are not included in the list.

COLOR

The property of color of a sedimentary aggregate is conditioned by four factors: (1) the summary or total effect of the colors of the component mineral grains and (or) rock fragments, (2) the color of the interstitial cement or matrix material, if present, (3) the presence of mineral pigment either coating the grains or disseminated throughout the cement or matrix, and (4) the degree of fineness of the grains. The last factor (4) is important in the case of very fine-grained sediments; the smaller the average particle size the darker the color of the aggregate. Pigmentary materials which may determine rock coloration include bituminous and carbonaceous material, oxides of iron, finely divided glauconite, ferrous sulphide, and so forth. Accurate color determination is a most difficult problem because of subjective errors and the lack of a generally accepted standard scale of colors.

Color of crinoidal sediment.—The crinoidal limestone of this report is light buff gray in color and has a slight greenish cast, locally, where glauconite is relatively more abundant. The predominant light gray color is imparted by the calcite grains of crinoidal origin, by the quartz grains, and by the calcite cement, pigmentary material being absent. The superimposed buff coloration of slight intensity is caused by the presence of limonite particles (Pl. I, Table II).

MASS CHEMICAL COMPOSITION (CHEMICAL ANALYSIS)

The property of mass chemical composition, evaluated in the form of a partial or complete chemical analysis of a sediment, is an important aspect of petrographic study and must be considered if the complete story is to be told for a deposit. This property may be summarily evaluated for a given sediment by the use of one or more adjectives which describe the dominant chemical composition; for example, siliceous, calcareous, magnesian, ferruginous, carbonaceous, and phosphatic.

A chemical analysis is a valuable tool in the study of fine-grained sediments, the mineralogy of which often defies determination by ordinary methods. From chemical analyses the quantitative mineral composition of a sediment often may be determined by recasting the chemical constituents in terms of the mineral species known to be present. This method of study also provides information about the MgO content and the calcite-dolomite ratios of limestones. In the case of sediments used for commercial purposes study of the property of mass chemical composition permits quantitative determination of the presence of deleterious and desired constituents. The organic content of sediments, important in the study of source beds of petroleum, likewise may be evaluated from chemical analysis; determination of organic content and its significance have been described by Trask.[19]

Chemical composition of crinoidal sediment.—Following is a partial chemical analysis of a composite sample of the crinoidal limestone of this report, samples from several horizons in Bed 9 having been mixed.

SiO_2	10.04
Fe_2O_3	0.66
Al_2O_3	0.49
CaO	49.20
MgO	0.36
Ignition loss (mostly CO_2)	38.17
Alkalies	Trace
	98.92

The analysis was made in the chemical laboratory of the Kansas State Geological Survey. Any H_2O present is included under ignition loss; some FeO may be contained in the Fe_2O_3 category.

Most of the silica is in the form of the mineral quartz; a minor amount is combined in the minerals glauconite and tourmaline (see Table II). Most of the Fe_2O_3 is in the form of limonite and magnetite although some must be contained in the other minerals. The CaO is dominantly combined with CO_2 as calcite.

ACID SOLUBILITY (CHARACTER OF RESIDUE)

The chemical property of a sediment of being soluble or insoluble in acid (HCl) finds concrete stratigraphic expression in the form of insoluble residue studies of limestones and dolomites. The character of an insoluble residue and its per cent by weight of the total sediment constitute an important sedimentary property useful in correlation and other phases of stratigraphic research. Procedures for preparation of residues have been developed by many workers; that employed by the writer was formulated by Workman and has been described in the section on study methods.

Results of insoluble-residue studies may be plotted graphically utilizing a bar diagram, such as that employed by Ireland,[20] the width of each bar representing the thickness of the sedimentary section sampled and the length of each bar representing the residue's per cent by weight of the total sample. The length of each bar may be subdivided into sections designated by different symbols, which represent the different types of material composing the residue. These bars may be plotted on well strips or on strips representing sections sampled in the field, and correlation between strips may be accomplished by matching the percentages by weight of the residues and by matching the character of the material, compositional and textural.

[19] P. D. Trask, *Origin and Environments of Source Sediments of Petroleum*, pp. 18–66, 110–205. Gulf Publishing Company, Houston, Texas (1932).

[20] H. A. Ireland, "Use of Insoluble Residues for Correlation in Oklahoma," *Bull. Amer. Assoc. Petrol. Geol.*, Vol. 20, No. 10 (October, 1936), pp. 1086–1121.

The insoluble-residue content of the crinoidal limestone of this report constitutes 11 per cent by weight of the aggregate and is illustrated graphically in the histogram of Figure 2.

INTERSTITIAL CEMENTATION

Cementation of the grains of a sedimentary aggregate is both a process and a resultant; as the latter it constitutes an important property inherent in lithified sediments. An evaluation of this property must consider the *degree* of cementation, the *texture*, and the *composition* of the cement. As a property, cementation pertains only to sediments the detrital grains of which are for the most part in mutual contact. In cases wherein the major detrital particles are separated in the aggregate, the binding material is termed the *matrix* (rather than cement), an example of which is provided by a well known rock type in which quartz sand grains are "floating" in a finely crystalline limestone matrix, the quartz grains evidently having been deposited together with a lime mud.

The degree of cementation may vary all the way from an openwork condition exemplified by openwork gravels, in which the interstices are free of cement, to a "tight" or closed-work condition in which the cement completely fills the interstices. Degree of cementation may be roughly evaluated in terms of *good*, *fair*, and *poor*.

Cements are of two textural types: *crystalline* and *fragmental*. Crystalline-textured cements may be in the form of chemical precipitates, syngenetic or epigenetic, deposited in the interstices as either aggregates of individual crystals or as growths in optical continuity on the detrital grains of the same composition. Fragmental-textured cements are in the form of clayey or silty binding material which, at the time of accumulation of the sediment, collects in the interstices between the larger grains. An example of fragmental-textured cement is provided by certain basal Cherokee sandstone lenses in the subsurface of eastern Kansas, which consist of medium to coarse quartz sand cemented by clay, fine quartz silt, or other fine-grained binding material. This type of cement is an expression of extremely poor sorting as to size. A primary fragmental cement, such as fine-grained lime mud deposited in the interstices of quartz sand grains, may become crystalline-textured as a result of epigenetic recrystallization.

Cementing substances display a wide range of composition and may be grouped in the following categories: (1) clay minerals, (2) carbonate cements (including calcite, dolomite, siderite, barite, *et cetera*), (3) iron oxides, (4) aluminum oxides, (5) silicate cements (including the chlorites, zeolites, feldspars, epidote, *et cetera*), and (6) sulphide cements.

Cementation of crinoidal sediment.—The crinoidal limestone of this report displays good cementation by a crystalline-textured cement consisting of very finely crystalline calcite, represented by the dark gray to black-appearing material between the crinoid plates and quartz grains in the thin section photographs of Plate I, A, B. This sediment probably was deposited as an openwork fragmental sand later to be cemented by epigenetic calcite precipitation.

MASS TEXTURE OF AGGREGATE

Sedimentary rocks display two fundamentally different types of mass texture: (1) fragmental (clastic) texture, and (2) crystalline texture. In sedimentary rocks of crystalline texture, as in igneous rocks, the aggregate consists of a mosaic of interlocking crystals wherein the individual crystal particles acquire shape by crystallization in place. In fragmental-textured rocks, on the other hand, grain shape is detrital in origin, and the grains are not mutually interferent but are bound together by a cement or matrix.

Crystalline textures may originate in three ways. First, they may form by primary chemical precipitation in place, exemplified by travertine and certain evaporites. Second, and quantitatively more important, they may form by secondary recrystallization and cementation of original fragmental-textured sediments; examples of this mode of genesis are limestones, including finely crystalline and lithographic varieties originally deposited as clastic lime muds and coarsely crystalline varieties deposited as calcareous sands. Third, crystalline textures may originate in sediments of original fragmental texture by secondary precipitation (secondary growth) of cementing material on the surfaces of fragmental grains; this material, of the same composition as the detrital grains, is precipitated in optical continuity on them. In this way the voids of the fragmental sediment become filled with secondary growth material, and a mosaic of interlocking crystals (crystalline texture) is formed. This third genetic type of crystalline texture is well illustrated by what Tieje has termed *orthoquartzite*,[21] a quartz sandstone firmly and completely cemented by secondary quartz precipitated on and in optical continuity with the detrital quartz grains. This sedimentary type should be distinguished from truly metamorphic quartzite, which Tieje has designated as *paraquartzite*, formed by recrystallization under high temperature-pressure conditions.

Most sedimentary rocks, including limestones and dolomites, have been deposited as fragmental-textured aggregates. Thus crystalline texture is for the most part acquired, resulting from diagenetic and metamorphic changes, and is the ultimate destiny of most sediments. There is a tendency for crystalline-textured sediments to become more coarsely crystalline with the passing of time.

All of the eight fundamental attributes apply to fragmental-textured sediments; only the attributes of composition, size, shape, space arrangement, and fabric, however, pertain to crystalline-textured rocks. As indicated by Table III, description of the attribute of size involves two different sets of terminologies, one for fragmental and one for crystalline aggregates. It is usually difficult, if not impossible, to evaluate the attribute of grain shape in crystalline-textured rocks.

Mass texture of crinoidal sediment.—The crinoidal limestone of this report, now of coarse to very coarsely crystalline texture, was originally deposited as a fragmental coarse to very coarse crinoidal sand. Thus its crystalline texture

[21] A. J. Tieje, "Suggestions as to the Description and Naming of Sedimentary Rocks," *Jour. Geol.*, Vol. 29 (1921), p. 655.

has been acquired, resulting from calcite cementation and possibly recrystallization of the detrital calcite grains.

PACKING OF AGGREGATE

The property of *packing* represents the degree of density of arrangement of grains in the sedimentary aggregate. In other words, the greater the mass per unit volume of the uncemented sediment the higher the degree of packing. In the absence of cementation, the property of *porosity* is inversely proportional to packing.

Packing is a function of the attributes of size, shape, space arrangement, and fabric. It is independent of average or mean size but is inversely proportional to the sorting as to size; that is, the poorer the sorting as to size the higher is the degree of packing inasmuch as the finer grains tend to fill the interstices between the larger. Packing varies directly as the sorting as to shape, the more heterogeneous the shapes of the particles the lower the degree of packing. Another aspect of its relation to shape is that it is inversely proportional to the degree of sphericity; in other words, the more the grain shapes approach a spheroidal condition (class III) the poorer is the packing. The attribute of mutual space arrangement of grains in the aggregate also controls the property of packing (see discussion of this attribute). Finally, packing is related to the attribute of orientation or fabric in such a way that it varies directly as the tendency toward preferred orientation of the components; in sediments of random fabric there is poor packing.

POROSITY

Porosity is the property that pertains to the character and relative volume of the interstices of a sedimentary aggregate. Distinction is made between *total* and *effective* porosity. The latter is a measure of the interconnected voids, whereas the former considers all pore spaces, isolated as well as interconnected. In petroleum geology and ground-water hydrology effective porosity is the more important aspect. Total porosity may be defined as the ratio of the voids to the total bulk volume of the rock; this ratio, multiplied by 100, is the per cent total porosity.

The property of porosity is related to two other properties, to which it is inversely proportional: (1) the degree of packing and (2) the degree of cementation. Porosity, like packing, is a function of the grain attributes of size, shape, space arrangement, and fabric. It is important to note that porosity is independent of average or mean size; thus, with the same degree of cementation and packing (including the same relative sorting as to size) a fine sand may have the same porosity as a coarse sand.

Porosity of crinoidal sediment.—The crinoidal limestone of this report has an average effective porosity in the section sampled of approximately 7.2 per cent, the measurements having been made with a Clough porosimeter. The relatively low average effective porosity is due to the fact that calcite cementation has for the most part closed the voids between the otherwise highly porous aggregate of crinoidal and quartz grains.

PERMEABILITY

Permeability is the sedimentary property that governs fluid flow through a porous aggregate. In ground-water hydrology and petroleum geology permeability is important because it controls the flow of water, oil, and gas through porous rocks. Whereas porosity determines the amount of fluid present, permeability controls the amount recoverable. Permeability may be defined operationally as the volume of fluid of unit viscosity which passes through a unit cross-sectional area in unit time under a unit pressure gradient; it is expressed in terms of the *darcy*, with dimensions $cm.^2$ A thorough discussion of the theory and methods of permeability measurement has been presented in a paper by Fancher, Lewis, and Barnes.[22]

Permeability is dependent on effective porosity; therefore, it is controlled by the properties which condition porosity, that is, by the degree of cementation and the degree of packing (in turn controlled by the attributes of size, shape, space arrangement, and fabric). Unlike porosity, however, permeability is a function of the average or mean size of the grains as well as the sorting as to size; the smaller the mean size the lower the degree of permeability. Thus, permeability is a function of all the attributes which determine packing and porosity (see discussion of packing) and *in addition* is a function of mean size of the grains.

E. SEDIMENTARY STRUCTURES

Sedimentary structures comprise the large-scale or mass features of sedimentary deposits. Unlike attributes and properties, which depend on the character and interrelationships of individual components, structures involve the relationships of masses or aggregates of components. Structures, such as stratification, mud cracks, and so forth, are conditioned by time and space (vertical and lateral) changes of discrete character in the fundamental sedimentary attributes. For example, stratification may be the resultant of discrete vertical (time) changes in composition, and (or) grain size, and (or) fabric.

Structures are of two temporal types: (1) *syngenetic*, that is, structures formed approximately with and at the time of deposition of the sedimentary aggregate; and (2) *epigenetic*, those formed after deposition and burial. The following outline presents a classification of sedimentary structures based on their character and their time and mode of genesis.

I. SYNGENETIC STRUCTURES
- A. Stratification
  - 1. Parallel type
  - 2. Cross-stratification type
- B. Organic Structures
  - 1. Bioherms ("reefs") of various types
  - 2. Biostromes of various types

[22] G. H. Fancher, J. A. Lewis, and K. B. Barnes, "Some Physical Characteristics of Oil Sands," *Pennsylvania State College Bull. 12* (1933), pp. 64–167.

- C. Markings
  - 1. Imprints of rain drops, *et cetera*
  - 2. Tracks, trails, and burrows
    - a. Made by organisms
    - b. Made by inanimate objects
  - 3. Ice-crystal impressions
  - 4. Wave and current marks
    - a. Ripple marks (aqueous and aeolian)
    - b. Swash marks
    - c. Rill marks
    - d. Beach cusps
- D. Contemporaneous Deformation Structures
  - 1. Tensional features (mud cracks)
  - 2. Compressional features, including folded and faulted structures caused by surficial thrust action of ice and other surface agents
  - 3. Gravitational features, including structures formed by sliding and gliding of sediments down sloping surfaces

II. EPIGENETIC STRUCTURES

- A. Solution Structures
  - 1. Stylolites
- B. Deformational Structures
  - 1. Compressional
    - a. Cone-in-cone
    - b. Deformation resulting from recrystallization
    - c. Differential compaction structures
    - d. Folds, faults, *et cetera*
  - 2. Tensional
    - a. Faults, joints, *et cetera*

A discussion of stratification structure is presented in the succeeding section, and organic structures are treated under the section on paleontologic analysis. The other types of structures need not be reviewed here inasmuch as they have been covered in numerous papers and texts. The crinoidal limestone of this report does not display much development of structures other than stratification, there being no discernible organic structures, markings, contemporaneous deformation features, solution structures, and so forth. Epigenetic deformational structures, however, are represented, inasmuch as strata of the Grand Tower and other Devonian and earlier formations of the southeastern Missouri region have been severely fractured and tilted by faulting.

STRATIFICATION

Definitions.—A *stratum* is defined as a layer, greater than 1 cm. in thickness, that displays continuous (non-discrete) variation in lithologic character and is visually separable from other layers above and below, the separation being determined by a discrete change in lithology, a sharp physical break in lithology, or by both. A *lamina* is similar to a stratum except that it is less than 1 cm. in thickness. The term *bedding plane* denotes the "plane" of separation between laminae or strata; it may represent the contact of lithologically different strata or laminae, without a sharp physical break, or it may represent a physical break resulting from a period of non-deposition (diastem) or erosion.

The term *bed* is somewhat loosely employed but usually denotes a unit of sedimentary rock, of lower than member or formation rank, composed of two or more strata or laminae that manifest some degree of lithologic unity. The component layers may be of the same composition and texture or may consist of two or more related lithologic types.

There are several miscellaneous terms used in describing stratification structure. The term *band* is applied to a stratum or lamina conspicuous because it differs in color from adjacent layers; a group of layers displaying color differences is described as being *banded*. The term *seam* is reserved for layers of coal enclosed above and below by other types of sedimentary rock. The term *parting* denotes a lamina occurring between massive strata different in lithology from the parting itself. A *varve* is an annual layer containing a record of seasonal environmental variation in the form of lithologic change.

Types of stratification.—The types and origin of stratification have been described in a classic paper by Andree.[23] The following outline presents in modified form the essential features of Andree's classification plan.

- I. Physical Break (Diastem) *without* Apparent Lithologic Change
  - A. Cross-stratification type
  - B. Parallel-stratification type
- II. Lithologic Change *without* Apparent Physical Break (Diastem)
  - A. Parallel stratification type
    - 1. Gradational ("graded-bedding")
      - a. Cyclical (glacial-lake varves)
      - b. Non-cyclical
    - 2. Non-gradational stratification
      - a. Cyclical
      - b. Non-cyclical
- III. *Combined* Lithologic Change and Physical Break (Diastem)
  - A. Cross-stratification type
  - B. Parallel-stratification type
    - 1. Gradational ("graded-bedding")
      - a. Cyclical
      - b. Non-cyclical
    - 2. Non-gradational stratification
      - a. Cyclical
      - b. Non-cyclical

The *physical break* (diastemic) type of stratification, marked by sharp bedding-plane breaks, is the resultant of cessation of sedimentation with or without erosional scour. On the other hand, the *lithologic change* type of stratification results from discrete vertical (time) changes in one or more of the fundamental sedimentary attributes, from vertical changes in composition, and (or) size, and (or) shape, and (or) fabric, and so forth. In addition to being represented by vertical lithologic change in any one genetic type (preëxisting rock, chemical precipitate, organic, or volcanic), stratification may be represented by vertical change from one genetic type to another; for example, change from strata of preëxisting rock to organic origin, or from organic to chemical precipitate origin.

British geologists, especially E. B. Bailey,[24] have been more concerned with studies of graded-bedding than have geologists in the United States. The term *graded-bedding* should be used in describing sedimentary deposits each stratum

[23] K. Andree, "Ursachen und Arten der Schichtung," *Geol. Rundschau*, Vol. 6 (1915), pp. 351–97.

[24] E. B. Bailey, "New Light on Sedimentation and Tectonics," *Geol. Mag.*, Vol. 67 (1930), pp. 84–90.

of which displays a gradation in grain size from coarse below to fine above. Thus, a sandstone with graded-bedding may be composed of strata, each of which grades from coarse sand at the bottom to fine sand at the top, or from fine sand at the bottom to silt at the top, and so forth. The bedding planes between gradational strata of this type are marked by rather abrupt lithologic (size) change and may or may not be marked by sharp physical breaks (diastems).

Thickness of layers.—This important aspect of stratification must be evaluated in petrographic analysis. Stratified deposits may be classified in four descriptive categories on the basis of the thickness of their layers.

1. *Fissile*, if deposit consists of laminae less than 2 mm. in thickness
2. *Shaly*, if stratification is in form of laminae from 2 to 10 mm. in thickness
3. *Flaggy*, if layers are in form of strata from 10 to 100 mm. in thickness
4. *Massive*, if strata are greater than 100 mm. in thickness

Thickness differences of layers in a vertical section of a given deposit should be described, as should also lateral variation in thickness of individual layers. Deposits of certain environments are composed of layers of nearly identical thickness in vertical section; other environments produce layers of different thickness, which may alternate in cyclical pattern or may show random vertical thickness variation. The neritic environment of epeiric seas tends to produce layers of marked uniformity in thickness over large areas, whereas other environments, such as river channel, deltaic, and littoral, produce sediments the layers of which show little vertical similarity and lateral continuity in thickness.

Form of bedding surfaces.—Bedding surfaces may be even, that is, in the form of regular planes; or they may be irregular—undulating, ripple-marked, wave- and current-marked, hummocky, and so forth. If they are irregular, details of form and dimensions of features should be described. With relation to each other, bedding planes may be parallel, non-parallel, convergent, and so forth.

Initial dip.—This aspect of stratification represents the angle of slope of bedding surfaces at the time of deposition, the contacts between layers usually being approximately parallel with the surface of deposition unless subsequently altered by differential compaction or other deformational processes. Initial dip may vary in degree from horizontal or a low angle up to the angle of repose for the sedimentary material in question. It is of particular significance in description of cross-stratification structures such as those in dunes and deltas, in description of strata flanking reefs or bioherms, volcanic masses, and so forth.

Stratification of crinoidal sediment.—The crinoidal limestone described in this paper has parallel stratification of the type involving physical breaks (diastems) without apparent lithologic change. The crinoidal strata vary in thickness from flaggy to massive, averaging approximately 1 foot (massive), and tend to be uniform in thickness in their lateral extent. The bedding surfaces are even, and little if any initial dip is represented.

F. LITHOLOGIC TYPE

The stratigraphic record is characterized by recurrence in time and space of lithologic and environmental *types*. *Lithologic classification* is more fundamental than and should be prerequisite to stratigraphic classification; little attention, however, has been directed to this field and there is no unanimity of usage. As pointed out by Tieje, there is an urgent need for simple *common names* which may be used to designate lithologic types and to represent their mode of genesis and their environment of deposition. The new stratigraphy must include a break-down of the stratigraphic record into lithologic types and their varieties and must provide widely acceptable lithologic common names.

Some lithologic names already are in widespread use; examples are provided by the terms arkose, graywacke, novaculite, clay ironstone, chalk, bentonite, travertine, tufa, tripoli, and coal. It still remains true, however, that our three most commonly used stratigraphic terms, *shale*, *sandstone*, and *limestone*, have little lithologic meaning when standing alone. The term *shale* is only of structural significance; it tells nothing of the composition, grain size, *et cetera*, of the rock inasmuch as shales may consist of clay minerals, limestone, quartz siltstone, fine sandstone, and so forth. Similarly, the term *sandstone* is of little more than grain-size significance; it fails to indicate the composition, structure, *et cetera*, of the rock inasmuch as sandstones may have shaly structure and may be composed of quartz, glauconite, and other minerals or rock fragments. Furthermore, even limestones, such as the crinoidal limestone of this report, may represent sandstone deposited as fragmental-textured sandy sediment later to be altered to crystalline texture. The term *limestone*, like sandstone and shale, is of limited lithologic significance, referring only to composition. Thus, it seems that the three lithologic names most widely employed in stratigraphic description have diverse petrographic implications, one based on the attribute of composition, another on the attribute of grain size, and a third on stratification structure.

Crinoidal limestone constitutes a fundamental lithologic type especially characteristic of a number of mid-Paleozoic stratigraphic divisions of North America. Tester[25] has suggested that the common name *criquina* be assigned to this lithologic type. The crinoidal limestone described in this paper occurs in a finely crystalline limestone and sandstone *association* in the Devonian Grand Tower formation of Missouri and Illinois, and it displays quartzose, glauconitic and heavy-mineral varietal features. The same lithologic type occurs in numerous other stratigraphic associations; that is, it is associated with numerous other lithologic types. For example, crinoidal limestone occurs as thin beds in the Hamilton shale sequence of central and western New York where it likewise shows distinctive varietal features. The type is also prominent where it flanks and dips away from the reef cores in the reef division of the Traverse beds of eastern Michigan and in the reef division of the Onondaga limestone of western New York and Ontario; in this reef associationship it shows distinctive varietal characters in that it is very coarse-grained, contains an abundance of detrital bryozoan fragments, and so forth. Numerous other association examples could be cited.

[25] A. C. Tester, *Guide Book Ninth Annual Field Conference Tri-State Geological Association* (1941), p. 6.

III. Paleontologic Analysis

In their studies of fossil assemblages paleontologists have been primarily interested in taxonomy and the description of species. Little attention has been given the ecologic aspects of the faunal assemblage picture. Paleontologic analysis should include description of those features of the fossil population which provide information for environmental interpretations; such features are treated in the following sections (A-D).

At the present time there is urgent need for quantitative studies of fossil populations, not only with regard to ratios of different species and classes but also with regard to size distributions, growth form variations, and other morphologic features. Statistical methods and approaches that might be employed in this new field of *paleobiometry* are presented in the book *Quantitative Zoölogy* by Simpson and Roe.[26]

A. Biocoenosis and Thanatocoenosis

The work of marine biological stations has shown that marine organisms exist as ecologic communities, known to ecologists as *biocoenoses*. The biocoenosis concept also applies, of course, to fresh-water and terrestrial organic communities. A *biocoenosis* has been defined by Hesse, Allee and Schmidt as an "association of living things which inhabit an environmentally uniform division of the biosphere and correspond in the selection and number of species with the average external habitat conditions."[27] The members of a biocoenosis are dependent on each other and are in a state of biological balance which is self-regulating and fluctuates about a mean. Some biocoenoses lack a plant foundation and are therefore for the most part dependent on the outer world for their food supply; others form virtual closed systems and are nearly self-sufficient. The Devonian crinoid-coral community, a record of which is contained in the crinoidal limestone of this report, must have depended largely for its food supply on abundant pelagic life wafted to it by marine currents.

The paleoecologist must make a clear mental distinction between a biocoenosis or *life assemblage* and a thanatocoenosis or *death assemblage*. The term thanatocoenosis was introduced by Wasmund[28] and is applied to a death association of organic remains, the components of which were either brought together after death or remained on the scene of their former life. As shown by the diagram of Figure 6, all fossil assemblages have passed through the thanatocoenotic stage inasmuch as they represent remnants of ancient death assemblages. Thanatocoenoses are of three general types (see Fig. 6).

[26] G. G. Simpson and A. Roe, *Quantitative Zoölogy*. McGraw-Hill Book Company, Inc., New York (1939).

[27] R. Hesse, W. C. Allee, and K. P. Schmidt, *Ecological Animal Geography*, p. 137. John Wiley and Sons, Inc., New York (1937).

[28] E. Wasmund, "Biocoenose and Thanatocoenose," *Archiv. f. Hydrobiologie*, Bd. 17 (1926), pp. 1–116.

1. Those composed largely of biocoenotic elements, that is, remains of organisms which lived on the scene and thus record the local environment of deposition; the fossil assemblage of the crinoidal limestone which this paper describes illustrates this type
2. Those composed of both biocoenotic (local) elements and foreign elements brought in from other environments, the latter having been transported in either after death or before to die on the site of accumulation of the sediment
3. Those composed primarily of foreign elements derived from other environments and biocoenoses; this type tells little if anything about the environment of deposition of the sediment in which it occurs and is illustrated by black shale assemblages of planktonic graptolites and by assemblages of pelagic protozoan remains such as Globigerina ooze

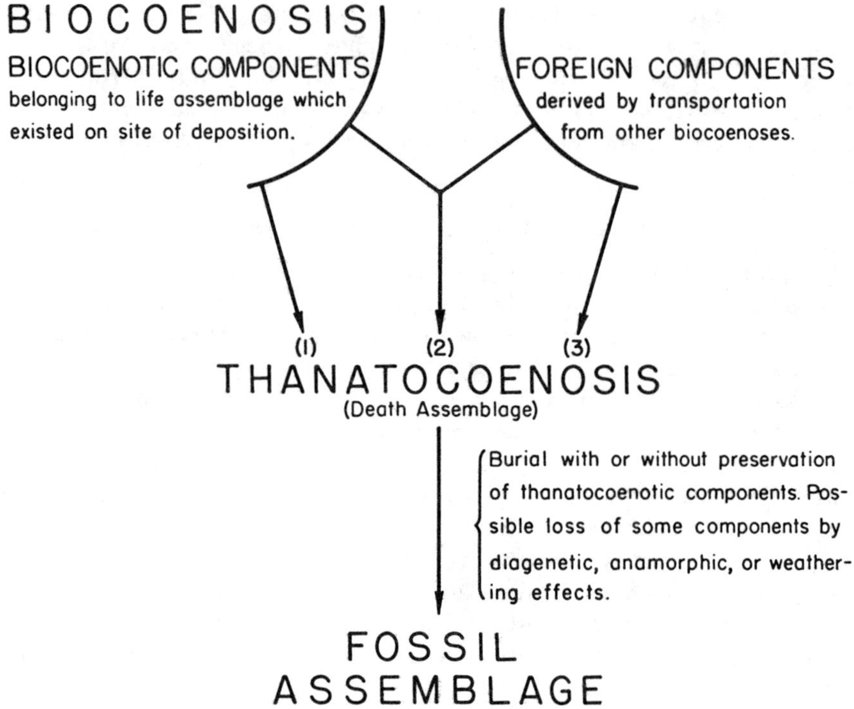

FIG. 6.—Diagram illustrating three thanatocoenotic types; and relationship between biocoenosis, thanatocoenosis, and fossil assemblage.

During life an organism is restricted to certain ecologic conditions and is thus representative of its environment, but after death its remains enter the mechanical realm or the chemical realm (are dissolved). In the mechanical realm the remains either stay in place (sedentary) or are transported in the same manner as are inorganic sediments. These transported remains or organic sediments may ultimately come to rest in deposits of environments to which the organisms are in no way related. As pointed out by Twenhofel,[29] many organisms live, die, and are buried under the same environmental conditions, but others die and are buried far from their place of life and may be entombed neither where they lived

[29] W. H. Twenhofel, *Principles of Sedimentation*, p. 153. McGraw-Hill Book Company, Inc. New York (1939).

nor died. In restoring the environment of a given deposit and its fossil assemblage a paleoecologist must isolate and consider the *biocoenotic species alone.*

B. CHARACTER OF BIOCOENOSIS

Component organisms of a biocoenosis have certain fundamental attributes which should be described and evaluated; these attributes also apply to fossil components of an extinct biocoenosis and should be considered in paleontologic analysis. The organic attributes are analogous to the previously described attributes of component grains of a sedimentary aggregate. The morphologic type (class, genus, and species) is more or less analogous to the sedimentary attribute of composition (mineral species, *et cetera* ; the attribute of growth form is somewhat analogous to the sedimentary attribute of component shape; and so forth. The organic attributes are treated in Table IV and in succeeding sections of this paper.

The character of the original life assemblage represented by fossil remains in the Grand Tower crinoidal limestone are described in Table IV. The crinoidal debris, although transported to some extent, undoubtedly was derived from the immediate environment and does not represent foreign material brought in from other environments; this is strongly indicated by the fact that some of the columnals have remained together in the form of undissociated stem sections and by the fact that crinoidal remains constitute the major portion of the deposit. Therefore, it seems that crinoids were part of the biocoenosis; in fact, they must have represented the dominant morphologic type.

The crinoidal limestone of this report does not lend itself readily to paleontologic analysis inasmuch as the coral, brachiopod, and other remains are poorly preserved in the coarsely crystalline rock. Furthermore, the fauna has not as yet been extensively sampled and studied. Table IV presents only a rough, preliminary picture of the life assemblage, and the paleontologic analysis phase of the present report is thus incomplete.

Morphologic composition.—The taxonomic position (class, genus, species, *et cetera*) of a given organism or fossil is its most important descriptive attribute and is rather analogous to the attribute of composition (mineral species, *et cetera*) of sedimentary components. When considered from the viewpoint of the aggregate of organisms of a biocoenosis, this attribute calls for a faunal and floral list naming the morphologic types present in terms of species and (or) genera and giving their relative abundance in the total population. Relative abundance may be estimated in the field by simple counting and ratio-computing methods and expressed in terms of the following arbitrary percentage classes: less than 1, 1–5, 5–15, 15–30, 30–50, 50–70, 70–90, and greater than 90. The morphologic types present in the fossil population of the illustrative crinoidal sediment are listed in Table IV simply in terms of genera; more extensive sampling and study at a later date will permit species designation.

Internal structure.—This attribute is important in taxonomic description but is not discussed here inasmuch as it is not particularly significant from the environmental viewpoint.

TABLE IV

CHARACTER OF BIOCOENOSIS REPRESENTED BY FOSSIL REMAINS IN ILLUSTRATIVE CRINOIDAL SEDIMENT

*Morphologic Type**	*Per Cent*	*Size*	*Growth Form*	*External Condition*	*Space Arrangement*	*Orientation*
TABULATE CORAL COLONIES *Favosites hemisphericus*	1–5	Diameter, 6–10 cm.; height, 4–12 cm.	Variable colony shapes	No signs of wear or breakage; poorly preserved	Tendency toward clustering of colonies along horizons	Upright, in life position; unmoved
INDIVIDUAL CUP CORALS *Heliophyllum sp.* Zaphrentids (several species) *Cystiphyllum americanum*	1–5	Calicular diameter, 2–5 cm.; height, 4–8 cm.	Variable cup shapes: conical, curved trochoid, irregular	No signs of wear or breakage; poorly preserved	Random; commoner along certain horizons	Random; upright to prone with axes trending in various directions
CRINOIDS Dissociated remains unidentifiable	70–90	Small size; columnal diameter, ½–3 mm.		Mostly dissociated to individual plates. Some columnals still together in form of stem sections	Omnipresent, random	
BRACHIOPODS *Atrypa reticularis* *Spirifer* (several species) Fragmentary evidence of other genera	1–5			Unbroken; no signs of wear; poorly preserved	Random; common along certain horizons	Probably unmoved and in life position
TRILOBITES *Phacops sp.*	–1	Length approx. 5 cm.		Very poorly preserved		
PELECYPODS AND GASTROPODS (fragmentary evidence)	–1					

* Extensive faunal collecting and study will expand this list and permit species designation.

Size.—Size is an attribute of an *individual* of a species as well as of a species *population* (aggregate) and is significant in environmental interpretation. Size and its evaluation are presentable in terms of a size-frequency distribution, which carries with it the aspects of *range* of variation and *average* size of a given species. Large range denotes poor size sorting and small range good sorting, these measures being ecologically significant.

Dwarf species are indicative of abnormalities of environment, with regard to such factors as salinity, temperature, turbidity, dissolved gases (H_2S *et cetera*), and others well summarized in Shimer's paper on dwarf faunas.[30] As pointed out by Shimer, there are two types of dwarf faunas: (1) faunas in which the individuals are of smaller size than that which the species normally exhibits and are thus stunted; this is the resultant of an abnormal, unfavorable habitat, and (2) faunas in which all the individuals are small but of the normal size for the species represented; in this case selective factors have weeded out the large species.

The second type of dwarf fauna defined above involves another aspect of the attribute of size; that is, that one must consider not only the size distribution of each species in the aggregate but also the size distribution of the aggregate or biocoenosis itself. For example, some biocoenoses, such as those associated with reefs in the Onondaga limestone of western New York, are composed almost entirely of large, heavy-shelled species, whereas other biocoenoses, such as those represented in the calcareous shales of the Hamilton group, tend to be composed largely of small, thin-shelled forms. Thus the *aggregate size* concept is one which is extremely significant to the ecologist.

Growth form.—The attribute of growth form is important in ecologic studies because it may reflect the impress of environment on the morphology of organisms. Certain taxonomic groups, such as corals, show much more individual variation in growth form than do other groups. For example, the Paleozoic rugose coral genera *Heliophyllum* and *Cystiphyllum* display an amazing amount of growth form variation which is to a large extent indicative of the environmental setting in which the corals thrived. Colonial corals likewise show environmental morphologic differences, well exemplified by the genus *Favosites* which abounds in and adjacent to the Onondaga limestone reefs of western New York. The same species of *Favosites*, occurring both in the reef cores and in the detrital beds flanking and dipping away from the reefs, tends to show contrasted growth forms in these different but adjacent environments. In the reef cores, where upward growth was unhindered, the colonies tend to display a high domal outline, the height in some cases exceeding the basal diameters. In the crinoidal limestone flanking beds, however, where upward growth was smothered by organic detritus swept from the reefs, the colonies stand as broad, low tabular bodies, ranging up to several feet in diameter and but a few inches in thickness.

Like the attribute of shape of sedimentary particles, growth form is an attribute of an *individual* as well as of an *aggregate* of individuals belonging to a species. In the latter case, the factors of sorting, range of variation, and average condition enter the growth-form picture. Some species may in the aggregate show good sorting as to growth form; that is, they have a small range of variation. On the other hand, certain species evince poor sorting and have a wide range of form; in this case it is important that the paleoecologist record the average or most typical condition as well as the range of variation.

Other organic attributes.—Three other biocoenotic attributes of importance in environmental interpretation are treated in Table IV; they are: (1) external con-

[30] H. W. Shimer, "Dwarf Faunas," *Amer. Naturalist*, Vol. 42 (1908), pp. 472–490.

dition of the shells and other hard structures with respect to wear, breakage, healing, and so forth, (2) space distribution or arrangement within the aggregate of the different morphologic types, and (3) orientation (fabric) of the shells with respect to the bedding planes and compass direction. These ecologically significant attributes have been largely ignored by paleontologists and call for detailed study and description of the type carried on by sedimentologists on the analogous sedimentary attributes.

Population and fossil density.—The density concept, employed by ecologists working with living populations, is a valuable tool in environmental restoration and may be applied to ancient biocoenoses represented by fossil remains. The population density of a biocoenotic area is a measure of the total mass of animal substance present and is determined primarily by the available food supply. Each area has the greatest mass of life that it is capable of producing and supporting. Two environmentally different areas having different organic assemblages but quantitatively similar food supply conditions will ordinarily have population densities of the same order of magnitude.

The population density of an ancient environment is difficult to ascertain from fossil evidence inasmuch as some environments do not allow total incorporation within the stratigraphic record of remains of their plant and animal life, the remains being transported elsewhere or destroyed in place. Some rock types almost devoid of fossil remains may have been associated, in the environment of deposition, with a high population density; thus, *fossil density* is not an accurate measure of *population density* but in most cases presents a crude quantitative picture of the life assemblage. Fossil density is expressed in per cent and represents the proportion of the total rock aggregate which is composed of organic remains. The crinoidal phase of the Grand Tower limestone has a fossil density of approximately 85 per cent.

Class and species densities.—In biocoenotic analysis population density should be distinguished from class and species densities, the latter two representing a measure of the number of taxonomic classes and species in the biocoenosis. Whereas the amount of available food determines the population density, class and species densities are determined primarily by other ecologic factors, such as salinity, temperature, and character of the substratum. Optimum conditions with respect to these factors favor speciation and permit high species and class densities. High densities imply poor environmental sorting as to the taxonomic classes and species represented, whereas low densities imply good environmental sorting. These densities may be evaluated in terms of the number of species and classes present in the organic assemblage under consideration.

Stratigraphic application of density concept.—A survey of the faunal assemblages of the stratigraphic record strongly suggests that application of the concept of densities to the study of faunal assemblage types may provide information pertinent to environmental restoration. The stratigraphic record affords numerous examples of interesting density relationships. For example, certain of the coral biostromes in the Hamilton group of western and central New York show high fossil (population) density, low class density, and medium to high species density, in that they are composed almost entirely of numerous species of corals. Examples of such Hamilton coral biostromes are provided by beds in the Centerfield member at several localities and by a bed in the Windom shale member in the vicinity of Leicester and Geneseo, New York.

A further illustration of the density concept is provided by the *Cystiphyllum* beds in the reef division of the Onondaga limestone of western New York and Ontario; these biostromes are composed almost exclusively of but one species of the rugose coral genus *Cystiphyllum* and may be described as having high fossil (population) density and low class and species densities. On the other hand, certain calcareous beds in the Hamilton shales of western New York display low fossil density, and high class and species densities. Finally, some deposits, such as the Jeffersonville limestone in the Falls of the Ohio district, display high densities from all three points of view.

C. ORGANIC STRUCTURES

Organic structures are of two types: (1) bioherms or "reefs" and (2) biostromes. The essence of these structures is that they consist for the most part of untransported remains of sedentary organisms. The illustrative crinoidal sediment of this report shows no organic structure (bioherms or biostromes), even though composed largely of organic remains, because the crinoidal debris is not in place but has been transported and deposited as a mechanical sediment.

The term *bioherm* has been defined by Cumings and Shrock so as to include "reeflike, moundlike, lenslike, or otherwise circumscribed structures of strictly organic origin, embedded in rocks of different lithology."[31] Bioherms vary greatly in size and shape and as to types of constituent organic remains. They may be composed in part or entirely of algal colonies, coral colonies, stromatoporoid colonies, crinoid remains, brachiopods, and so forth. The Tepee Buttes in the Pierre shale represent small bioherms composed of mollusc shells, the pelecypod *Lucina*. Thriving oyster bioherms are situated at the mouth of the Newport River, North Carolina. As illustrated by Laudon and Bowsher,[32] crinoid bioherms may be of great size.

Some organic structures of biohermal type are small, ranging down to a few feet or even a foot or less in diameter and height. These miniature bioherms consist of masses of organic remains embedded in lithologically different rocks and may be composed of bryozoans, crinoids, brachiopods, small corals, gastropods, and other fossil types; some are composed of but one fossil type whereas others represent a hodge-podge composite of several types. Some miniature bioherms are globular in shape, in which case they are known as "ballstones."

[31] E. R. Cumings, "Reefs or Bioherms," *Bull. Geol. Soc. America*, Vol. 43 (1932), p. 333.

[32] L. R. Laudon and A. L. Bowsher, "Mississippian Formations of Sacramento Mountains, New Mexico," *Bull. Amer. Assoc. Petrol. Geol.*, Vol. 25 (1941), pp. 2126–29, 2140–44.

The term *biostrome*, according to the definition of Cumings,[33] includes "purely bedded structures, such as shell beds, crinoid beds, coral beds, *et cetera*, consisting of and built mainly by sedentary organisms, and not swelling into moundlike or lenslike forms. . . . " The bedded character of a biostrome thus distinguishes it from the lenslike bioherm. Many of the so-called "reefs" of the stratigraphic literature, including the famous coral "reefs" of the Falls of the Ohio region, are not truly reefs (bioherms) but are rather of the biostromal or bedded type. Biostromes may be composed of the remains of individual and colonial corals, algae, brachiopods, pelecypods, and so forth. Most coal beds belong in this category. Many organic deposits, including some shell beds, coquinas, and crinoidal limestones, are not true biostromes because of the fact that the organic remains represent transported detritus, not in position of original growth.

D. BIOCOENOTIC TYPE

The concept of biocoenotic type is analogous to the previously described concept of lithologic type. Just as a sedimentary deposit is an assemblage of sedimentary components representing a response to external conditions and to each other, so a biocoenosis stands as an assemblage of coexisting and ecologically related animal and plant species likewise representing a response to external conditions and to each other. At the present time there is need for a breakdown of fossil faunal and floral assemblages into generalized types. Paleontologists have long been concerned with taxonomic classification but have given little attention to classification of assemblages.

The crinoidal limestone of this report contains a *crinoid-coral-brachiopod* biocoenotic type which occurs in Devonian rocks in many regions. This does not imply that different manifestations of this type contain the same genera and species but only that similar morphologic classes are represented. The crinoid-coral-brachiopod type is commonly associated in mid-Paleozoic rocks with the crinoidal limestone or "criquina" lithologic type. Thus there is an environmental relationship between the life and lithologic assemblages.

Certain biocoenotic types, such as the one cited above, recur vertically and laterally in the stratigraphic record, and it would seem that they existed wherever their ecologic requirements were satisfied and no migration barriers obtained. From a narrow point of view a biocoenosis represents an acutal community or life assemblage inhabiting a restricted area at a given time; an assemblage thus restricted represents a *faunule*, that is, a local concrete expression of a biocoenotic type. From a broad viewpoint a biocoenosis is different in that it may recur in time and space in somewhat different forms. The different faunules differ with respect to the component species represented, with respect to ratios of species, to size and growth form of the species, and so forth. In spite of this variation, however, they retain a degree of unity and distinctiveness, which makes it possible to classify them as belonging to the same parent biocoenotic type. For example, dif-

[33] Cumings, *op. cit.*, p. 334.

ferent tide-pool localities of the New England coast have different *faunules;* these, however, have much in common and may be classified together as belonging to the tide-pool biocoenotic type.

IV. Sedimentary Petrogenesis

The interpretive study of *genesis* of sedimentary deposits is divisible into the various fields here outlined, and discussed in subsequent sections of this paper.

A. SEDIMENTARY ECOLOGY (Environmental Analysis)
   1. *Syngenetic phase,* dealing with the surficial environment of deposition and accumulation of a deposit
   2. *Progenetic phase,* dealing with the surficial environmental history of the components of a deposit *prior* to their incorporation in the deposit; this treats their history from the source site to the site of final deposition
   3. *Epigenetic phase,* dealing with the subsurface environmental history of the deposit subsequent to the time of deposition and incorporation in the stratigraphic record

B. SEDIMENTARY DYNAMICS (Processes and Resultants)
   1. *Syngenetic phase,* dealing with the surficial sedimentation *processes,* inorganic and organic, which are involved in the deposition and accumulation of a deposit and which determine its attributes, properties, and structures
   2. *Progenetic phase,* dealing with the gamut of sedimentation processes and resultants affecting the components of a deposit during their trek from point of origin to point of final accumulation and incorporation
   3. *Epigenetic phase,* dealing with the subsurface processes and resultants affecting the attributes, properties and structures of the deposit subsequent to the time of deposition and incorporation in the stratigraphic record; this includes the processes of diagenesis anamorphism, and weathering

Four genetic controls.—The components of the sedimentary record owe their existence in a deposit to the functioning of four controls, the sequential operation of which is prerequisite to the presence of a component, inorganic or organic, in a sedimentary rock. These four controls are expounded in Figure 7.

There is a tendency on the part of stratigraphers to overlook one or more of these four controls. It is commonly supposed, for example, that similar sedimentary environments (control II) obtaining in different areas will lead to the genesis of similar rock types in these areas; that this supposition does not logically follow is indicated by the following illustration. It is a well known fact that silica present in marine waters is absorbed on clay particles during deposition of a mud or shale. If this be so, then it appears that the most favorable conditions for true silica precipitation would obtain during a period of an abundant dissolved silica supply and clear, clay-free water; such conditions must have obtained during the deposition of the great Clear Creek chert section of the Devonian of the Mississippi Valley region. The question arises as to what would have happened in Clear Creek time if suspended clay had been available (control I) in the marine waters, all the other environmental factors (control II) remaining the same (weak currents *et cetera*). Would not a shale section have been deposited under the same environmental setting, the abundant silica being absorbed and deposited with the clay? Thus it would seem that by holding the environmental control constant and varying the supply control two entirely different rock types might be generated.

Another case in point is the accumulation, under similar environmental con-

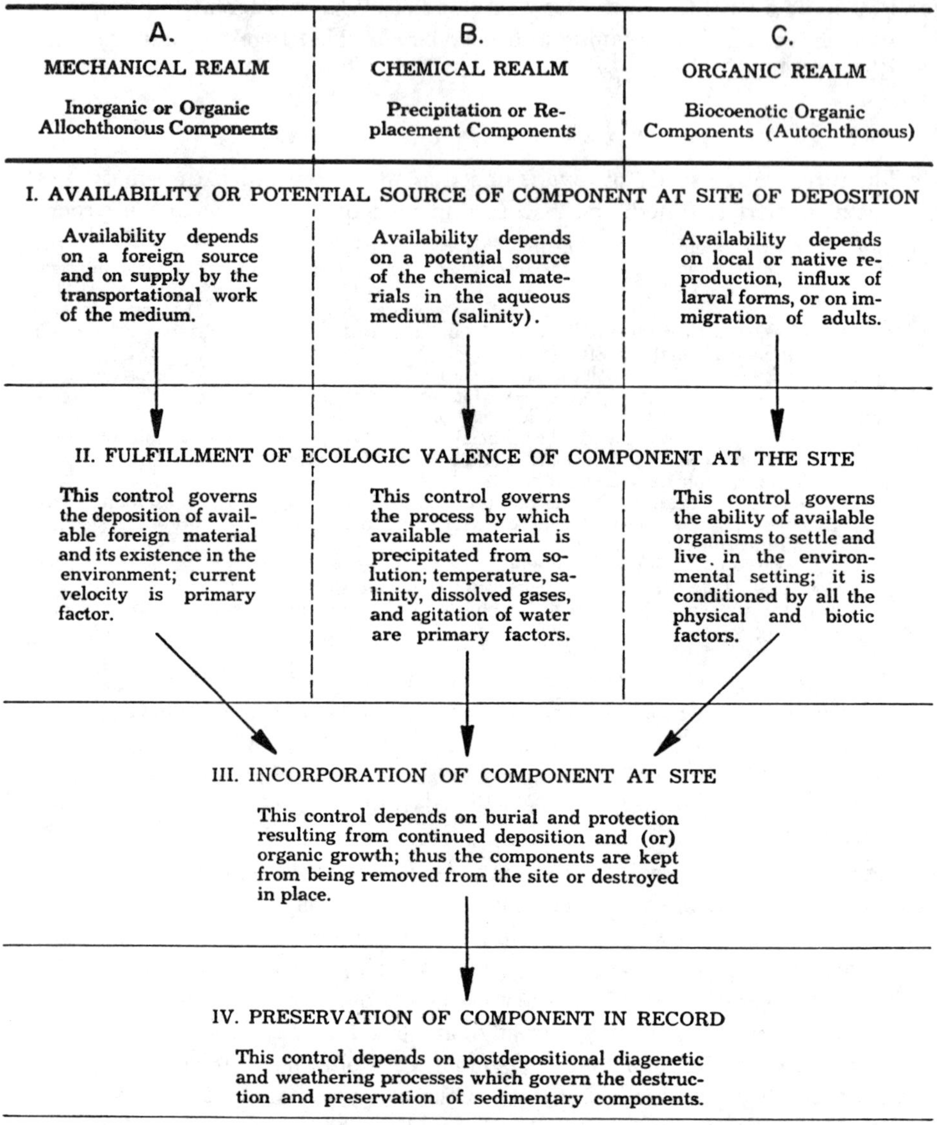

FIG. 7.—Four genetic controls governing existence of sedimentary components in a deposit.

ditions, of a coarse quartz sandstone and a crinoidal limestone, both of which might be cross-stratified and in general indicative of strong current conditions; in the one case terrestrial quartz sand is provided and in the other detritus from flourishing crinoidal growth. This citation provides a clear-cut illustration of a fundamental but seldom recognized sedimentational law: *the sea will tend to pro-*

duce, under given environmental (valence) conditions, sediments of similar textural character regardless of differences as to source and type of material being supplied. The expression *similar textural character* implies that the fundamental attributes, properties, and structures will be similar, excepting those which are functions of mineral composition.

A concrete stratigraphic expression of the quartz sandstone-crinoidal limestone relationship cited above is provided by the basal portion (Croneis' Bed 9)[34] of the Grand Tower limestone of central and eastern Missouri and southwestern Illinois; this stratigraphic division is believed to vary locally from practically pure crinoidal rock, to quartzose crinoidal limestone, to crinoidal quartz sandstone, to quartz sandstone (Dutch Creek sandstone). The particular phase of Bed 9 described in the present paper is from the Little Saline fault complex of eastern Missouri and represents a quartzose crinoidal *phase* of this system of continous lithologic variation. Had crinoids failed to flourish on the scene, a quartz sandstone like the Dutch Creek would have been the resultant. This fact indicates that the supply or availability control (I) is the determinative one in the Bed 9 system of lithologic variation; the environmental control (II) probably was rather constant throughout the area of basal Grand Tower deposition.

The supply or availability control likewise is functional in the organic realm and, as indicated in Figure 7, is dependent on local reproduction, influx of larval forms, or immigration of adult forms. Just as there may be barriers to the dispersal of sediments in the sea, so may there be barriers to the dispersal of organic species, and these may be of several types. Actual geographic barriers probably are most effective in limiting organic migration and distribution, but others also may operate, such as areas with unfavorable conditions as to temperature, salinity, and other physical factors as well as biotic factors. Furthermore, current conditions on which larval forms are largely dependent for their distribution, may rule out the introduction of species to ecologically favorable areas where they would thrive if the availability control were functional. Paleontologists, in their studies of distribution of fossils and correlation, commonly fail to consider the possibility of the effects of these positive or negative barriers. Two separated but *ecologically similar* areas, on the same time horizon, may have different biocoenotic populations (different species and genera) because of availability (control I) differences alone. In spite of these superficial differences the two populations should be morphologically and structurally analogous inasmuch as the same rule cited above for sediments applies equally to organisms: *under given environmental conditions the sea will tend to produce organic assemblages of similar nature (similar biocoenotic type)* regardless *of differences as to species and genera available.*

Control III, that of *Incorporation,* is of vital significance in that it governs the induction of components into the stratigraphic record; if it fails to function, the components are either removed from the environment of deposition or destroyed on the site. Increased current velocities during times of storm may remove com-

[34] Croneis, *op. cit.*

ponents from the environmental scene and thus prevent incorporation. There are many environments, teeming with life, in which the conditions are such that organic remains are not incorporated and protected. As pointed out by Twenhofel,[35] wherever bottom conditions favor the existence of organisms which eat mud and bore into shells, all remains are apt to be broken and ultimately reduced to powder. Organisms which accomplish this work of destruction include certain boring sponges, annelids, fishes, crabs and other crustaceans, and some bivalves, echinoids, and holothurians. Incorporation is best accomplished in environments in which deposition or growth of organisms is extremely rapid or in which the bottom is unfavorable for scavengers and borers, the latter being true of environments of deposition of black shales. The low fossil (population) densities of most black shales is due to unfavorable environmental conditions for life (control II) and not to the incorporation control, which in this case operates favorably.

The *Preservation* control (IV) is the last in the sequence which conditions the existence of a sedimentary component. This control depends on the various diagenetic, weathering, and anamorphic processes by which components may be destroyed or altered in a subsurface environment. The preservation control is of vital importance to the paleoecologist engaged in restoring ancient life environments inasmuch as the character of a fossil assemblage is partially dependent on its diagenetic history. Fossil remains undestroyed by passage through the surficial and transitional realms (Fig. 8) are either obliterated in the subsurface realm or preserved by one of the several fossilization processes. Thus differences in the fossil assemblage of the same bed at two separate localities may not be due entirely to initial differences in the biocoenoses (controls I and II), or to incorporational differences (control III), but also may be due to different post-depositional (preservational) histories of the two areas.

A. SEDIMENTARY ECOLOGY (ENVIRONMENTAL ANALYSIS)

ENVIRONMENTAL REALMS

Environmental progression.—Through their cycle of existence the materials of the stratigraphic record progress through different environmental realms, and it is important that the paleoecologist understand and evaluate the impress of these realms on the sedimentary material he studies. The surficial or gradational realm, although having but brief jurisdiction over sedimentary materials, is nevertheless the most important in shaping the character and destiny of deposits. If an ancient surficial environmental setting is to be accurately restored from the study of a sedimentary rock, the student must first obtain a clear picture of the sediment as it appeared at the time of deposition and of its associated organic community. To do this he must isolate and describe the influence of each post-depositional environmental realm through which the sediment has passed. Primary and secondary lithologic characters, easily confused, must be clearly differentiated. It

[35] W. H. Twenhofel, "Environment in Sedimentation and Stratigraphy," *Bull. Geol. Soc. America*, Vol. 42 (1931), pp. 419–23.

is perhaps a truism that in some cases the rock materials studied in the field and laboratory have little in common with the sedimentary assemblage as it appeared at the time of deposition; similar primary assemblages may, as a result of different environmental histories, go to form contrasted rock types.

Figure 8 portrays in diagrammatic fashion the character and interrelationships of the various environmental realms, due cognizance being taken of the fact that nature draws few sharp boundaries.

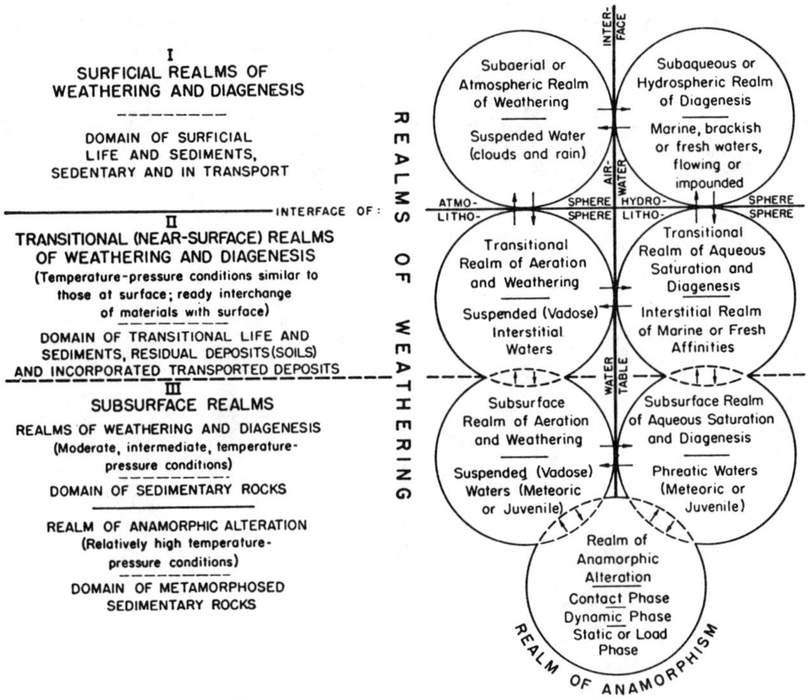

FIG. 8.—Environmental realms through which sedimentary materials may pass. Diagram has no depth connotation. High-temperature conditions, resulting in anatexis or melting in intrusive aureoles, cause sedimentary materials to leave the realms pictured and to enter the realm of fusion to form igneous rocks.

The environmental subdivision into realms is accomplished on a two-fold basis: (1) on whether the medium (interstitial or free) is dominantly water or air; and (2) on the temperature and pressure conditions. In realms II-A and III-A an interstitial *aerial* medium obtains (Zone of Aeration), these intergrading realms corresponding with Van Hise's "Belt of Weathering."[36] In realms II-B and III-B an interstitial *aqueous* medium is represented (Zone of Saturation), these realms corresponding to Van Hise's "Belt of Cementation."[37]

[36] C. R. Van Hise, "A Treatise on Metamorphism," *U. S. Geol. Survey Mon.*, Vol. 47 (1904), pp. 409–560.

[37] *Ibid.*, pp. 562–655.

Subsurface environmental factors.—The nature of subsurface environments and their impact on incorporated sediments comprise a little explored field of investigation in which detailed studies are urgently needed. The combination of environmental factors which influence a sedimentary componènt existing in a subterranean environment differs in certain respects from that which affects a component in a surficial environment. The subsurface factors are as follows.

1. Temperature conditions
2. Motion of interstitial water or air medium (sealed or stagnant *versus* free circulation conditions)
3. Chemical character of interstitial medium
   a. Salinity and dissolved gas content (O_2, CO_2, *et cetera*), if the medium is water. *pH*-value
   b. Chemical character of suspended water and gaseous character, if interstitial medium is air
4. Pressure of water or gaseous medium (hydrostatic pressure and air or gas pressure)
5. Stress conditions to which the rock components are subjected
6. Petrographic character of sedimentary material (attributes, properties, structures). Mineral composition, grain size, shape, fabric, *et cetera;* packing, cementation, porosity, and permeability
7. Character and activities of subsurface organisms, if present

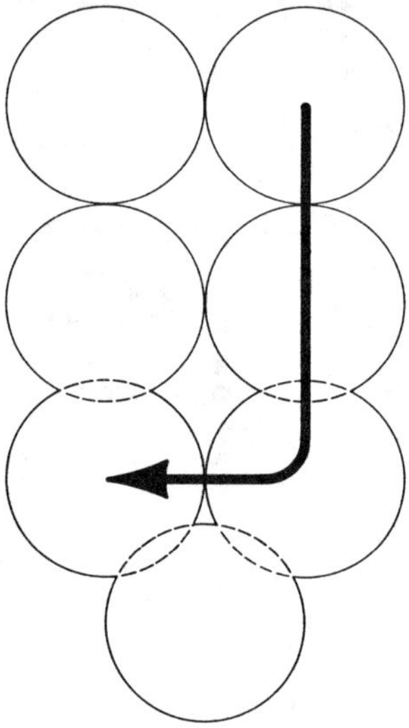

FIG. 9. Diagram of environmental history of crinoidal sediment, Bed 9, Grand Tower limestone, Ozora, Missouri.

Environmental progression (history) of crinoidal sediment.—Figure 9, patterned after Figure 8, pictures the environmental history of the illustrative crinoidal sediment, Bed 9, Grand Tower limestone of Ozora, Missouri. The surficial environment of deposition and accumulation of this sediment is described in succeeding sections of the paper, and the subsurface environmental history subsequent to deposition and burial is described in the section on Sedimentary Dynamics, in connection with the discussion of the epigenetic processes and resultants (diagenesis and weathering) which have affected the crinoidal sediment in the subsurface realms.

INTERPRETIVE RESTORATION OF SURFICIAL ENVIRONMENTS

Prerequisite to the analysis and reconstruction of an ancient surficial environment is a thorough understanding of the primary sedimentary assemblage and its associated organic community. Too often, paleontologists attempting to restore an ancient environment have had an inadequate understanding of the character of the sediment itself and have based their environmental interpretations solely on analogies between recent and fossil plants and animals and assemblages. A paleoecologist must be a sedimentologist as well as a biologist.

A cursory examination of the stratigraphic literature serves to suggest that environmental interpretations have followed no logical plan. Statements are made that a rock type is a beach deposit or a lagoonal deposit, or that it probably was laid down in deep water or under rough-water conditions, and so forth. These interpretations are valuable insofar as they are backed by thorough description and understanding of the rock type with regard to which they are made, but they present only a small part of the environmental picture. The following sections of this paper present a rigorous environmental breakdown into fundamental factors inherent in an environmental setting and provide a logical and organized plan of approach.

The concept of *surficial environment* has two different aspects: (1) an *ecologic aspect*, which pertains to the sum total of factors which influence an organism or sedimentary component existing at any point on the earth's surface, and (2) a *geographic aspect*, which pertains to the regional or space environmental relations of that point to the source and depositional provinces in general.

ECOLOGIC FACTORS INHERENT IN SURFICIAL ENVIRONMENTS

The factors listed in the following outline pertain only to subaqueous environments. Study of a subaerial environment would entail the listing and consideration of a related set of analogous factors, in which case the factor of salinity would drop out of the picture, the character of pelagic life would be replaced by that of flying organisms, and so forth. The factors which influence the existence of an organism or sedimentary component anywhere on the earth-water interface are as follows.

- I. Factors Pertaining to the Substratum
  - A. Composition and textural character
  - B. Inclination
- II. Factors Operative on Substratum
  - A. Movement (transportation) of sediment on substratum
    - 1. Individual
    - 2. Collective
  - B. Character and activities of benthonic life
- III. Factors Pertaining to the Medium
  - A. Motion of the medium
    - 1. Agitation (turbulence)
    - 2. Bottom currents (velocity)
  - B. Dissolved gas content
  - C. Salinity

D. Character of suspension load (composition, grain size, volume, turbidity, *et cetera*)
E. Temperature
F. Light intensity
G. Hydrostatic pressure

IV. Factors Operative in Medium above Substratum

A. Movement (transportation) of suspended sediment
B. Character and activities of pelagic life

Most of the ecologic factors are *dynamic* and in the present oceans show periodic, seasonal and diurnal, as well as non-periodic variations in time and space. For each factor, such as current velocity, it is important in restoring an ancient environment to evaluate both the *range* of variation and the *average* value. Most of the factors are more or less interrelated and represent dependent variables. Three independent variables, *temperature*, *salinity*, and *hydrostatic pressure*, determine the character and distribution of physical properties in the ocean. The paper by Fleming and Revelle[38] on physical processes in modern oceans should be consulted in the interpretive reconstruction of physical processes and factors obtaining in ancient seas.

Character of substratum.—This factor is of prime importance to the existence of many benthonic creatures. It includes the composition and textural attributes, such as the average and degree of sorting as to grain size of the bottom sediment and also includes properties such as the degree of cementation and lithification and the degree of packing and porosity. Composition of the substratum is important in the case of a subaerial soil substratum devoted to agricultural purposes.

Inclination of substratum.—This factor is significant in regions of marked relief of the surface of deposition, for example, delta environments, and reef margins. Sedimentary particles and unattached living organisms may exist on a highly inclined substratum if the slope does not exceed their maximum angle of repose.

Movement of sediment on substratum.—Tractional transportation of bottom sediment, whether in the form of individual or collective motion of the particles, is significant in the existence of benthonic creatures. It is well known that areas of shifting sands (collective movement) are extremely unfavorable for the development of bottom communities and are apt to be barren, organic deserts. This is true of most cross-stratified sandstones, such as parts of the Beauvais sandstone of Missouri, which bear but a sparse benthonic fauna. On the other hand, the fact that *individual movement* of sand grains does not ordinarily inhibit benthonic animal growth is evinced by the abundant bottom fauna of many parallel-stratified sandstones and crinoidal limestones, in the genesis of which the individual type of grain movement must have occurred.

Character and activities of benthonic and pelagic life.—In the categories of these two factors are included the biotic aspects of the environment. Of these, food supply is most important, a sufficient amount of organic food being an indispensable condition for the habitability of an area by animals. Organic associations and

[38] R. H. Fleming and R. Revelle, "Physical Processes in the Ocean," *Recent Marine Sediments, A Symposium*, Amer. Assoc. Petrol. Geol. (1939), pp. 48–141.

interrelationships and the struggle for existence also are part of the biotic picture, as exemplified by the failure of bivalves, mussels and clams, to thrive in an area dominated by starfish. The population density of the environment also should be considered. It is impossible for the paleoecologist to evaluate these biotic factors in other than a general way inasmuch as the zoöplanktonic food supply and organic interrelationships can not be interpreted from paleontologic evidence.

Motion of medium.—This factor includes directive currents and random agitation or turbulence and is of great significance especially in controlling the size character of sediments in an environment. Only *bottom current* conditions can be evaluated in the restoration of an ancient environment; they may be designated roughly in terms of centimeters per second with the use of Hjulstrom's size-velocity curves for erosion and deposition. From the study of the range of size of particles of a sediment it is possible to make a rough estimate of the range of variation in current velocity in the depositional environment. Current *direction*, as well as velocity, is a significant aspect which sometimes may be determined from a knowledge of regional stratigraphic relations, average direction of cross-stratification, and so forth.

Currents exert a powerful influence over an organic community in that they serve as a transporting agent for food, provide a means for immigration and emigration of larval forms, and in that they in some cases have a strong impact on growth form. Experiments performed on branching corals, for example, have shown that in shallow, strongly agitated water short stumpy branches are formed whereas quiet-water conditions permit the growth of long, slender fragile branches, both growth forms being manifested by the same species and even by parts of the same individual placed in the contrasted environmental settings.

Dissolved gas content.—This factor of an aqueous environment usually does not constitute a strong limiting factor inasmuch as oxygen, the most vital gas, is absent or nearly so in but few marine environments. In the stagnant depths of the Black Sea, however, and in many Norwegian fiords closed by a bar, the evolution of hydrogen sulphide has combined nearly all of the available oxygen, and such conditions probably existed in tracts of ancient epeiric seas. This factor may be evaluated in terms of highly aerobic, aerobic, anaerobic, and highly anaerobic. Highly aerobic environments, such as those of tide pools and shallow current-swept banks, commonly have a large organic population (density) and might be expected to produce dark sediments; on the contrary, light-colored sedimentary deposits are the resultant because of nearly complete oxidation of the organic matter. The converse relationship obtains in the case of anaerobic (reducing) environments.

Salinity.—The extent of salinity variation that can be endured by animals is different for different species. Some types, such as reef corals, are influenced by slight changes in salt content, whereas others can withstand wide variation. Present-day reef corals can endure a salinity range of from 30 to 40‰, higher concen-

TABLE V

ECOLOGIC ANALYSIS OF MARINE ENVIRONMENT REPRESENTED BY CRINOIDAL PHASE, GRAND TOWER LIMESTONE, OZORA, MISSOURI

Ecologic Factors		*Evaluation and Evidence*
Factors pertaining to substratum	Composition and textural character of substratum	Fragmental. Very coarse to medium sand composed of calcareous crinoidal débris and quartz. Admixture of very fine sand and silt composed of quartz, heavy minerals, *et cetera*
	Inclination of substratum	Rather level or horizontal surface maintained, as indicated by evenness of bedding surfaces, lateral uniformity of strata, and lack of noticeable cross-stratification
Factors operative on substratum	Movement of sediment on substratum (individual or collective motion of particles)	Individual movement (rolling, sliding, saltating) of larger components—medium to coarse crinoidal sand and quartz sand. Movement intermittent, effective during storms and times of strong current action. Movement indicated by introduction of quartz sand and by distribution of crinoidal débris in strata of uniform thickness. Lack of cross-stratification and bedding irregularities suggests no collective movement
	Character and activities of benthonic life	Densely populated crinoid-coral plantation. Dominantly small crinoids, with admixture of tabulate corals, cup corals, brachiopods, pelecypods, *et cetera*
Factors pertaining to medium	Motion of medium over bottom (turbulence and directive currents)	Variable current action. Poor size sorting indicates rather wide range of current velocity from strong to weak. Generally mild currents and wave agitation (turbulence) suggested by little worn and incompletely dissociated crinoidal débris. Intermittent strong action during storms. Periods of calm allow deposition of silt and fine sand from suspension. Hjulstrom's size-velocity curves suggest 30 cm. per second current necessary for moving coarse sand (crinoidal débris and quartz) and 0.2 cm. per sec. necessary for silt deposition
	Dissolved gas content	Aerobic conditions, varying with current strength, evinced by density of benthos with its high O_2 requirements
	Salinity	May have been normal marine to slightly concentrated (30 to 40‰) as suggested by ability of living corals to withstand high but not low salinities. Density of shell life indicates large-scale turnover of *Ca* salts
	Character of suspension load (composition, grain size, volume, and turbidity)	Very fine sand and silt, mostly quartz and heavy minerals. Small volume (low turbidity). Intermittent suspension during times of strong current action. Current velocity moving coarse sand would keep fine sand and silt in suspension
	Temperature	At least moderately warm, with probable seasonal change. Warmth suggested by coral abundance and shallow water conditions
	Light intensity	Probably rather strong, because of coral abundance and large phytoplankton requirements
Factors operative above substratum	Movement of suspension load	Very fine sand and silt, moved intermittently during times of stronger currents and turbulence, must have settled to bottom during periods of calm
	Character and activities of pelagic life	Unknown. Large volume of microplankton, larval forms, *et cetera* suggested by large planktonic food demands of benthonic community and by reproduction

tration being more favorable than lower. Thus corals do not thrive off river mouths where marine water is freshened and salinity is abnormally low; they are known to thrive, however, in parts of the Red Sea where salinity is higher than in normal marine water. In addition to conditioning animal existence, the factor of salinity is important in serving as the source of chemically precipitated sediments as well as of organically precipitated shells, tests, and so forth.

Character of suspension load.—The composition, grain size, and volume of suspended material affects organisms and bottom sediments. Turbidity of the medium may limit organic growth and make it impossible for certain species to inhabit an area. The suspension load constitutes a potential supply for bottom sediments, effective when the wind or water currents and turbulence fall below the critical velocity and contributing to building of such deposits as loess and marine shales.

Temperature.—This factor of the medium, of maximum importance to marine zoögeography, may be evaluated for an ancient environment only by uncertain analogies between living and fossil plant and animal types. For example, present-day reef corals flourish only in relatively warm water (mean temperature greater than 60°F.); fossil coral-reef structures of Paleozoic rocks suggest, by analogy, that the Paleozoic marine waters in which they were formed were relatively warm. In modern oceans water temperature varies with location (latitude and longitude) and with seasons, and there are indications that such was also the case in ancient seas. Deposits of halite, gypsum, and other evaporites suggest warmth of ancient water bodies.

Light intensity.—This usually does not function as a direct limiting factor for animal existence but ordinarily determines the development of plant life. Inasmuch as animals ultimately are dependent on plants for their food supply, this factor affects animals indirectly. Sedimentary deposits are affected by this factor insofar as they are composed of plant remains.

Hydrostatic pressure.—This factor does not exert a strong controlling influence on either organisms or sediments in relatively shallow water bodies such as those of epeiric seas. Pressure increases one atmosphere for every 10 meters of depth, and, except at great depths, plays an unimportant rôle in the life of marine animals other than certain fishes.

ECOLOGIC VALENCE CONCEPT

The concept of ecologic valence, expounded in the text by Hesse, Allee, and Schmidt,[39] is of paleontologic and stratigraphic import. Coexisting organisms and sedimentary particles are governed by the same set of afore-described ecologic factors; they are not tied hard and fast to unalterable values of these conditioning factors but rather can tolerate a range of variation. The composite of the factorial ranges constitutes the ecologic valence of an organism or sedimentary component.

[39] Hesse, Allee, and Schmidt, *op. cit.*, pp. 20–21.

For example, the valence of a coral species might be designated as follows: salinity 30–40°/₀₀ (parts per mille), mean temperature 60–85°F., current velocity less than 100 cm. per second, dissolved oxygen greater than $4.0^{ml}/L$, light intensity greater than some measurable value, and so forth. Although such valence determinations are theoretically feasible, experimental studies have in general been made for only one or two measurable factors affecting a living species. Thus the valence concept is little more than a valuable aid to proper mental orientation in environmental studies.

If a species is to reside in a given environmental setting, that environment must satisfy its ecologic valence; that is, the latitude of variation of each of the factors must lie somewhere within the range tolerated by the species. Furthermore, the nearer the environment approaches the species' optimum for each factor, the more flourishing will be its existence. The more the factors depart from optimum values, the higher the degree of adaptation the species population must make and thus the less it will flourish. Forms that can not survive extremes must perish. In addition to being conditioned by physical factors of their environment, members of an organic community are also dependent on biotic factors, on plant and animal relationships, the struggle for food and energy, and so forth.

The factor for which a species has the narrowest range of tolerance acts as the limiting factor in its existence, regardless of the fact that the species may withstand a wide range with respect to the other factors. A counterpart of this rule is that the selection of species in a given environment is to a large extent determined by that habitat factor which most nearly approaches the viability minimum, this being known in ecologic circles as the *law of the minimum*. For example, there may be sufficient oxygen, favorable temperature, and abundant food supply in a salt-water pond, but the high salt content or salinity will permit the existence of only a few euryhaline animals.

An organic community or biocoenosis represents an assemblage of species whose ecologic valences overlap or are roughly coincident. Thus the valence of a biocoenosis itself is a composite of the valences of the component species. Unless altered by biotic factors a community type tends to be coextensive in time and space with the maintenance of similar physical conditions. There are numerous illustrations indicating that communities in widely separated parts of the world may be similar if similar ecologic conditions obtain. Any permanent time or space change in the physical or biotic factors of an environment bring about a change in the composition and character of a biocoenosis. Some members of the community drop out, others thrive better than before, foreign species may enter; growth forms of some species may undergo alteration, size distributions may change, and so forth. In these ways the community strives toward equilibrium with its changing environment.

Sedimentary components, like organisms, are governed by the factors of their environment and have an ecologic valence. The influence of the different factors on organisms and sedimentary grains, however, differs markedly. Such factors as temperature and light intensity, of importance to the existence of organisms, have little direct influence on sedimentary components. Of significance to chemically precipitated sediments and readily soluble components are the factors of salinity, temperature, motion or agitation of the water, and dissolved gas content. On the other hand, little soluble and relatively inert clastic components, such as quartz, are dependent primarily on the factor of motion of the medium (current velocity and turbulence). A lithologic type tends to show but little character variation throughout time and space provided there is maintenance of similar factorial conditions. Any permanent time or space change in the environmental factors causes lithologic changes. When the range of current velocity of the environment of a clastic deposit, such as quartz sand, is shifted in the direction of higher velocities, the sand assemblage undergoes a change; small grains out of equilibrium with respect to the new size-velocity range are swept away, and larger grains may be added to the deposit. In these and other ways the sedimentary community, like the organic community, strives to approach equilibrium with its changing environment.

GEOGRAPHIC ENVIRONMENTAL FACTORS

In the geographic analysis scheme here propounded the various geographic factors are classified in three categories: (1) physiographic factors, (2) diastrophic factors, and (3) gradational factors. These factors pertain to the regional or space relations of a given environmental setting and to the source and depositional provinces in general.

Depositional province and subprovince.—As here employed, the term *depositional province* denotes the areal extent of a region of continuous and related sedimentation, such as the Eastern Interior basin during Middle Devonian time. Within the confines of such a province several distinct but environmentally related lithologic types may be undergoing simultaneous sedimentation, the restricted areal extent of any one lithologic type constituting the *subprovince* of that type.

Source province and subprovince.—The concept of source province, as here employed, embraces the areal extent of a positive region undergoing reduction and contributing preëxisting rock components for the building of sedimentary deposits. During Middle Devonian time Ozarkia and Appalachia were the dominant source provinces in eastern North America. The term *source subprovince* denotes a restricted area of the general source province and embraces the areal extent of all preëxisting rocks, igneous, sedimentary, and metamorphic, which contribute components via transportation routes for the building of the deposit in question.

Outline of geographic factors.—Following is an outline classification of the geographic environmental factors.

I. PHYSIOGRAPHIC FACTORS
  A. Depositional Subprovince
    1. Character of depositing medium
    2. Geographic place represented
      a. Location
      b. Areal extent
      c. Shape and degree of enclosure
    3. Attitude and topographic relief of depositional surface
    4. Depth of bottom below water surface, in case of subaqueous environment
    5. Distance from shore or boundary of source province
  B. Source Subprovince
    1. Geographic place represented
      a. Location
      b. Areal extent
      c. Shape or outline
    2. Attitude and topographic relief of erosional surface
    3. Elevation above sea-level
II. DIASTROPHIC FACTORS
  A. Depositional Province and Subprovince
    1. Structural type (trough, basin, shelf, *et cetera*)
      a. Location
      b. Areal extent
      c. Shape or outline
    2. Rate and magnitude of diastrophic movement
      a. Orogenic subsidence
      b. Epeirogenic subsidence
  B. Source Province and Subprovince
    1. Structural type (dome, arch, geanticline, shield, *et cetera*)
      a. Location
      b. Areal extent
      c. Shape or outline
    2. Rate and magnitude of diastrophic movement
      a. Orogenic emergence
      b. Epeirogenic emergence
III. GRADATIONAL FACTORS
  A. Depositional Subprovince
    1. Agents and media of transportation
    2. Modes of transport
    3. Rate and volume of transport
    4. Distance and direction of transport from boundary of source province
    5. Position of depositional surface with regard to base level or profile of equilibrium
    6. Rate and volume of deposition, accumulation, and incorporation
    7. Continuity of sedimentation (diastems *et cetera*)
  B. Source Subprovince
    1. Lithology of source rocks
    2. Climate and type of weathering
    3. Agents and media of transportation
    4. Modes of transport
    5. Rate and volume of transport
    6. Distance and direction of transport
    7. Position of erosional surface with regard to base level or grade (stream age)
    8. Rate and volume of erosion, land reduction and removal of sediment

PHYSIOGRAPHIC FACTORS

Character of medium.—The various types of depositional media are as follows.

I. AQUEOUS
  A. Impounded Water Body
    1. Without oceanic connections (lake)
    2. Marine
  B. Running Water
    1. Confined surface (streams)
    2. Open surface (sheet wash *et cetera*)
II. AEOLIAN
III. GLACIAL

Geographic place.—The geographic place represented by the source or depositional subprovince may be described with little difficulty for present-day environmental settings. It is difficult and often impossible, however, to evaluate this factor for ancient environments which are not so self-evident and must be interpreted from our slim knowledge and understanding of the stratigraphic record.

Geographic place names of impounded water bodies include, for example, estuary, fiord, epeiric shelf sea, epeiric interior sea, lake delta, and marine swamp. Geographic places represented in running water environments include river flood plain, valley bottom, delta, piedmont slope, and alluvial fan. Geographic place designations also may be applied to aeolian and subglacial environments.

The location, areal extent, and shape and degree of enclosure of the source and depositional province and subprovince can best be described by means of a paleogeographic map such as that of Figure 10.

Depth of water.—In an impounded water medium the physiographic factor of *depth* is important inasmuch as it conditions such ecologic factors as character of the substratum, temperature of the medium, hydrostatic pressure, salinity, dissolved gas content, and light intensity. A picture of depth conditions in present-day impounded water bodies, epeiric seas, continental shelf seas, lagoons, and lakes, has been presented by Barrell.[40] Table VI is modified from one in Barrell's paper.

TABLE VI

AVERAGE DEPTH CONDITIONS IN PRESENT-DAY IMPOUNDED WATER BODIES
(Modified from table by Barrell)

Type of Water Body	*Distances from Shore, in Miles*				
	5	10	20	80	100
	Depths of Water, in Feet				
Stormy epeiric seas	55	70	90	110	110
Quiet epeiric seas	35	50	70	90	90
Wide lagoons	15	15	—	—	—

Quiet epeiric seas are represented by the shelf sea east of the Mississippi delta and by the Persian Gulf. The latter is an epeiric sea with a central unfilled portion below wave base. Stormy epeiric seas are represented by Lake Erie and the North Sea, in both of which full depth of water is attained within 100 miles from shore, as far as wave action is concerned.

Ancient epeiric seas typically were very shallow water bodies. Inasmuch as such shallow-water features as ripple marks, cross-stratification, and mud cracks often occur at the same horizon over areas covering thousands of square miles, their development must not have been restricted to near-shore environments but

[40] J. Barrell, "Rhythms and the Measurements of Geologic Time," *Bull. Geol. Soc. America*, Vol. 28 (1917), pp. 776–85.

evidently could take place anywhere on the epeiric sea bottom. Widespread marine silts and sands of the ancient epeiric seas indicate the effectiveness of wave action in agitating the bottom material and working it by oscillatory action to great distances. From the foregoing evidence, and others, depths of 20 to 50 or 100 feet may be regarded as typical of Paleozoic epeiric seas.

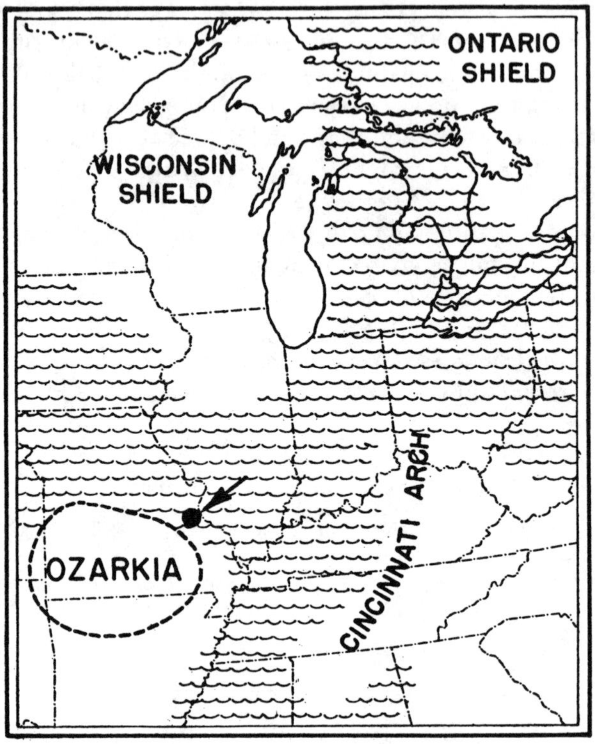

FIG. 10.—Paleogeographic map of time represented by Grand Tower limestone deposition. Dot and arrow indicate position of subprovince of crinoidal phase of Grand Tower.

Physiographic (regional) environment of crinoidal sediment.—A physiographic analysis of the Middle Devonian regional environmental setting in which the illustrative crinoidal limestone was deposited is presented in outline fashion.

I. PHYSIOGRAPHIC FACTORS
- A. Depositional Province and Subprovince
  1. *Character of depositing medium.* Impounded water body (marine)
  2. *Geographic place.* Depositional province—epeiric sea occupying Eastern Interior basin (see paleogeographic map of Figure 10). Subprovince of crinoidal sediment situated near Ozarkian shore of epeiric sea in southeastern Missouri
  3. *Attitude and topographic relief of depositional surface.* A marine basin flat of little relief; possibly some shoal tracts
  4. *Depth.* Shallow water (neritic zone), estimated as being in the order of magnitude of 25–100 feet; more probably 25–50 feet

5. *Distance from shore.* Subprovince of crinoidal sediment less than 10 miles from eastern shore of Ozarkia, possibly much less

B. Source Province and Subprovince

1. *Geographic place.* Landmass of Ozarkia, a positive tract (dome) probably connected with land to southwest (see paleogeographic map, Figure 10). That the Ozarkian land area of Middle Devonian time represented the source province is suggested by regional stratigraphic relationships (Fig. 11). Middle Devonian sediments of preëxisting rock origin decrease in volume or entirely disappear outward from Ozarkia northward, northeastward, eastward, and southeastward
2. *Attitude and topographic relief.* A gently sloping surface with small erosional relief
3. *Elevation above sea-level.* Low to moderate

The subprovince of the crinoidal sediment must have been rather localized, occupying a small tract off the northeastern shore of Ozarkia (Fig. 10), inasmuch as the lower Grand Tower crinoidal limestone phase grades laterally into quartz sandstone southeastward and eastward (Dutch Creek sandstone and Hoing sandstone) and northwestward (sandstone phase of Mineola formation). The exact extent of the subprovince of this lithologic type is difficult if not impossible to determine because of structural complexities and paucity of outcrops and well logs and samples.

DIASTROPHIC FACTORS

Diastrophism is the major geologic process, the functioning of which produces surface irregularities and sets the counteracting process of gradation in operation. The sedimentational processes are but part of the major process of gradation which tends toward the development of a plane surface. If it were not for diastrophism, the earth's surface equilibrium would not be disturbed by rising and sinking movements, and sedimentation therefore would not function; thus diastrophism is the ultimate cause of the development of sedimentary deposits.

It follows from the foregoing reasoning that the character of a sedimentary deposit (attributes, properties, structures) is largely governed by the character and degree of diastrophism obtaining in the source and depositional provinces during its genesis. Diastrophic conditions vary between two extremes. Although there are all stages of intergradation, the conditions in a given area may be classified as belonging to one or the other of two fundamental types: (1) *orogenic*, and (2) *epeirogenic.* Conditions may be intermediate between these two extremes, and the sediments will then show mixed characteristics.

Orogenic type of diastrophism and sedimentation.—The orogenic type of diastrophic movement is characterized by large-scale and rapid movement. *Positive* orogenic movements result in rapid and high elevation of a region, attended by erosional dissection and rapid reduction; reduction, however, does not keep pace with emergence. *Negative* orogenic movements result in rapid subsidence of a region (trough or basin), attended by rapid accumulation of thick sections of sediments. The orogenic type of diastrophism and resulting sedimentation may in turn be subdivided into two contrasted types: (1) the tensional or block-fault type exemplified by diastrophic movements and sedimentation in the Basin-and-Range province during Tertiary time and in the area east of the Appalachians during Triassic and Jurassic time, and (2) the compressional type, exemplified by the Alpine orogeny with its flysch and molasse phases of sedimentation. Thick sections of orogenic sediments accumulate in the troughs and are eroded from the rising geanticlines.

Sediments deposited by continental glaciers and by streams draining therefrom commonly show orogenic characteristics. Such sediments are *pseudo-orogenic* in character; that is, they are deposited in regions of epeirogenic movements of the earth's crust under glacial conditions.

Epeirogenic type of diastrophism and sedimentation.—This is characterized by small-scale and slow movement, unattended by pronounced faulting and folding. *Positive* epeirogenic movements result in slow elevation of a region, dome or arch; inasmuch as erosion and reduction almost keep pace with uplift, high mountains are not formed. Examples of positive epeirogenic tracts are the Ozark dome and the Cincinnati arch, which underwent slow emergence during part of Paleozoic time. *Negative* epeirogenic movements result in slow subsidence of a region, basin or trough; inasmuch as sedimentation and accumulation almost keep pace with subsidence, more of a structural than a topographic basin or trough is formed. An example of a negative epeirogenic tract is the Eastern Interior basin during much of Paleozoic time.

Diastrophic environment of crinoidal sediment.—The outline which follows describes the regional Middle Devonian diastrophic conditions under which the Grand Tower crinoidal limestone phase (Bed 9) was deposited.

II. DIASTROPHIC FACTORS

- A. Depositional Province and Subprovince
  1. *Structural type.* A structural basin (Eastern Interior basin) occupying southeastern Missouri, southern half of Illinois, western Tennessee and Kentucky, and Indiana. Subprovince of crinoidal sediment situated near western margin of basin, where it joins the Ozarkian positive province
  2. *Rate and magnitude of diastrophic movement.* Slow epeirogenic subsidence of intermittent and differential character
- B. Source Province and Subprovince
  1. *Structural type.* A structural dome (Ozark dome); see map of Figure 10
  2. *Rate and magnitude of diastrophic movement.* Slow epeirogenic emergence of intermittent character

GRADATIONAL FACTORS

Gradation is the major geologic process by which the irregularities of the earth's surface, produced by diastrophism, tend to become smoothed to produce a plane surface. The array of gradational factors are listed in the foregoing outline of geographic environmental factors and are discussed in the following paragraphs. Those pertaining to epeiric sea environments, including equilibrium profile and wave-base control, are admirably discussed and illustrated in Barrell's classic paper on "Rhythms and the Measurements of Geologic Time."[41] A picture of probable depth conditions of the epeiric seas may be gleaned from Barrell's paper, and it expounds the often forgotten law that ***subsidence is the ultimate basis of deposition, accumulation, and incorporation of sediments.***

[41] Barrell, *op. cit.*, pp. 776–809.

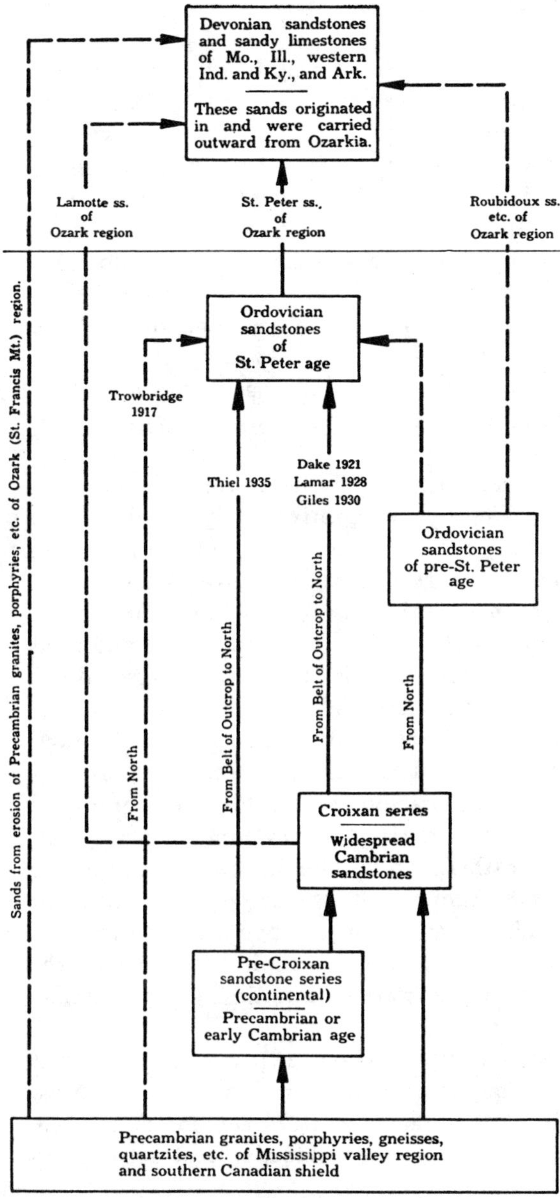

S.W. ILL. AND W. KY.	SOUTHEASTERN MISSOURI	CENTRAL MISSOURI
	Basal ss. beds of St. Laurent ls.	Sandstone phase of Callaway ls.
	Beauvais sandstone	Auxvasse Creek ss.
Sandstone beds of basal Grand Tower ls. Dutch Creek ss. (Pendleton and Hoing ss.)	Sandy limestone phases of Grand Tower ls.	Sandstone phase of Mineola fm.

MO., ARK., & SOUTHERN ILL.	IOWA, MINN., WISC., & ILL.
Thebes ss. St. Peter ss. Everton ss.	St. Peter ss.

MISSOURI, ARKANSAS	IOWA, WISC., AND ILL.
Sandstone beds of Powell, Cotter and Jefferson City formations	Sandstone beds of Shakopee formation
Roubidoux ss. Gunter ss.	New Richmond ss.

MO., ARK., & SOUTHERN ILL.	IOWA, MINN., WISC., & ILL.
Lamotte ss.	Madison ss. Devils Lake ss. Jordan ss. Mazomanie ss. Franconia ss. Dresbach ss. Eau Claire ss. Mount Simon ss.

The vast supply of sand reworked through formations listed above was generated for the most part during the late Precambrian or early Cambrian, a time of continental peneplanation and drastic chemical decay followed by a thorough mechanical attack by rivers, wind, and littoral agencies, forming river, dune, beach, littoral, and neritic sands.

MAIN AVENUES OF SEDIMENTATION (solid arrow)

POSSIBLE ALTERNATIVE OR ACCESSORY AVENUES WHICH MAY HAVE FUNCTIONED (dashed arrow)

FIG. 11.—Flow diagram showing probable progress of sedimentation of early Paleozoic sands (quartz and associated stable minerals) in Mississippi Valley district. Diagram shows possible sources of terrestrial (preëxisting rock) sand material in crinoidal phase (Bed 9) of the Grand Tower limestone.

Lithology of source rocks (source analysis).—Genetic study of a sedimentary rock containing allochthonous components derived from preëxisting rocks must include an interpretive synopsis of the probable source rock or rocks which yielded these allochthonous components to the transporting agents. Krynine's pioneering work[42] in this field might well serve as a guide for students concerned with problems of source analysis. Source studies should if possible treat the *ultimate* as well as the *immediate* source of the preëxisting rock components. The ultimate source of most of the stable component grains is, of course, an igneous rock, and its determination may not be an easy task, especially if the components have progressed through a number of sedimentary cycles. A flow diagram (Fig. 11) helps in obtaining a mental picture of the history or lineal descent of sedimentary components and presents both the probable source rocks by stratigraphic divisions and the probable areas of derivation.

The rock type or types which have served as source material for the building of a sedimentary deposit may be revealed by interpretive study. Study of the mineral composition of a sediment usually shows a certain suite of mineral species which individually or collectively may be suggestive of the source rock or rocks. An understanding of probable mineral associations is helpful in diagnosis, and an understanding of the regional stratigraphy and petrography of all potential source rocks is prerequisite to a sound source analysis. The more diversified the rocks of the source subprovince, the more varied is the character of the descendant deposit; few deposits derive their components from a single preëxisting rock type. Source rocks of mineral species commonly occurring in sediments are described in the mineral outlines of Milner's text on sedimentary petrography.[43]

Climate and type of weathering.—Certain detrital mineral species are highly indicative of the climate and type of weathering obtaining in the source province from which they were derived. It is important to note the degree of alteration of minerals. An abundance of fresh feldspar grains in a deposit is indicative of mechanical weathering under rigorous desert, glacial, or tropical monsoon conditions of high relief; whereas altered feldspar grains are indicative of chemical weathering in humid areas of little relief.

Weathering of the chemical type is accomplished by the processes of solution and katamorphic alteration—carbonation, oxidation, hydration, and hydrolysis. Mechanical weathering is the resultant of abrasive work of running water, ice, and wind, and the disintegration work of plants and animals, frost action, and insolation.

Gradational analysis, environment of crinoidal sediment.—Gradation conditions in the Middle Devonian regional environment of the crinoidal phase (Bed 9) of the Grand Tower limestone are here described in outline fashion.

[42] Krynine, *op. cit.*, pp. 78–87.

[43] H. B. Milner, *Sedimentary Petrography*, 3d ed., pp. 229–357. Nordeman Publishing Company (1940).

III. GRADATIONAL FACTORS

A. Depositional Subprovince

1. *Agents and media of transportation.* Transport by weak to moderate marine currents
2. *Modes of transport.* Tractional transport of crinoidal débris and medium to coarse quartz sand; suspension transport of silt and very fine sand composed of quartz, tourmaline, *et cetera*
3. *Rate and volume of transport.* Probably slow rate and small volume except during storms
4. *Distance and direction of transport from boundary of source province.* Unknown. Probably less than 50 miles in a direction somewhere between northeastward and southeastward, depending on location of river mouth from which preëxisting rock components were derived
5. *Position of depositional surface with regard to base level.* Depositional surface generally at base level or wave base—somewhat above during storms and below during periods of calm
6. *Rate and volume of deposition, accumulation and incorporation of sediment.* Small volume and relatively slow rate indicated by dominant organic source and admixture of subordinate amount of derived preëxisting rock material
7. *Continuity of sedimentation.* Accumulation intermittent with diastems represented; this is suggested by discrete vertical changes in faunal content and by sharp physical breaks in stratification without apparent lithologic change

B. Source Subprovince

1. *Lithology of source rocks.* Sand and silt material of terrestrial (preëxisting rock) origin for most part derived from St. Peter sandstone of Ozark region (Fig. 11). This is suggested because quartz grains and heavy mineral suite (Pl. I), including tourmaline, rutile, magnetite, and glauconite, are similar to those of St. Peter. St. Peter source also suggested by stratigraphic relationships in Ozark region which indicate the St. Peter sandstone was undergoing erosion during Middle Devonian time. As shown by Figure 11 there are three other possible sources of quartz and heavy-mineral grains in crinoidal rock: (1) Ordovician sandstone beds of Powell, Cotter, Jefferson City, Roubidoux, and Gunter formations of Ozark region, (2) Cambrian Lamotte sandstone of Ozark region, and (3) sands from erosion of pre-Cambrian granites and porphyries of St. Francis Mountain region, this last source being improbable because of high degree of rounding of material
2. *Climate and type of weathering.* Unknown. Warm climate suggested by abundance of corals in adjacent seas. Humid conditions with chemical weathering suggested by presence only of stable derived minerals in crinoidal sediment; lack of unstable minerals may, however, be due to the fact that the derived grains had been reworked through several sedimentary cycles
3. *Agents and media of transportation.* River currents
4. *Modes of transport.* Tractional transport of medium to coarse quartz sand; suspension transport of silt and fine sand
5. *Rate and volume of transport.* Probably slow rate and small volume except during floods
6. *Distance and direction of transport.* Unknown. Probably less than 100 miles in a direction somewhere between northward and eastward, depending on location of rivers and drainage
7. *Position of erosional surface with regard to base level or grade.* Can not be interpreted from available evidence. Large volume of quartz sand in Middle Devonian beds (Fig. 11) equivalent to the Grand Tower limestone crinoidal phase suggests that at this time the Ozarkian source province was not at grade but was in youthful or early maturity stage of erosional cycle
8. *Rate and volume of erosion, land reduction and removal of sediment.* Relatively slow rate and small volume under epeirogenic conditions

B. SEDIMENTARY DYNAMICS (PROCESSES AND RESULTANTS)

The dynamic aspect of sedimentary petrogenesis is strongly related to the environmental aspect inasmuch as it deals with the processes and resultants (changes) by which sedimentary deposits and their components tend to become adapted to, or to be brought into equilibrium with, their changing environments. The dynamic aspect is subdivided into two fields of study: (1) sedimentation, which pertains to the processes active on the surface, and (2) subsurface (epigenetic) processes and effects.

SEDIMENTATIONAL ANALYSIS

The term *sedimentation*, as here employed, refers to the dynamical aspects of gradation; that is, it pertains to processes active on the earth-water or earth-air interface which bring about leveling of surface irregularities by distribution and reallocation of surficial materials. The field of sedimentation is divisible into three phases.

1. *Reduction* of topographically positive areas above base level of erosion by *initial* erosion (beginning of new sedimentary cycle); removal of material in solution or by mechanical transporting agencies
2. *Transportation*, including *temporary* erosion and deposition in transit; sedimentary particles ordinarily come into motion (are eroded) and come to rest (are deposited) many times in their trek from source site to site of final accumulation and incorporation
3. *Accumulation* of sediment in topographically negative areas below base level of deposition; *final* deposition and incorporation

TRANSPORTATION

Mobility of sedimentary components and organisms is of major importance in the environmental picture and plays a significant rôle in sedimentary petrogenesis. Reconstruction and description of the environment and genesis of a sedimentary deposit must include an analysis of the components with regard to both their probable *mode* of transport and deposition and the *agents* and *media* responsible for their transportation and deposition. Such a transportational analysis is exemplified by Figure 13.

A sedimentary deposit may be heterogeneous with respect to the transportational histories of its components; that is, the deposit may represent an assemblage of components of different transportational histories, different in regard to the agent and (or) the mode of transportation and deposition. Such heterogeneity is typical of many deposits of the stratigraphic record. On the other hand, some deposits are highly homogeneous in this respect.

Agents and media of transportation.—These may be outlined as follows.

- I. SUBAERIAL AGENTS (Air Medium)
  - A. Gravity directly (rock avalanche *et cetera*)
  - B. Air currents and turbulence (wind)
  - C. Organisms
- II. SUBAQUEOUS AGENTS (Marine, Brackish, or Fresh-Water Medium)
  - A. Gravity directly (submarine mud-flow *et cetera*)
  - B. Water currents and turbulence
    - 1. In impounded body (currents generated by convection, tides, or wind)
      - a. Lake
      - b. Marine body
    - 2. In running water body
      - a. Stream or river currents
      - b. Sheet wash
  - C. Organisms
- III. SUBGLACIAL AGENTS (Ice Medium)
  - A. Ice currents (lamellar flow)

Motion of the medium.—In painting an environmental picture the paleoecologist must consider and evaluate the factor of *motion of the medium*. According to Fleming and Revelle,[44] the motion of a water particle may be divided into two

[44] Fleming and Revelle, *op. cit.*, p. 77.

parts: first, local agitation involving random or turbulent and reciprocating or wave motion; and second, mean or current motion. Turbulent eddies and wave motion alone do not result in net transport; current motion, however, is characterized by a certain uniformity of velocity and, when combined with turbulence, results in net transport in a definite direction.

It is possible to make a rough quantitative estimate of the velocity of motion of the medium by consulting Hjulstrom's size-velocity curves for erosion and deposition previously mentioned. The crinoidal limestone of this report contains transported components up to 3 mm. in mean diameter, which indicates that the mean velocity of the medium during its deposition must have at times been in the order of magnitude of 30 cm. per second, at least. Among other size classes,

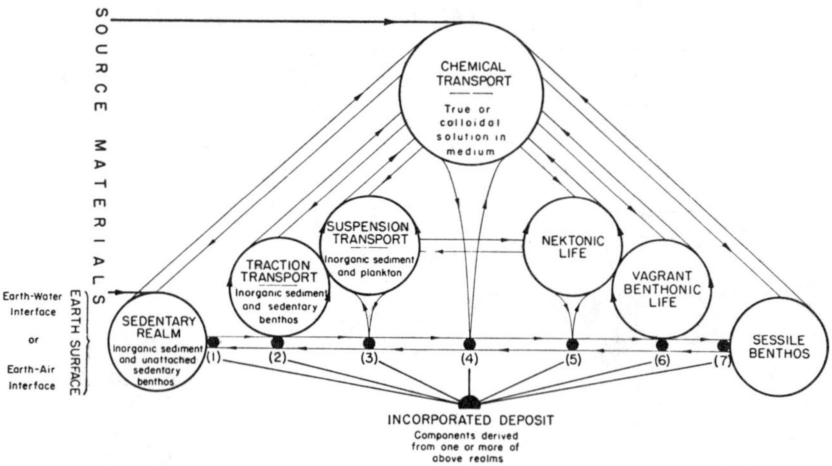

FIG. 12.—Mobility realms of sediments and organisms; modes of transportation and self-locomotion (migration).

coarse silt particles also occur in the limestone, and this indicates that at other times the velocity must have been as low as 0.2–0.4 cm. per second to allow their deposition from suspension. This combination of sizes suggests that the turbulence and current conditions were variable and fluctuated widely in time.

Modes of transportation and locomotion.—The modes of transportation and the mobility realms of sediments and organisms are pictured in the diagram of Figure 12 and outlined in the following paragraphs. The lines and arrows connecting the various realms emphasize the possibility of interchange or transfer of materials. For example, some pelagic creatures are at certain times *nektonic* or free-swimming and at other times *planktonic*, in suspension and dependent on movements of the water. Other examples of transfer are provided by fine sand grains in traction transport, which enter the suspension realm during times of increased current velocity; and pebbles in the sedentary realm entering the realm of traction transport under more rapid current conditions. On death, organic materials leave the life realms, and the remains either enter the mechanical (traction or suspension) realm or the chemical realm (are dissolved), or they may remain *in situ* and go to form sedentary organic deposits.

Physical (mechanical) realms of transport.—In these realms mobility of components is dependent on motion of the medium (fluid mechanics) and mechanical laws. The relationship between size and mode of transport in an aqueous medium is suggested by Table VII. If the current velocity is insufficient to move a given component, whether an unattached living benthonic organism or a lifeless sedimentary component of inorganic or organic origin, that component is said to be *sedentary*. Examples of deposits composed largely of sedentary components include residual soils, lag gravels, and oyster biostromes.

TABLE VII

RELATIONSHIP BETWEEN SIZE AND MODE OF TRANSPORT*

Mode of Transport	*Size of Component*
Gravity traction fall (landslides, *et cetera*)	Massive blocks Boulders Cobbles
Traction in water currents (6th-power law)	Pebbles Granules Coarse sand
Inertia suspension in water currents (6th-power law)	Medium sand Fine sand
Viscous suspension in water (Stokes' law)	Coarse silt Fine silt
Colloidal suspension or solution in water	Clay Colloidal particles
True solution in water	Ions

* Relationship between size and mode of transport, as indicated by spacing of size classes at right, is generally used and varies with current conditions. For example, under certain conditions cobble and boulder-size components may be carried in traction transport, coarse sand components may be carried in suspension, *et cetera*.

The realm of *traction transport* likewise includes unattached living benthonic organisms and lifeless sedimentary components of inorganic or organic origin. Traction movement may occur solely in the form of *individual motion* of particles (rolling, sliding, or saltating), or may occur in the form of collective motion (dune or ripple phase, smooth phase, or antidune phase). Examples of deposits composed for the most part of traction-transported components include coarse quartz sandstones and crinoidal limestones, and stream and beach gravels.

The realm of *suspension transport* carries living planktonic organisms and lifeless sedimentary grains of inorganic or organic origin. Suspension may be in the form of inertia suspension (6th-power law) or viscous suspension (Stokes' law). Deposits composed largely of suspension-transported and deposited grains include loess, clay shales, *Globigerina* ooze, and so forth.

Chemical realm of transport.—In this realm mobility is dependent on motion of the medium and chemical laws pertaining to true and colloidal solution and precipitation. Organic precipitation brings about transfer of material from the chemical to the life realms. Inorganic deposition may be in the form of primary chemical precipitates, such as gypsum, halite, and other evaporites, or chemical replacements, such as phosphorite and some cherts and dolomitic limestones, or may be in the form of colloidal precipitation (coagulation and flocculation). As indicated in Figure 12, chemical precipitation may occur directly on the substratum or may occur up in the medium, in which case the precipitated grains enter the suspension realm of mechanical transport and either settle to the bottom or are carried away to be deposited elsewhere.

Life realms of self-locomotion (migration).—In these realms mobility is for the most part independent of motion of the medium, although influenced thereby, and is dependent on self-locomotion. In the *nektonic realm* of swimming creatures death brings about transfer of the organic remains to the mechanical transportation realm, and the remains either settle to the bottom and become incorporated or are carried away in suspension or traction transport. An example of a deposit composed largely of organic components from the nektonic realm is provided by fish-bone breccias.

In the *vagrant benthonic realm* locomotion is by crawling or walking, jumping (saltating), wriggling, or burrowing. On death, the remains enter the mechanical realm and are either incorporated on the death site or carried away in traction transport to be incorporated elsewhere. Trilobite and gastropod beds represent types of deposits composed of components from this realm.

The realm of *sessile or attached benthonic* creatures includes certain algae, corals, pelecypods, crinoids, and so forth. On death, the organic remains may enter the mechanical realms to form such lithologic types as cross-stratified crinoidal limestones or may remain *in situ* to form such sedentary organic deposits as algal and coral reefs, and brachiopod biostromes.

Erosion, transport, and deposition.—Mobility or transportation involves the operation of three processes: *erosion*, *transport*, and *deposition*. As pointed out by Hjulstrom,[45] erosion and deposition are integral parts of transportation inasmuch as individual particles usually come into motion (are eroded) and come to rest (are deposited) many times in their progress from source to place of final accumulation and incorporation. Rolling and saltation are the usual modes of traction transport, and sliding usually is of minor importance. According to Hjulstrom, size of the particles and roughness of the substratum strongly influence the nature of the motion. Large particles are more readily eroded by currents than small particles, partly because of their larger surface area and partly because of the fact that they stand up higher in the water where the motion is stronger.

Turbulence has a strong influence on sedimentary particles in that upward

[45] F. Hjulstrom, "Transportation of Detritus by Moving Water," *Recent Marine Sediments, A Symposium*, Amer. Assoc. Petrol. Geol. (1939), p. 13.

streams may cause them to pass easily from saltation movement to the suspension realm of transport, the substratum thus being eroded. Saltation may be depicted on the diagram of Figure 12 by a figure-8 circuit of the traction and suspension realms. Suspended particles have their own individual settling velocities; the position of the particles in the water is determined by the net effect of two opposing factors: the turbulent motion which causes the particles to rise and the force of gravity (settling velocity) which makes them sink. The factors influencing settling velocity have been studied by Wadell and include size, shape, and specific gravity of the particles and the specific weight, temperature and viscosity of the medium through which they settle. Particles of silt and fine sand size ordinarily are carried in suspension and belong in the traction realm only under very slow current conditions. Suspension transport does not greatly affect the particles, except by solution, inasmuch as there is little mutual contact.

If turbulence or velocity decrease is marked and sudden, the suspension and traction loads are deposited together, and sorting is then poor. Loads are deposited from both modes of transport on the basis of the size, shape, and specific gravity of the particles.

In the realm of chemical transport the colloidal phase carries such substances as ferric hydroxide, aluminum hydroxide, silica, certain clay minerals, and organic colloids. Colloidal particles do not settle in response to gravity, and their suspension does not depend on movements of the medium. They are precipitated (deposited) through flocculation or coagulation. In the true solution phase of chemical transport, erosion occurs by solution and deposition by precipitation, these processes resulting from changes of temperature or pressure, by changes of concentration, or by chemical reactions.

Effects of transportation.—In the case of autochthonous components genesis and deposition are simultaneous and synonymous. Allochthonous components, however, are more complex in that transportation intervenes between genesis and final deposition and institutes chemical and physical changes by the processes of solution, abrasion, chipping, and splitting. In this way the component parameters of size, shape, roundness, and surface texture are affected. The effects of transportation on sedimentary particles have been described by Russell.[46]

Because of the selective action of the transporting agent, a sedimentary assemblage in transport undergoes local and progressive sorting as to size, shape, and mineral composition, the latter being due to differences in specific gravity, hardness and resistance to solution of the different minerals. The effect of sorting is caused by *selective erosion* and *selective deposition* by water and air currents. Sorting as to size is dependent on current velocity and follows approximately the pattern of Hjulstrom's size-velocity curves for erosion and deposition.[47]

Transportational analysis of crinoidal sediment.—The transportational story

[46] R. D. Russell, "Effects of Transportation on Sedimentary Particles," *Recent Marine Sediments, A Symposium*, Amer. Assoc. Petrol. Geol. (1939), pp. 32–45.

[47] Hjulstrom, *op. cit.*, p. 10.

of the crinoidal phase (Bed 9) of the Grand Tower limestone is a fascinating one and is set forth in the tabular diagram of Figure 13. It is probable that the columnals and other crinoidal débris, which constitute the greater part of the rock and range up to 3 mm. in mean diameter, were carried only in traction transport; during times of lessened current velocity, periods of calm, they were sedentary. The same was true for the medium to coarse quartz sand grains (Pl. I); their space distribution in the rock aggregate coincides with and has the same random

TRANSPORTATIONAL TYPES OF COMPONENTS	TRANSPORTING AGENTS AND MEDIA (HISTORY)	MODES OF TRANSPORTATION (HISTORY)
Crinoidal debris (1/4 to 6 mm.)	Marine currents	Sessile Crinoids —Death / Dissociation→ Traction transport ↓↑ Sedentary
Medium to coarse quartz sand (1/4 to 1 mm.)	River currents → Marine currents	(Varying with current velocity) Traction transport ↓↑ Sedentary
Very fine to medium sand composed of quartz, tourmaline, limonite, etc. (1/16 to 1/4 mm.)	River currents → Marine currents	Sedentary ⇄ Suspension transport ⇄ Traction transport; Sedentary → Traction transport; Traction transport → Sedentary (Varying with current velocity)
Silt composed of quartz, glauconite, etc. (1/256 to 1/16 mm.)	River currents → Marine currents	(Varying with current velocity) Suspension transport ↓↑ Sedentary
Benthonic corals, brachiopods, etc. (See Table IV)	Unmoved by marine currents (autochthonous)	Sedentary benthos

FIG. 13.—Tabular diagram illustrating transportational analysis and history of components of the illustrative crinoidal sediments (see Pl. I).

pattern as that of the crinoidal débris. On the other hand, the medium to very fine sand and silt, consisting of quartz, tourmaline, *et cetera* (Pl. I, Table I), evidently were carried in suspension transport during storms and times of stronger current action. The reason for this belief is two-fold: (1) a current velocity (Hjulstrom curve) sufficient to move the larger quartz and crionoidal sand grains would carry the fine sand and silt in suspension, and (2) the fine sand and silt have an interesting space distribution in the rock aggregate, occurring in the form of patches, stringers, and bands (Pl. I, a, b) which suggests that they were deposited during times of quiet water, buried in the crinoidal débris, and not picked up with the return of stronger currents.

ACCUMULATION AND INCORPORATION

Little attention and study has been directed toward that phase of sedimentation pertaining to final deposition, accumulation and incorporation of sedimentary deposits. The concepts of base level, wave base, and profile of equilibrium thus far have had but little application to stratigraphic problems. An outstanding contribution in this field, as it relates to accumulation of sediments in epeiric sea environments, has been made by Barrell in the paper, previously cited, "Rhythms and the Measurements of Geologic Time."

Accumulation controlled by base level and subsidence.—The most important concept to be gleaned from Barrell's paper is that the rate and volume of permanent accumulation and incorporation of sediments in epeiric seas are determined not so much by the rate of supply as by the rate of subsidence of the surface of deposition. Sediment deposited at a given point ordinarily is only a small fraction of that which is carried past. Thus, in the long run, *accumulation is a function of subsidence*, which, as a rule, is differential in character.

Under orogenic conditions there tends to be large-scale incorporation of great thicknesses of sediments. This is due to relatively rapid and continuous subsidence, fewer diastems and disconformities, and larger supplies of sediments. In contrast to this epeirogenic conditions tend to bring about small-scale incorporation of relatively thin bodies of sediments, this resulting from slow and discontinuous subsidence, with many diastems and disconformities and smaller volumes of sediment supplied.

Wave base, as defined by Barrell, is that depth at which wave action ceases to be strong enough to erode bottom sediments. Above wave base the coarser bottom sediments, medium to coarse sands and gravels, are transported and deposited. These shallow-water sediments, in traction transport, frequently are eroded and deposited by wave and current action and are gradually moved toward an area wherein they may become permanently deposited and incorporated below wave base. Below wave base the bottom sediments are comparatively undisturbed and may be permanent, unless the bottom is brought above wave base. The finer sediments, including clays, silts, and fine sands, carried in suspension transport, are only deposited below wave base; products of this environment include clays, siltstones, shales and fine-grained limestones.

SUBSURFACE (EPIGENETIC) PROCESSES AND RESULTANTS

As a result of their dynamic impact, the subsurface environmental factors, previously described, control the operation of processes which may alter or obliterate certain components of a sedimentary aggregate in an attempt to keep it in equilibrium with its changing subsurface environments. In addition, new components, epigenetic in origin, may be added to the sedimentary aggregate. Thus the attributes, properties, and structures of a deposit may undergo alteration after burial. These processes, as outlined in Table VIII, operate with different degrees of rigor in the different subsurface environmental realms through which sedimentary materials may pass; the realms are portrayed in diagrammatic fashion in Figure 8. Each realm is characterized by the relative competence and net results of the processes obtaining therein, some processes being of major importance and others of a secondary or minor degree of importance (Table VIII). The relative operational importance of the processes in any individual case depends to a large extent on the character of the rock materials affected.

TABLE VIII

SUBSURFACE PROCESSES BY WHICH SEDIMENTARY COMPONENTS ARE ALTERED OR OBLITERATED AND NEW COMPONENTS ADDED IN SUBSURFACE ENVIRONMENTAL REALMS

Processes	*Realms of Weathering*	*Realms of Diagenesis*	*Realm of Anamorphism*
KATAMORPHIC ALTERATION. Mineral changes through processes of carbonation (desilication), oxidation, hydration, *et cetera*	MAJ	MIN	
ANAMORPHIC ALTERATION. Simple recrystallization without mineral change		MIN	MAJ
ANAMORPHIC ALTERATION. Recrystallization with mineral change through processes of dehydration, silication, deoxidation		MIN	MAJ
SOLUTION. Partial or complete removal by being dissolved	MAJ	SEC	MIN
CHEMICAL PRECIPITATION. Deposition from solution in aqueous medium. Precipitation in optical continuity on grains or as discrete crystal entities	SEC	MAJ	MIN
METASOMATISM. Mineral change by chemical replacement by solutions	MIN	MAJ	SEC
MECHANICAL INTEGRATION. Compaction of particles, including welding under high pressure conditions	MIN	SEC	MAJ
DIASTROPHIC MECHANICAL DISINTEGRATION. Breakage by diastrophic stress beyond elastic limit; granulation, brecciation	MIN	SEC	MIN
GRADATIONAL MECHANICAL DISINTEGRATION. Breakage by gradational stresses resulting from action of wind, water, ice, insolation, and plants and animals	MAJ	SEC	

MAJ—Major importance. SEC—Secondary importance. MIN—Minor importance.

DIAGENESIS

The term *diagenesis* has been involved in a diversity of geologic usage. Twenhofel[48] includes under the term all modifications of sediments under conditions of temperature and pressure normal to the surface or outer portion of the crust, and under environmental conditions which do *not* result in weathering or

[48] W. H. Twenhofel, *Treatise on Sedimentation*, 2d ed. (1932), p. 108. The Williams and Wilkins Company, Baltimore, Md.

destructive changes. By stipulating "under normal conditions of temperature and pressure" he automatically excludes most changes occurring in the high temperature-pressure anamorphic realm, either under deep-seated conditions or in shallower zones by dynamic or contact metamorphism. By excluding "katamorphic changes having for their objective the disaggregation of the sedimentary materials," he rules out most of the changes occurring in the atmosphere and interstitially down to the water table, that is, changes occurring in the zone of aeration or Van Hise's Belt of Weathering.

The essence of diagenesis and the extent of its domain are subjects which demand geologic attention. It seems that diagenesis is not a process or resultant in itself but is rather an integration of processes (Table VIII) and resultants by which sedimentary rocks and their components become adapted for existence in and to changes of an *aqueous medium*, free or interstitial, under moderate conditions of temperature and pressure, and in an environmental setting in which anamorphic and katamorphic alterations play a minor or incipient rôle. Temperature-pressure conditions strongly influence diagenesis and vary from those obtaining at the surface to those which cause the gamut of anamorphic alterations to be set in operation. Inasmuch as diagenesis does not necessarily or in all cases include cementation and lithification, Van Hise's term *Belt of Cementation* is not strictly applicable to this realm. Lithification is just an episode which may or may not, but ordinarily does, occur as a result of the series of changes which constitute diagenesis.

As indicated by Table VIII, chemical precipitation and metasomatism or replacement are the major diagenetic processes; aqueous solutions are the dominant operative agents. There are three potential sources of the precipitated and replacing materials: (1) from the weathering realms by solution and removal, (2) from igneous emanations, and (3) from solution and removal within the diagenetic realm itself. Metasomatism commonly is selective in its attack and may allow replacement of certain constituents of a sediment without affecting the remainder. Aragonite, as a simple example, is much more prone to the influence of metasomatising solutions than is calcite. Of secondary importance in diagenesis are the processes of mechanical integration or compaction, diastrophic disintegration or breakage, and solution and removal. Katamorphic and anamorphic alteration processes are of minor importance.

Realms of diagnesis.—Diagensis does not function uniformly throughout its sphere of influence but rather shows some differentiation into realms, in which its processes operate with different degrees of rigor and thereby produce somewhat different effects. Diagenesis in the *surficial* or *hydrospheric realm* (I-B) is influenced by the free mobility of the aqueous medium and by transportation of sedimentary materials. Among the processes which may occur is metasomatism, including primary dolomitization, silicification, sideritization, and so forth.

Transitional or *near-surface diagenesis* (II-B) is poorly understood, but the few studies that have been made indicate that it is capable of producing marked changes in the incorporated sediments. For example, Twenhofel's study[49] of the diagenetic changes occurring in the deposits of Lake Mendota near Madison, Wisconsin, have shown that there is a marked gradation within a 10-foot section of sediment, from a deep black soup-like mixture filled with carbonaceous matter on the lake bottom down to a white wet mud in which there is little carbonaceous matter 10 feet below bottom. It is suggested that microörganisms are responsible for this progressive downward dissipation of the organic materials. Many near-surface marine sediments are known to support a teeming population of microscopic and macroscopic organisms, and it is believed that most marine sediments of slow accumulation make many passages through intestinal tracts, these passages producing attritional effects, eliminating organic matter, and causing other chemical alterations. Fossil remains also may be nearly or completely destroyed by this action.

The transitional diagenetic realm is one of maximum change and adjustment to subsurface aqueous conditions. Materials most prone to redistribution are unstable minerals, dispersible colloids, and chemically reactive substances, and include colloidal silica, calcium phosphate, calcium carbonate, ferrous carbonate, iron disulphides, and several other compounds. In addition to organic processes, others are active, such as compaction, metasomatism, and incipient anamorphic alteration, the latter including the abundant development of fine white mica in some clays and shales soon after deposition. The results of some marine dredging have shown that in many cases lithification is accomplished in this transitional realm.

The *subsurface diagenetic realm* (III-B) differs from the transitional in that it lacks the ready interchange of interstitial solutions with surface waters and in that temperatures and pressures are slightly to considerably higher. Thus the processes differ in their competence; the activities of organisms practically cease, compaction becomes more effective, and stress conditions are initiated that may cause fracturing of certain of the components. Reducing conditions tend to prevail in this realm, especially if circulation is sluggish, and highly oxidized substances are likely to suffer a degree of reduction. Metasomatism may cause great changes, such as secondary dolomitization, phosphatization, and silicification of limestones. Some components may be dissolved and obliterated; others may undergo simple recrystallization; concretions may form. Many little understood alterations occur which deserve investigation, such as the formation of sedimentary feldspars, rutile, leucoxene, and glauconite. Thus many of the attribute parameters and properties of the primary sedimentary assemblage may be changed, textures and structures partially or completely destroyed and new ones instituted.

The British geologist Smithson[50] is pioneering in the study of the mineralogic

[49] W. H. Twenhofel, "The Physical and Chemical Characteristics of the Sediments of Lake Mendota, A Fresh Water Lake of Wisconsin," *Jour. Sed. Petrology*, Vol. 3, No. 2 (1933), pp. 68–76.

[50] F. Smithson, "The Alteration of Detrital Minerals in the Mesozoic Rocks of Yorkshire," *Geol. Mag.*, Vol. 77, No. 2 (1941), pp. 97–112.

and textural changes wrought by subsurface diagenesis, these changes being designated as "intrastratal changes" by some workers. Smithson's recent studies have shown that mineralogic variations within certain British strata, variations formerly thought to be due to initial distributive differences in sedimentation, are actually the result of differential post-depositional diagenetic action. He points out that there are three possible factors which may cause differences in the heavy-mineral assemblages in different localities: (1) the existence at the time of deposition of two or more contemporaneous distributive provinces, (2) differences in the age of the rocks, (3) areally differentiated mineralogic changes subsequent to deposition. The importance of the latter of these factors is coming to be more and more realized; failure to take it properly into consideration may lead the petrographer to erroneous conclusions, particularly on questions of provenance. Smithson has shown that mineralogic changes in Triassic and Jurassic rocks in Yorkshire have been particularly active in certain belts and have led to the general impoverishment of the detrital suite through the decomposition of garnet, staurolite, possibly monazite, and so forth. Outward from these belts the mineral assemblage tends to increase in richness. Other changes such as the regional development of authigenic brookite and of outgrowths on zircon also have occurred.

WEATHERING

Weathering and *diagenesis* are complementary. Taken together, they comprise the gamut of changes which rock materials suffer under the moderate, non-anamorphic temperature-pressure conditions of the outer portions of the lithosphere. Minerals formed under weathering and diagenetic conditions are in general few in number, chemically simple, and with low specific gravities.

Weathering, however, rather than denoting adaptation to totally aqueous conditions, stands for the sum total of processes and resultants by which rock components become adapted for existence in and to changes of an *aerial medium*, atmospheric or interstitial, under the influence of oxygenated and slightly acid meteoric (vadose) waters. All of the processes obtaining in the weathering realms occur also under diagenetic conditions; the relative importance of these processes, however, differs greatly in the two cases. Katamorphic alteration by carbonation (desilication) and oxidation is the dominant process under weathering conditions and of minor import below the water table in the diagenetic realms where these processes are limited, as a rule, by the small amounts of CO_2 and O_2 present. Removal by solution is important in the weathering realms and usually of secondary importance under diagenetic conditions where the complementary process of precipitation from solution dominates the scene. Gradational mechanical processes, such as spalling, exfoliation, splitting and abrasion, are important in the surficial environments of weathering. Plants and animals play a significant rôle in the chemical and mechanical changes; organic acids are generated and further the weathering processes.

Realms of weathering.—Weathering, like diagenesis, does not function uniformly but shows a vertical differentiation. *Surficial weathering* (I-A of Fig. 8) takes place under the full sweep of the atmospheric gradational agencies and is influenced by pronounced temperature changes and climatic vagaries. Surficial plants and animals play important direct and indirect rôles. The *transitional weathering realm* (II-A) is one of disaggregation, of maximum change and adjustment to surface climatic conditions; here the stage is set for reduction of the surface by erosion. That there is a differentiation of processes and products within this realm is indicated by the development of the soil profile, due partly to the selective redistribution of mineral and organic matter by vadose water migrating through the soil. The most potent climatic factor influencing transitional weathering is the amount of rainfall and the resulting moisture content of the soil. The *subsurface realm* (III-A), extending down to the water table, is for the most part protected from surficial climatic vagaries; thus the weathering processes differ in competence, and plant and animal activities are of little consequence.

REALM OF ANAMORPHISM

As indicated by Table VIII, this realm, with its dynamic, contact, and static phases, permits the functioning of several of the processes common to the realms of diagenesis and weathering but is differentiated by the dominance of anamorphic alteration, occurring as simple recrystallization without mineral change or as recrystallization with mineral change through the processes of silication, dehydration, deoxidation, and so forth. Environmental adaptation results in the production of pyroxenes, amphiboles, chlorite and micas, and other platy or columnar silicate minerals from the carbonates, hydrates, oxides and other components of sedimentary rocks. In this way dense crystalline slates, schists, and gneisses are formed and environmental equilibrium approached. The contact phase of anamorphism furnishes high temperature and pressure conditions for the acceleration of anamorphism in the more deep-seated regions and for introducing the anamorphic realm to regions nearer the earth's surface where diagenesis or weathering normally dominates the scene. Thus the anamorphic realm has no depth restriction.

In the high temperature-pressure environment of this realm, deformation is largely by chemical processes, in contrast to the dominantly mechanical deformation of the diagenetic and weathering realms; mechanical work nevertheless occurs, in the form of further compaction, welding, and granulation. Precipitation and cementation here are of far less consequence, and metasomatism occurs but differs in character from that of the other realms.

REALM OF FUSION

Once a sedimentary rock becomes fused, the resultant material is a magma, and the crystalline rock which forms by cooling has the characters of an igneous rock. Thus, sedimentary identity is partially or completely lost, depending on how extensive the melting is, whether by such partial processes as *lit-par-lit* injection or by anatexis or complete regeneration (palingenesis). When this environmental stage is reached, the rock materials leave the sedimentary domain and thus are no longer pertinent to the subject of the present paper.

EPIGENETIC PROCESSES AND CHANGES AFFECTING CRINOIDAL SEDIMENT

Diagenesis.—The outstanding effect of diagenesis on the Grand Tower limestone crinoidal phase (Bed 9) has been cementation by epigenetic precipitation of calcite in the interstices between the crinoid plates and quartz grains (Pl. I). Lithification by this process may have been partially or completely accomplished soon after burial and in the early or transitional stage of diagenesis (Fig. 8).

The *MgO* content of the crinoidal limestone, treated in the section on chemical composition, is only 0.36 per cent, which is much lower than it should be judging from results of chemical analyses of modern crinoids. In a paper by Clarke and Wheeler[51] the *MgO* content of modern crinoids from various localities in the Atlantic and Pacific was described as being from 2.49 to 5.13 per cent, averaging approximately 4.25 per cent; the calcium carbonate of the plates was found to be in the form of the mineral calcite, not aragonite. Clarke and Wheeler also found from analyses of crinoids from Ordovician to Eocene in age that the remains of these fossil forms contained considerably less *MgO*, nine of the ten crinoids analyzed averaging only 0.75 per cent *MgO*. According to their reasoning, it is possible that the ancient crinoids may have been deficient in magnesia, but it is more probable that the loss is due to alteration during diagenesis, perhaps to the infiltration and precipitation of calcium carbonate in the somewhat porous crinoidal plates. This obviously would lower the apparent proportion of magnesia. Thus the magnesia deficiency in the composite analysis of the Grand Tower crinoidal limestone, compared with analyses of remains of modern crinoids, probably is due at least in part to diagenetic infiltration of calcite in the crinoid plates and to the presence of interstitial calcite cement and other minerals (mainly quartz), which reduce the relative proportion of *MgO* in the chemical analysis. Part of the magnesia deficiency may have been caused by solution and leaching during diagenesis, but calcium carbonate should then have been removed to a greater extent than magnesium carbonate.

Diagenesis has brought about a change in the mass texture of the crinoidal sediment from a fragmental-textured sand, at the time of deposition, to a crystalline-textured limestone. This change is due largely to interstitial precipitation of calcite, binding the aggregate so as to form a calcareous mosaic of crystals, each crinoid plate representing an individual calcite crystal.

The thin-section photomicrographs of Plate I show the crinoid plates to be irregularly fractured. It is probable that this is the result of diastrophic breakage,

[51] F. W. Clarke and W. C. Wheeler, "The Composition of Crinoid Skeletons," *U. S. Geol. Survey Prof. Paper 90-D* (1914), pp. 33–37.

occurring during diagenesis when the rock was subjected to stress conditions which attended faulting of the region. Intense faulting is known to have occurred in the Little Saline Creek area, in the vicinity of Ozora, soon after Middle Devonian time.

Weathering.—One notable effect of weathering on the illustrative crinoidal rock is the local katamorphic alteration of magnetite grains to limonite, producing slightly yellowish patches in the rock. Although it is possible that the limonite grains may have been deposited as such, it is more probable that they have been formed by epigenetic alteration (oxidation and hydration) of magnetite grains, some of which remain unaltered in the deposit; this alteration probably was accomplished by vadose waters in the belt of weathering.

Another probable effect of weathering on the crinoidal rock is the increase in porosity of the rock aggregate as a result of solution and removal of calcite locally along minute channel ways in the rock. As described under porosity, the crinoidal limestone in the section sampled has an average effective porosity of 7.2 per cent; much of this probably is due to solutional removal in the belt of weathering.

Reprinted from:
BULLETIN OF THE AMERICAN ASSOCIATION OF PETROLEUM GEOLOGISTS
VOL. 37, NO. 5 (MAY, 1953), PP. 940–960, 7 FIGS.

REGIONAL STRATIGRAPHIC ANALYSIS OF "UPPER CRETACEOUS" ROCKS OF ROCKY MOUNTAIN REGION[1]

W. C. KRUMBEIN AND F. G. NAGEL[2]
Evanston, Illinois, and Salt Lake City, Utah

ABSTRACT

Lithologic and thickness data on the "Upper Cretaceous" rocks of the western states were collected in the field and from the literature. The data were arranged into isopach-lithofacies maps showing average lithology of the total interval, and of the Dakota-Colorado and Montana-Laramie groups. Vertical variability maps of the number of sandstones in the units were also prepared.

The isopach-lithofacies maps indicate a general thickening and increasing sand content in all units from east to west. The facies lines are roughly parallel with the isopachs, and suggest a general westerly source for most of the sediments, although local sources were operative at times. The number of sandstone beds is greatest along an intermediate belt, indicating a major intertonguing of sand and shale in the general area of the average shoreline trend.

Detailed study of sand and shale lithology in sections from Arizona, New Mexico, and southwestern Colorado indicates that most of the sands are relatively clean subgraywackes, decreasing in angular components northeastward. The shales are mainly silty, and the association of lithologic types suggests unstable shelf conditions over most of the area.

INTRODUCTION

Just a century ago, in 1853, Meek prepared the first columnar section of Upper Cretaceous rocks in the western interior of the United States (Hall and Meek, 1856). In the ensuing century a vast amount of published material has accumulated on Upper Cretaceous stratigraphy in that area. In the past quarter century, moreover, the outcrop literature has been supplemented by a large number of well records, which make available a truly three-dimensional view of this large body of rocks.

Much of the literature on the Upper Cretaceous is descriptive rather than interpretive. Most writers were concerned with relatively small areas, such as quadrangles or mining districts, but their publications commonly afford specific data for regional analysis. About 1,000 publications were examined in the preparation of this study, of which about 300 yielded information that could be used directly.

Relatively few papers on the Cretaceous deal with regional aspects of stratigraphic investigation, but some threads of continuity can be followed from the earliest studies. As long ago as 1876, Meek established his "lower series" (Dakota, Fort Benton, and Niobrara) and "upper series" (Fort Pierre and Fox Hills). He did not include the Laramie group, which was then considered to be Tertiary. In 1876 White applied the term "Colorado group" to the Fort Benton and Niobrara, and in 1889 Eldridge proposed "Montana group" for the Fort Pierre and Fox Hills. These groupings are in current use.

[1] Based on a dissertation for the degree of Master of Science, submitted by the junior writer to the Graduate School, Northwestern University, 1951. Manuscript received, October 6, 1952.

[2] Department of Geology, Northwestern University, and Ohio Oil Company. The writers are indebted to their colleagues at Northwestern University, especially E. C. Dapples and L. L. Sloss, for helpful suggestions and critical reading of the manuscript. The junior writer expresses grateful acknowledgment to The Ohio Oil Company for permission to publish this report.

Among outstanding publications, which carried on the regional tradition, are White's monograph (1891), on the Cretaceous of North America, Stanton's monograph (1893) on the Colorado formation, Lee's studies (1915) in the Four-Corners area, Bartram's paper (1937) on the Rocky Mountain area and the contributions of Sears, Hunt, and Hendricks (1941) on the San Juan area, of Spieker (1946) on central Utah, and of Pike (1947) on the San Juan basin. In 1944 Reeside published a comprehensive series of isopach maps on the Cretaceous. A recent symposium on sedimentary facies includes papers by McKee (1949) and Spieker (1949) in Arizona and Utah.

The present paper is designed to follow in this tradition by integrating published information on the Upper Cretaceous in the form of isopach and lithofacies maps. The area under consideration extends from Canada to Mexico, and from western Kansas to eastern Nevada. Figure 1 shows the area covered, and indi-

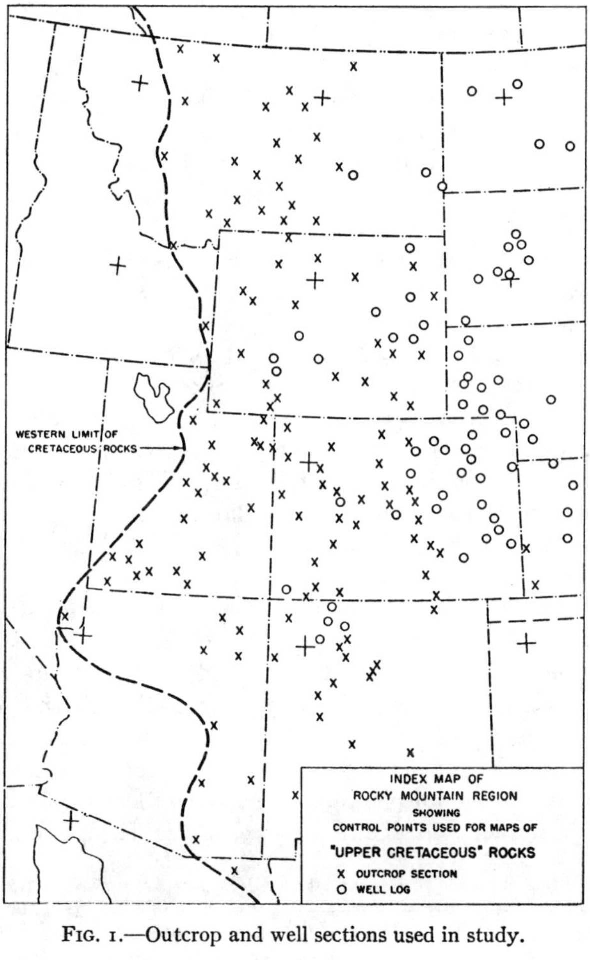

FIG. 1.—Outcrop and well sections used in study.

cates control points. The project was started as an experimental study under the supervision of the senior writer, and was expanded into a Master's thesis by the junior writer.[3] The chief purposes of the paper are to present the broad pattern of thickness and lithologic variation in the Upper Cretaceous and to introduce methods for expressing the vertical variability, as well as the lateral variation in the section. These maps provide a basis for broad interpretation of tectonic and environmental conditions during accumulation of the rocks.

SELECTION OF STRATIGRAPHIC UNITS

The stratigraphic unit selected for study comprises all the rocks from the base of the Dakota sandstone to the top of the Laramie and equivalent beds. This unit has generally been considered as comprising the Upper Cretaceous, although recent work by Cobban and Reeside (1951) and Cobban (1951) shows that the Upper-Lower Cretaceous boundary lies above the Dakota in parts, at least, of the Rocky Mountain province.

The present study is based on an "operational stratigraphic unit," whose base is the relatively easily recognized base of the Dakota sandstone and whose top is the generally accepted upper limit of the Cretaceous. In terms of Rocky Mountain Cretaceous stages, the major unit extends from the base of the Dakotan to the top of the Laramian, and the sub-units are separated at the division between the Coloradoan and Montanan.

The problem of locating the exact upper and lower boundaries of the major unit had to be solved arbitrarily in some localities. As one example of boundary selection, the upper limit in the Wasatch Plateau (Spieker, 1946) was placed approximately at the middle of the North Horn formation. Selection of the lower boundary was generally more satisfactory, although in the absence of typical Dakota sandstone the decisions had to be based on somewhat arbitrary grounds. In western Montana, for example, the Colorado shale rests conformably on the Kootenai; and in the Wasatch Plateau the Dakota is not recognizable as such in the thick Indianola section. During the compilation stages of this study, a correlation network of measured sections and wells was set up to carry the selected limits over the map area.

In most instances the limit selected is believed to be consistent with the regional picture. Seldom does an arbitrary choice involve more than a few per cent of the total thickness of the major unit. Such adjustments, though questionable for detailed studies of local areas, exert only a slight effect on the broader regional aspects. Thickness is more frequently affected than facies, and the net effect is probably a slight increase or decrease in facies and isopach rates of change.

The selection of the middle boundary, between Coloradoan and Montanan, depended more on faunal evidence than on recognizable rock units. Fortunately,

[3] The complete thesis is on file in Northwestern University Library. The present paper greatly condenses the full discussion presented in the thesis, and only touches upon the detailed lithologic analysis of samples from the Black Mesa area, which forms an essential part of the study.

the top of the Coloradoan is nearly everywhere marked by an adequate assemblage of fossils. In summary, therefore, the major mapping unit used in this study has a lower surface based as nearly as possible on rock correlation, and transgresses time planes. The upper surface is as nearly as possible a time-rock surface and is believed to be essentially parallel with time planes. The middle division is based on biostratigraphic data and is believed to be essentially time-parallel. To the degree in which these objectives have been met, the upper sub-unit (Montanan-Laramian) is a time-rock unit, and the lower sub-unit (Dakota-Coloradoan) is an operational unit with time-rock top and rock-unit bottom.

Three sets of maps were prepared, which included first an isopach-lithofacies map of the total interval and both sub-units; and second, a net-sand-isopach and number-of-sands map of the total interval and of the sub-units. These six maps furnish the basis for the interpretive discussion.

METHOD OF PREPARING MAPS

The base map used in this study was the Kelly 1:2.5 million base, which shows townships and permits accurate location of wells and outcrops. Reliance was placed mainly on published material, which resulted in relatively open control on the maps. The total map area (Fig. 1) is approximately 700,000 square miles, and the control consists of 206 outcrop or well sections. This is one control point per 3,500 square miles. For the map scale and stratigraphic unit used, one point per 2,500 square miles would be more satisfactory for discerning regional trends.

The density of control is not evenly distributed over the area, and in some places, notably Colorado and parts of Wyoming and Utah, the control approaches optimum spacing. In other areas, notably North Dakota, the spacing is very wide. In part, selection of control points (where a choice is available) should be based on relative rates of facies change, which concentrates control in areas of rapid change and spreads it out in more uniform areas.

The data at each control point were evaluated for completeness and quality. Most of the published well logs and measured sections provided adequate data, but in some areas incomplete or generalized sections had to be fitted into the framework. A consistent policy of terminology was followed as far as possible. Emphasis in the study was placed on the sand-shale ratio and on the number of sandstone units present. In evaluating the data, conglomerate, sandstone, volcanic tuff, and agglomerate were included as coarse clastics. Fine clastics included claystone, mudstone, siltstone, and all varieties of shale. Shaly sandstone was counted as three-fourths coarse and one-fourth fine clastic, whereas silty sandstone and sandy siltstone were taken as coarse clastics.

In counting the number of sandstones, an operational procedure was adopted to avoid subjective errors as far as possible. In tallying the number of sandstone beds at each control point, only those sandstones (and conglomerates) more than 10 feet thick, and separated from similar strata by at least 10 feet of intervening non-sand, were counted. Thus, two 50-foot sandstone beds separated by a 3-foot

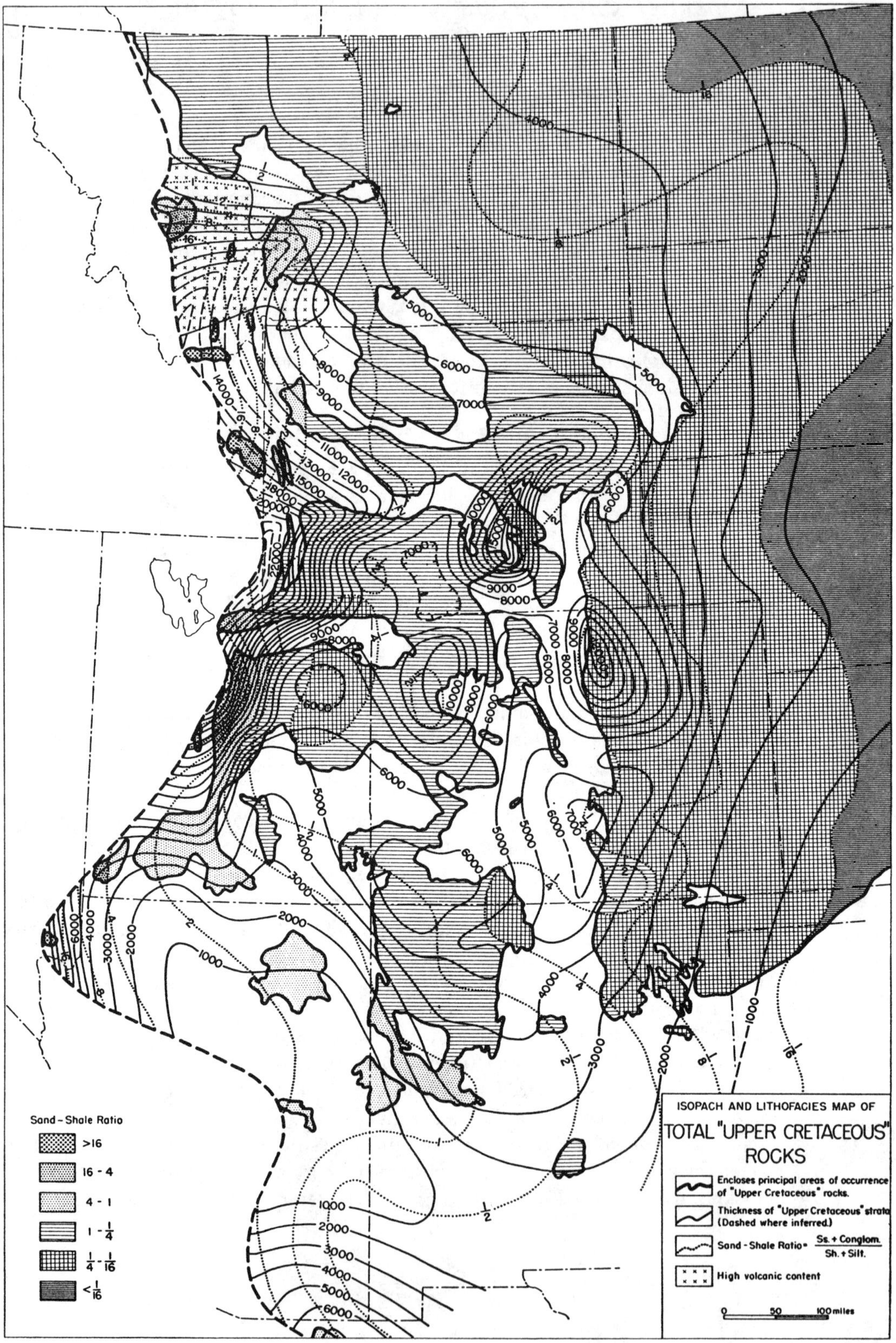

FIG. 2.—Isopach and sand-shale ratio of total operational unit.

shale parting were counted as one sandstone; whereas two 10-foot sandstone beds separated by 15 feet of non-sand were counted as two. Thin sands, less than 10 feet thick, finely interbedded with shales, were not counted at all.

The non-clastics (limestone, coal, *et cetera*) in the section were included in the total isopach interval. Their thickness was recorded, and a clastic ratio computed, but the values were not included on the maps. With few exceptions the content of non-clastics was significantly less than 10 per cent of the total section, so that emphasis was placed on the sand-shale ratio for analysis.

The first stage of mapping included preparation of isopach maps for the total unit and the two sub-units. A definite attempt was made to avoid undue reconstruction in the maps, inasmuch as the intention was to show the present distribution of thickness and facies. Some compromise was necessary, however, inasmuch as the Upper Cretaceous rocks have been wholly or partly eroded from many parts of the region. The procedure adopted was to indicate with heavy lines the present areas of occurrence of the rocks, but to extrapolate thickness lines across areas where the Upper Cretaceous was reduced or completely removed since Miocene time. Justification for such extrapolation is based on Lee's demonstration (1915) that the deposits originally extended across the present Rocky Mountains.

In the far western part of the region, erosion has been so profound that it is now almost impossible to differentiate the effects of pre-Miocene and post-Miocene denudation. Hence, extrapolation of lines in this area is largely conjectural, although it is felt to be a reasonable interpretation. The resulting isopach maps were used as base maps for gross lithofacies maps. In similar manner, the facies lines were extrapolated, but the map patterns are shown only in areas of present occurrence of the rocks.

ISOPACH AND LITHOFACIES OF TOTAL "UPPER CRETACEOUS"

Figure 2 is a combined isopach-lithofacies map of the total operational unit. The heavy lines with superimposed patterns indicate areas where rocks of this unit are now present, either in outcrop or beneath cover. The heavy dashed line on the west is tangent to presently known limiting outcrop areas; it represents the line beyond which there is no control in the map area.

The isopachs, drawn on 1,000-foot intervals, indicate that the unit is thickest at its western edge and thins toward the east, although several locally thickened areas complicate the primary wedge-like configuration of the sedimentary body. The thickest western accumulations exceed 20,000 feet; the general linearity of this thickest belt, and the rapid rate of change eastward from it indicate the presence of a Cretaceous orthogeosyncline. The general trend of isopach lines suggests that this element extended southeastward across Arizona into northern Mexico. Presence of volcanics in western Montana implies eugeosynclinal conditions there.

Intracratonic depositional basins are suggested by the thicker areas farther east. Most prominent of these is in southern Wyoming, in the site of the present Hanna Basin, where the unit is 16,000 feet thick. In north-central Colorado, east of the Front Range, is another basin with about 12,000 feet of Upper Cretaceous rocks. A third well defined basin lies in northwestern Colorado, with thicknesses exceeding 10,000 feet. A possible fourth basin lies in southeastern Colorado. It is suggested by a single control point at Florence, but erosion has stripped out most of the critical area.

In three areas the unit is notably thin. These lie in the present Washakie basin of southern Wyoming, parts of the Uinta basin of northeastern Utah, and much of eastern Arizona, where the rocks are less than 1,000 feet thick. It is believed that part, if not all, of the thinning in these areas is a function of post-depositional erosion. In the northern Uinta basin, for example, Cretaceous beds are overlain by Tertiary with angular unconformity. Deep drilling within the basin may disclose original thicknesses of the rock.

There seems to be no evidence for regional thickening of the stratigraphic unit in the present Big Horn, Powder River, San Juan, or Black Mesa basins. Evidently, these represent Tertiary events.

The lithofacies pattern in Figure 2, based on the sand-shale ratio (Krumbein and Sloss, 1951, p. 271), yields an essentially simple pattern. Coarse clastics predominate in the western part of the area (ratio values greater than 1.0); elsewhere finer clastics are prevalent. With few exceptions the sand-shale ratio increases from east to west, and the rate of change is gradual until it approaches the western orthogeosyncline. The ratio lines have been contoured on a geometric basis, which brings out the exponential rate of change, inasmuch as each westward increasing contour value doubles the proportion of coarse clastics.

Several anomalies in the pattern deserve comment. The high shale area in northeastern Utah results from a combination of depositional and erosional factors. There is a large proportion of shale in the lower sub-unit, and the upper more sandy phases of the Montana-Laramian were stripped off, reducing the sand-shale ratio value. Two salients of dominant coarse clastics extend eastward from the western margin. One of these projects into west-central New Mexico, and the other is associated with the volcanic area in Montana. The volcanic area is particularly noteworthy. Lava and other extrusive materials occur near the base of the Upper Cretaceous in the extreme western part, and rise progressively in the section eastward, continuing into the Tertiary. Pyroclastic materials also occur notably in the northern San Juan basin, although no actual flows are known.

The map of the total unit is of interest in showing the over-all features of the Upper Cretaceous. In essence, it answers three questions: what is the geographic distribution of the rocks? what is the regional pattern of its thickness variations? and, what is the average proportion of coarse and fine clastics within it? The map has principal value as an orienting device, and is suitable for very broad environmental and tectonic interpretation. If greater detail is desired, the basic

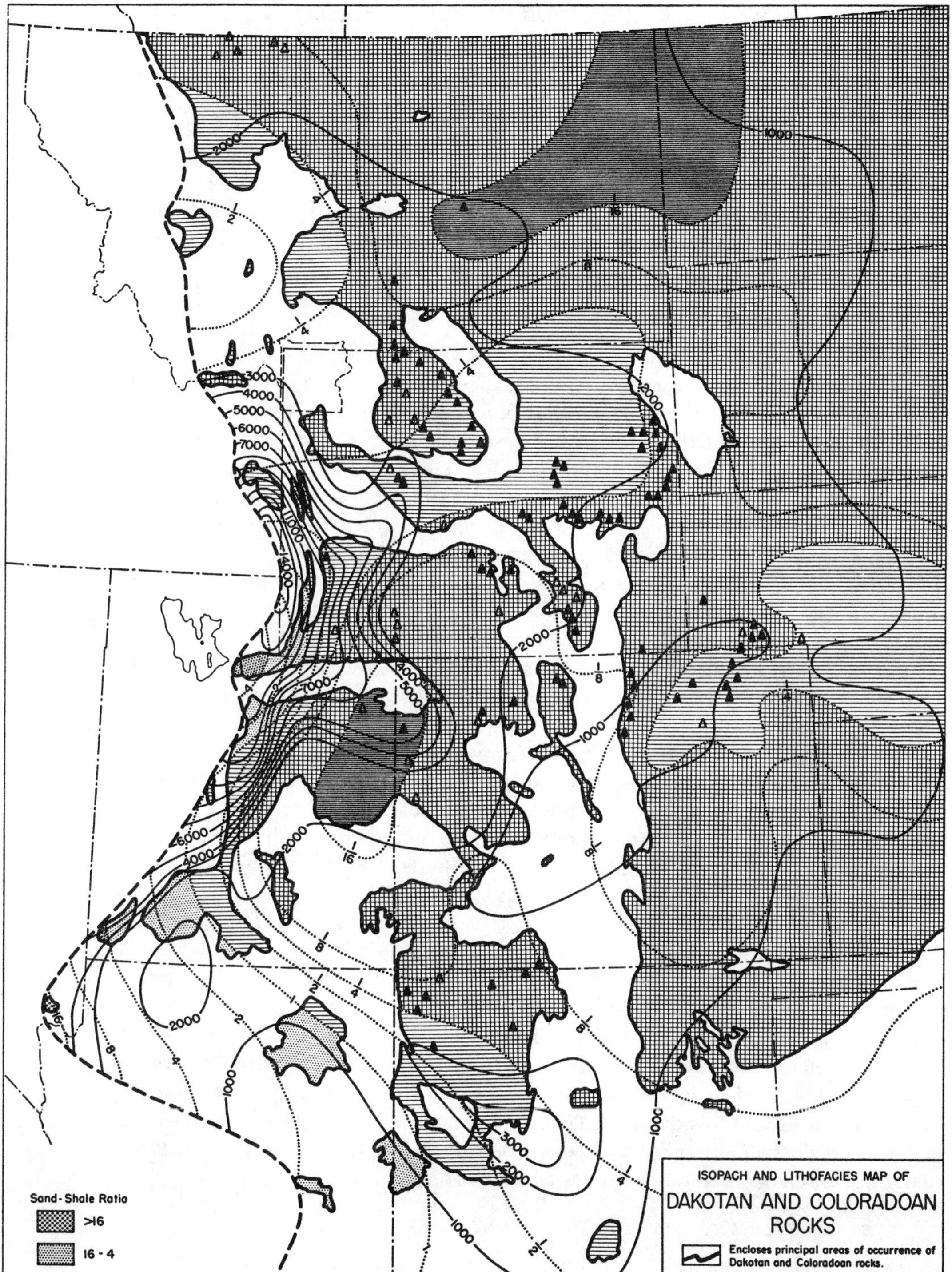

FIG. 3.—Isopach and sand-shale ratio for lower sub-unit.

data would be reassembled to show the distribution of specific lithologies (types of sandstone, marine *vs.* non-marine shale, *et cetera*) in the unit.

In the present study, two extensions of the basic map were prepared. The first involves the isopachs and gross lithology of the Dakota-Coloradoan and the Montanan-Laramian, and the second shows the distribution of sandstone bodies within the two sub-units. The first extension contrasts the coarse and fine clastics of the upper and lower part of the major unit, and the second shows how the sand bodies are distributed through each sub-unit.

ISOPACH AND LITHOFACIES OF DAKOTAN-COLORADOAN ROCKS

Figure 3 shows the present distribution and gross characteristics of the lower part of the "Upper Cretaceous" section. In general, the rocks are relatively evenly distributed over the Rocky Mountain region, with the thickest parts in the west-central part of the map area. These western thickened parts represent the orthogeosynclinal belt. A prong of thick accumulation extends eastward along the general area of the Uintas. This accumulation and its relation to an inferred extra-cratonic source suggest an exogeosyncline. Thickening in west-central New Mexico suggests development of a mild intracratonic basin. The thin area in the belt trending northeast-southwest through Colorado does not appear to be due to post-depositional erosion. Evidently, this area remained slightly higher, or subsided less rapidly than surrounding areas.

There is no regional evidence of unusual thickening of the Dakotan-Coloradoan sub-unit in the present Hanna Basin, northwestern Colorado, or other areas where the complete unit displayed stronger development. Evidently, all the intracratonic basins shown in Figure 4 developed in late Upper Cretaceous time.

The sand-shale ratio in Figure 3 shows that the Dakotan-Coloradoan rocks are overwhelmingly dominated by fine clastics. Coarse clastics comprise more than 50 per cent of the section in a relatively narrow western belt. Over most of the region shales are four or more times as abundant as sand, and in eastern Montana the gross proportion of sand is less than $\frac{1}{16}$ that of shale.

Two variations in the pattern have interesting implications. The tongue of slightly greater sand content (ratio between $\frac{1}{4}$ and 1) which extends eastward across Wyoming to the Black Hills, and the area of higher sand-shale ratio in northeastern Colorado and western Nebraska, may represent the influence of local eastern source areas. According to Crowley (1951) the Newcastle sand in the Powder River Basin received detritus from the Black Hills. Hunter (1950) concluded that local conglomerates in the Frontier formation indicate uplift in the Big Horn Mountains during Coloradoan time. Lack of control in the Powder River basin does not permit closer definition of the high sand area; it may actually not extend completely across the basin as a continuous sheet.

ISOPACH AND LITHOFACIES OF MONTANAN-LARAMIAN ROCKS

Figure 4 shows the areal distribution, isopachs, and sand-shale ratio of the upper stratigraphic sub-unit. This unit is conspicuously thicker than the lower

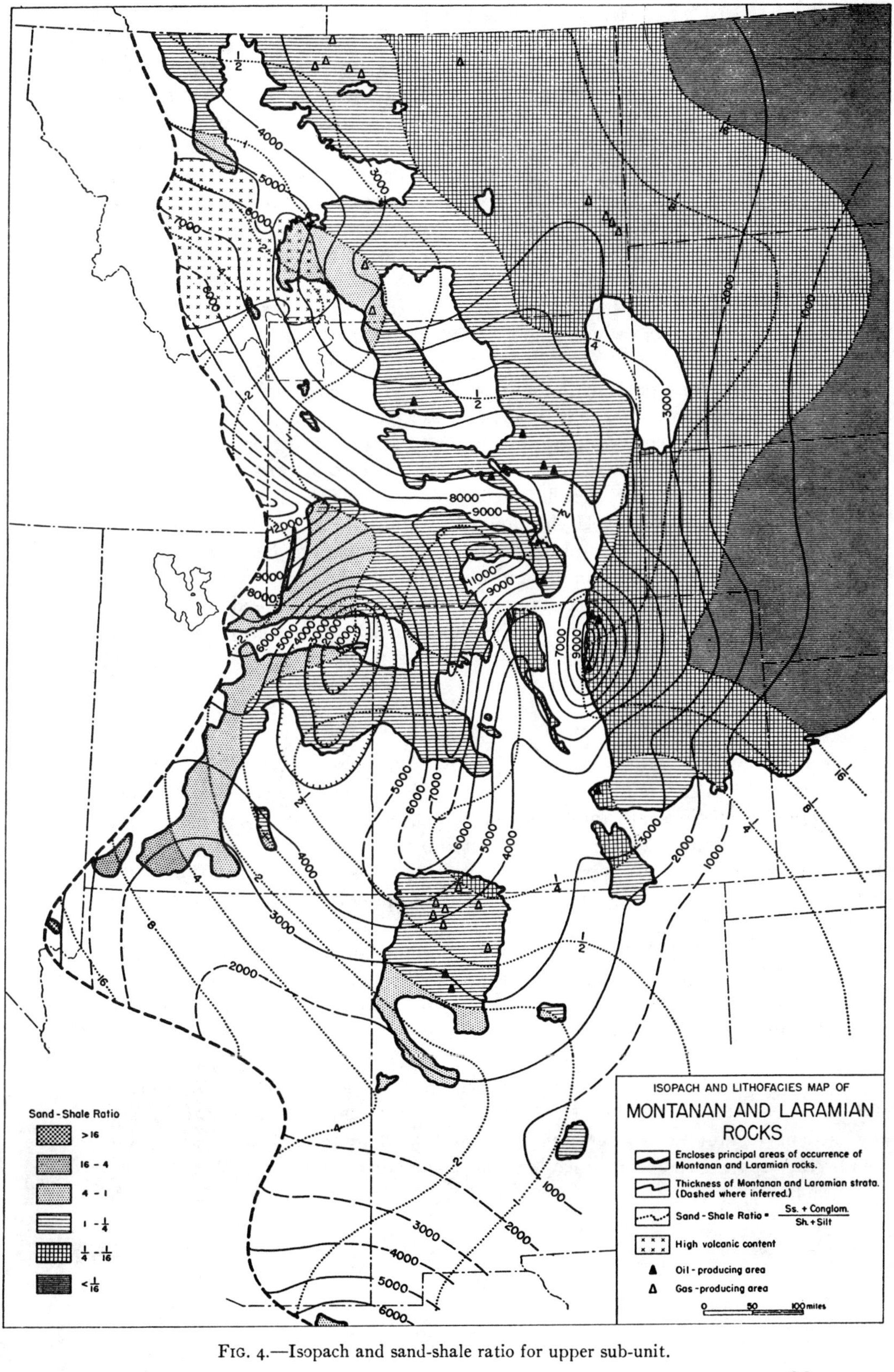

FIG. 4.—Isopach and sand-shale ratio for upper sub-unit.

sub-unit, and shows more diversification in its isopach pattern. Concentrations of isopach lines in southern Wyoming and northern Colorado indicate the presence of two well developed intracratonic basins. There is some evidence of thickening in northwestern Colorado, but none in the southern part. The conspicuous thinning in northeastern Utah apparently reflects pre-Miocene erosion. The thin area in eastern Arizona and central New Mexico may represent either non-deposition or post-depositional erosion.

The geosynclinal element along the western margin of the map area is apparently restricted to southwestern Wyoming and adjacent areas, although the pattern may be misleading because of sparse control. The volcanics in western Montana suggest eastward extension of eugeosynclinal conditions there in Upper Cretaceous time.

The sand-shale ratio pattern of Figure 4 shows a marked increase in sand content since earlier Upper Cretaceous time. Sand exceeds shale in a broad, more continuous belt along the western margin of the map area, and the pattern shows a relatively gradual and regular decrease eastward. Three eastward trending lobes of high sand content may be seen, however. One is associated with the volcanics in Montana; one in Wyoming-Utah may reflect uplift in the Uintas in late Cretaceous (Walton, 1944); and the broad tongue in Arizona and New Mexico suggests general eastward migration of sand over a broad shelf area.

The large area of low sand content in northeastern and central Colorado is probably an erosional phenomenon. The dominantly sandy Laramie and Fox Hills formations, at the top of the section, have been eroded, with a consequent decrease in over-all sand-shale ratio.

JOINT INTERPRETATION OF FACIES MAPS

Consideration of Figures 2, 3, and 4 jointly affords some clues to the tectonic framework of "Upper Cretaceous" sedimentation. It should be emphasized that only average conditions are indicated, inasmuch as the main operational unit and the two sub-units contain rocks deposited over a considerable span of time. Only strongly developed or persistent features are shown; minor variations tend to be masked or averaged out, and can be discerned only from larger-scale studies of thinner units.

The three isopach-lithofacies maps have several points in common: (1) the isopach and lithofacies lines exhibit a rough regional parallelism, suggesting a fairly uniform relation of sedimentation to contemporaneous tectonism; (2) isopach and facies lines exhibit a general north-south trend, suggesting a similar orientation of the regional tectonic elements; (3) general thickening from east to west, combined with increasing coarse-clastic content, indicate a principal source of Upper Cretaceous sediments from the west; (4) areas in which the rocks are now absent exhibit no strong influence on the main isopach and lithofacies trends: the Upper Cretaceous once blanketed the entire region; (5) the isopachs and lithofacies lines appear to intersect the western limit of deposits at high angles, suggesting that the rocks once continued farther westward, and were not abruptly terminated in a large eroding land mass; and (6) there is no apparent correlation between present intermontane basins (Big Horn, North Park, Black Mesa, *et cetera*) and configurations of the isopach and lithofacies lines. This suggests that these basins may accordingly be interpreted as Tertiary phenomena.

The preceding generalizations tend on the whole to support earlier conclusions regarding Upper Cretaceous rocks. The following brief account of tectonic and sedimentary events appears to be substantiated by the maps.

The tectonic framework of the "Upper Cretaceous" consisted mainly of an orthogeosynclinal belt in the west, of which some portions of the eugeosyncline are still preserved. The miogeosyncline appears to have been partly compounded of overlapping or *en échelon* subsiding areas rather than one major linear belt. The remainder of the region was cratonic, and consisted mainly of shelf areas. The lithologic associations suggest unstable shelf conditions for the most part. Within the main cratonic area were three or four intracratonic basins, shown mainly by isopach lines. The basins in southeastern Wyoming and northwestern Colorado are reflected by an increase in coarse clastics (Fig. 4). The well defined basin in north-central Colorado and the doubtful one in south-central Colorado are not evident in the lithofacies map patterns.

During Dakotan-Coloradoan time, the principal source area lay in the west and southwest. Rocks comprising this unit were deposited as an almost continuous sheet across the area. A few island sources in the east apparently contributed to sedimentation. In the ensuing Montanan-Laramian interval, the main source area still remained in the west, and volcanics invaded the orthogeosynclinal belt in the northwestern part of the area. Local sources of sediment played an increasing part toward the close of the interval. For example, Walton (1944) showed that the Current Creek formation in northeastern Utah was derived from a source in the Uinta Mountains. The San Juan volcanic area in southwestern Colorado supplied much andesitic débris to the northern San Juan Basin (Zapp, 1949). There is evidence that sands in the Pierre in northern Colorado were at least partly derived from a source in the Front Range area (Griffiths, 1949). The broad environmental picture during Montanan-Laramian time, as clearly developed by various authors (Lee, 1915; Sears *et al.*, 1941; Spieker, 1946; Pike, 1947) is that of a marine invasion from the east, marked by numerous oscillations as continental sediments and coal interfinger with the marine deposits in a complex time-transgressive manner. Pike (1947) shows the inferred shorelines during this oscillation in parts of New Mexico.

SANDSTONE VERTICAL VARIABILITY MAPS

Vertical variability maps are lithofacies maps which show the vertical arrangement of selected rocks in the stratigraphic column (Busch, 1950; Krumbein and Sloss, 1951, p. 411). Such arrangements can be shown in terms of position within the column, average thickness or spacing in the column, or number of occurrences in the column. In this study the number of sandstone occurrences in the stratigraphic unit was chosen for analysis, because of its importance in several connections.

Upper Cretaceous rocks in the Rocky Mountain region are composed dominantly of sand and shale. Hence, the number of discrete sandstone beds in the section at any one control point indicates approximately the degree of sandstone-shale intertonguing at that point. In a broad sense, most of the sandstones are non-marine and most of the shales are marine. The number of sandstones at a given location is therefore a rough indication of the number of marine advances and withdrawals across that area.

From an economic standpoint, the number of sandstones represents the number of possible reservoir beds to be drilled and tested. Likewise, each of the interjacent shales is a possible source rock. Moreover, in geophysical exploration the number and placement of sand-shale interfaces may influence quality and number of seismic reflections.

The sandstone maps were prepared by plotting the total thickness of coarse clastics and the number of sandstones at each control point. Net sand isopachs were first drawn to indicate the variation in total sand and conglomerate thickness within the stratigraphic unit. Contours of the number of sandstone beds were then drawn as an overlay on the isopachs. The "number of sands" contours thus represent the areal variation in the number of discrete sand bodies within the stratigraphic unit. There is no implication that these sands occur as continuous sheets, although certain zones of sands are widely distributed.

By mentally dividing the thickness value at any point by the number of beds, the average thickness of the sandstone beds at that point can be visualized. The values so obtained will be somewhat larger than the true figures, however, inasmuch as the isopachs show total coarse clastic thickness, whereas only sands 10 feet or more in thickness were tallied in the sandstone count.

NUMBER OF SANDSTONES IN "UPPER CRETACEOUS" ROCKS

Figures 5, 6, and 7 are the sandstone maps corresponding with the gross lithology maps of Figures 2, 3, and 4. As before, these maps show average conditions for the stratigraphic units, and do not show conditions at any specific time within the Upper Cretaceous. They are to be interpreted as indicating in a broad way the thickness of total coarse clastics within the units, and the regional tendencies for few or many discrete sands to be developed.

Figure 5 shows that the number of individual sandstones in the total "Upper Cretaceous" exceeds 25 along a belt extending southward from Montana through central Wyoming and Colorado to northwestern New Mexico. The maximum number of sands, about 45, occurs in the present Hanna Basin area. In the eastern part of the map area, there are less than five sands on the average. Net sandstone thickness increases fairly uniformly from east to west, which indicates that the sands thicken and coalesce westward, and thin and pinch out toward the east.

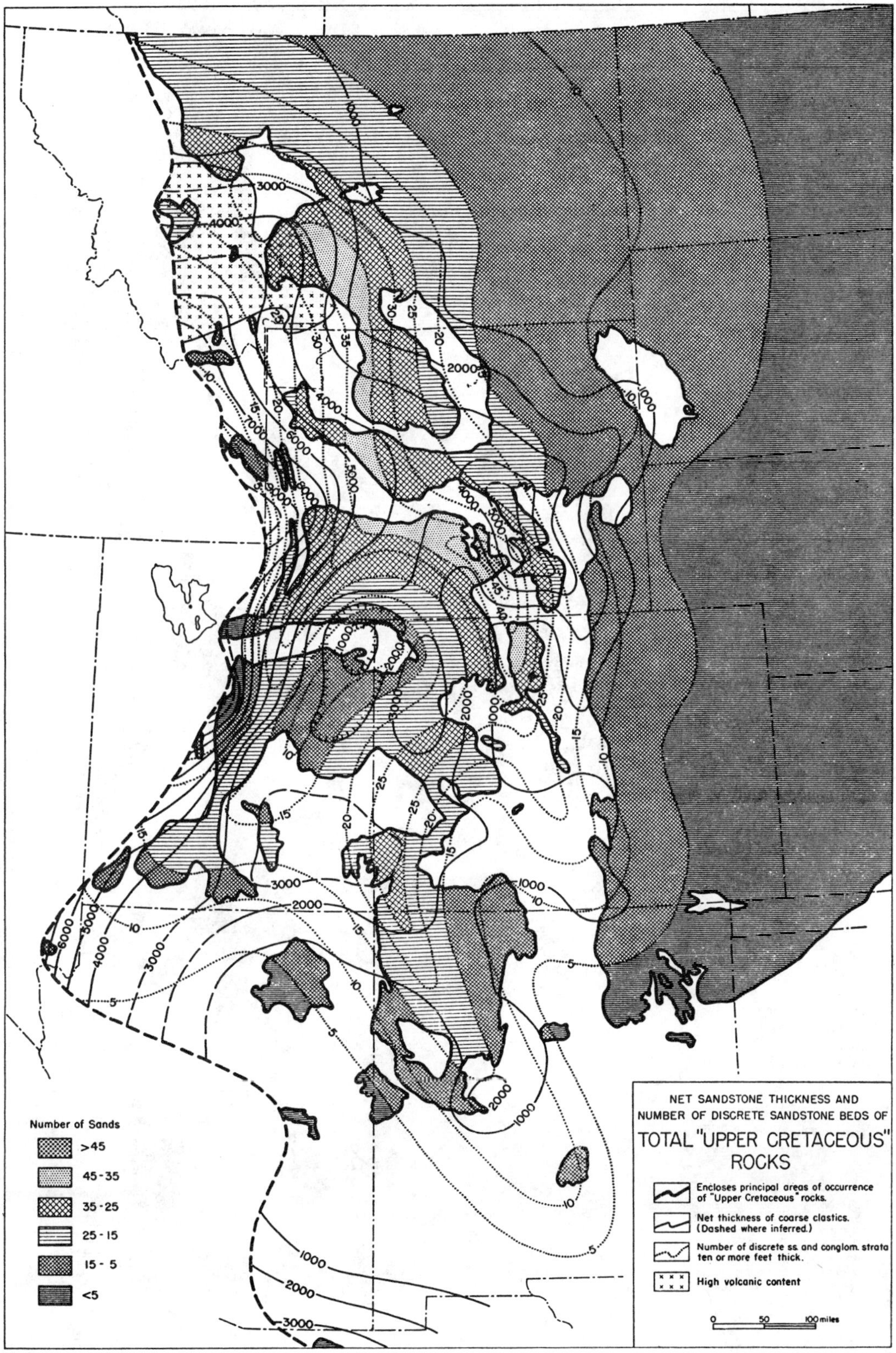

FIG. 5.—Sand isopach and number of sands in total operational unit.

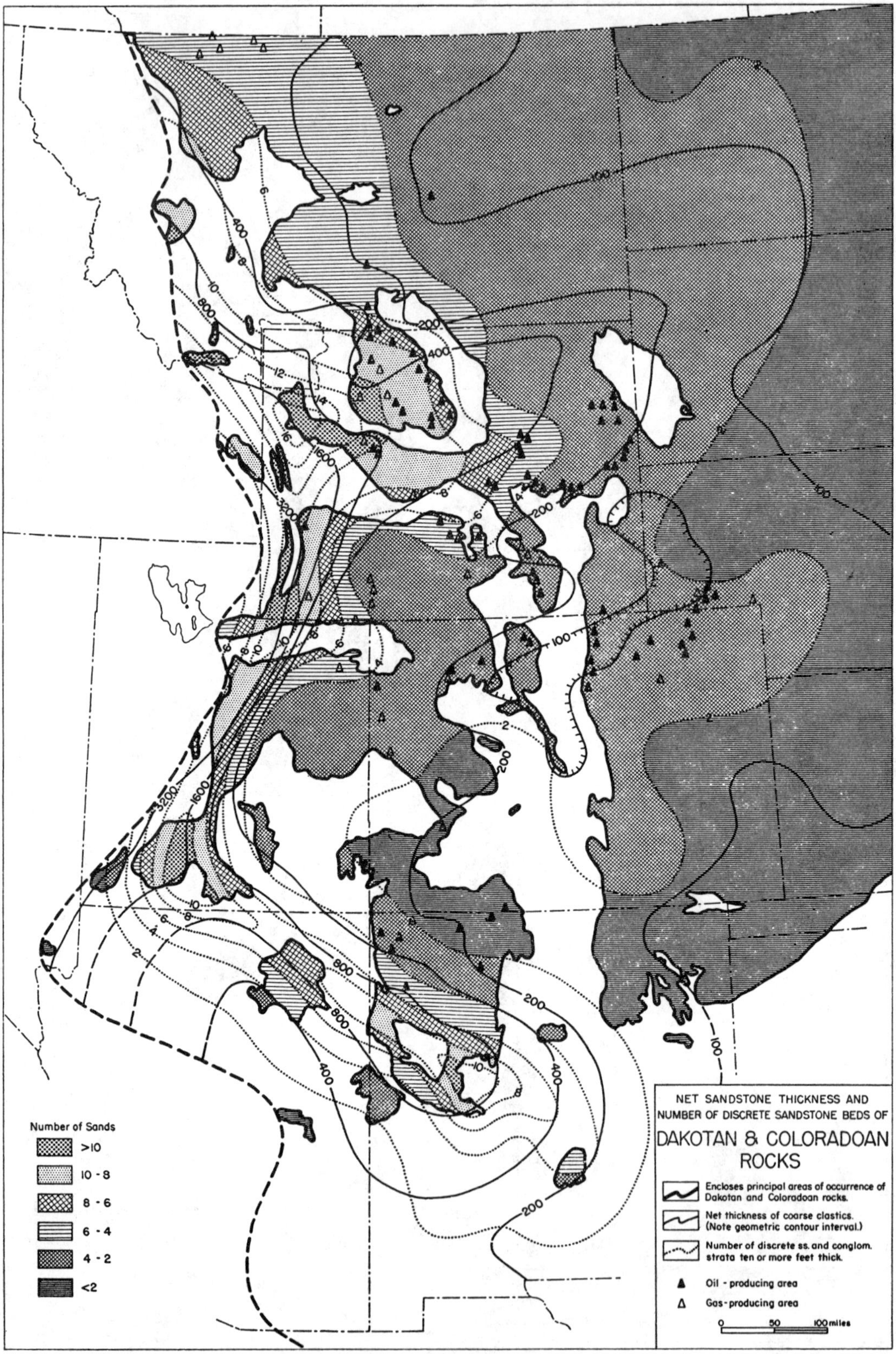

FIG. 6.—Sand isopach and number of sands in lower sub-unit. (Isopachs shown in geometric intervals.)

The pattern of sands shown on these maps is partly a function of the manner in which the data are assembled. By requiring shale or other sedimentary breaks in the section in order to define discrete sand beds, it follows that where the entire section is sand or conglomerate, it will be tallied as a single bed. In a sense this is what the maps attempt to show, in order to bring out the merging and coalescing of individual coarse clastic beds as the Upper Cretaceous is followed westward. As a result of this manner of tallying the data, most such maps have a central maximum belt, unless the western part has been eroded away, or if the map includes only part of the larger regional picture.

Merging and coalescing of sands occur near source areas, and this approach toward source can probably be even more effectively shown by superimposing a conglomerate-sandstone ratio on the map. In the present maps, it would show a significant increase in conglomerate in the vicinity of the western map limit.

The anomalous area of low number of sands in northeastern Utah is probably the result largely of Late Cretaceous and Tertiary erosion. In a "reconstructed map" of the total Upper Cretaceous, this anomalous area would probably not be included.

Figure 6, for the Dakotan-Coloradoan sub-unit, shows that six or more sandstone beds occur in a zone very near the western edge of the map area. This maximum band extends roughly southward from southwestern Montana through western Wyoming and central Utah into southwestern Utah. From there it swings southeast across northern Arizona into west-central New Mexico. The greatest number of sandstone beds, about 16, occurs in western Wyoming.

In general, net sandstone thickness increases from east to west, although there appears to be an area of abnormally low thicknesses in northern Colorado and western Nebraska (compare this with Figure 3). Along the eastern map margin, and extending as a lobe into southwestern Colorado, is a large area with two or less sandstones on the average.

In the Montanan-Laramian sub-unit (Fig. 7), a zone of 15 or more sands extends south from southern Montana through western Wyoming and western Colorado into northwestern New Mexico. Its course is deflected eastward in northeastern Utah, where discrete sands are few and net sand thickness is low. This feature apparently resulted from erosion of Montanan-Laramian rocks, as mentioned previously. The largest number of sands, about 35, occurs in the present Hanna Basin area. As with Figures 5 and 6, the sandstone thickness increases from east to west, the sandstone beds thicken and coalesce to the west, and they thin and disappear into a dominantly shale section on the east, with an average of five or less sands in the eastern area.

Comparison of the three sandstone maps discloses several interesting points: (1) the map patterns are linear and trend roughly north-south, paralleling the tectonic grain of the region; (2) individual sandstones are few in the eastern part of the region, increase to a maximum along a relatively narrow belt, and then decrease in number farther west. The narrow belt is the zone of maximum sand-

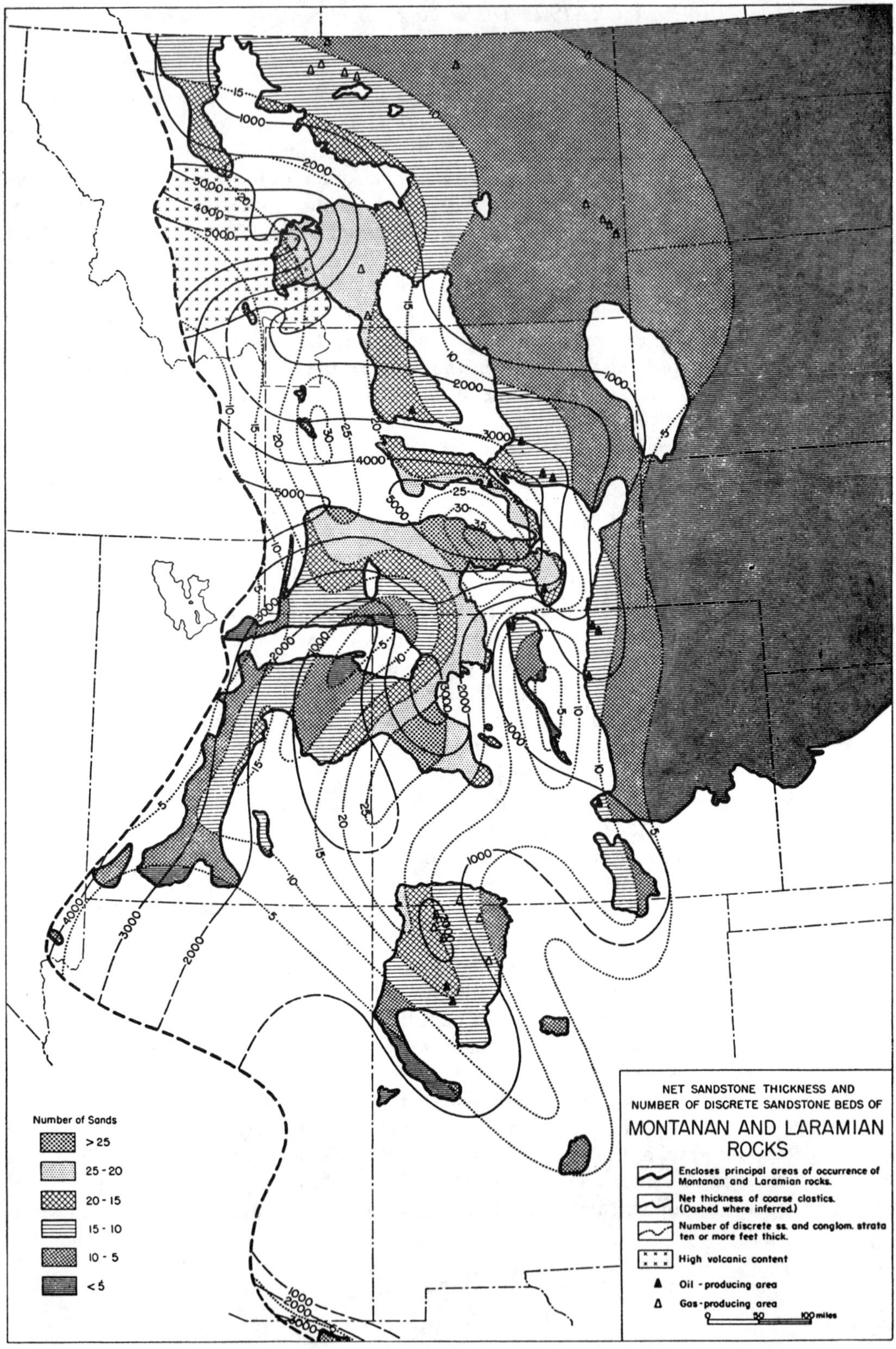

FIG. 7.—Sand isopach and number of sands in upper sub-unit.

shale intertonguing, and in a broad way it is also the average zone of maximum marine oscillation; (3) the number of sands in the Montanan-Laramian is considerably greater than in the Dakotan-Coloradoan; and (4) the belt of maximum sand-shale intertonguing migrated east and north during Upper Cretaceous time.

Figures 5, 6, and 7 permit some environmental reconstruction, much as Figures 2, 3, and 4 permit some tectonic interpretation. To the extent that the sands are dominantly non-marine, the zone of maximum interfingering is an approximate average shore line trend for the stratigraphic unit. Pike (1947, p. 94) shows strandline positions for the Mesa Verde-Mancos in New Mexico. Comparison of Figures 6 and 7 with Pike's map shows fairly good agreement between Pike's detailed strandlines and the writers' broader average zones.

If one wished to draw an "average shoreline" for the Upper Cretaceous or its sub-units, a first approximation would be had by drawing a line along the center of the maximum contour bands on the number-of-sands map. Figures 5, 6, and 7 are interesting in this connection in that they show how the map of the total Upper Cretaceous merges and blends the separate zones from the two sub-units. As with all facies maps, the magnitude of the stratigraphic unit controls the extent to which individual aspects are merged into broad average values or trends. Detailed study of shorelines or other specific features depends on mapping much thinner units, as was done by Pike.

SUMMARY OF FACIES MAPS

Several principal objectives were kept in mind in the original study represented by this paper. The first was to prepare a set of regional maps as complete as information in the public domain permitted; the second was to examine mapping techniques for vertical variability maps; the third was to achieve a broad regional picture of sedimentary and tectonic events; and the fourth was to see how the facies patterns relate to broad regional aspects of known oil occurrence.

In summary of the first several objectives, it may be stated that the study supports earlier interpretations of Upper Cretaceous rocks. Perhaps the present maps make these regional relations more explicit by showing them as part of a three-dimensional continuum. Obviously, there is much more that can be done with Upper Cretaceous facies maps. Detailed study of the kinds of sands present would probably show changes from a dominant graywacke-subgraywacke association in the orthogeosyncline to subgraywackes in the central part of the map area, perhaps with increasing proportions of quartzose sheet sands farther east. Distribution of coals, lignitic shales, and related swamp or lagoonal associations would show a decreasing content of such rocks eastward, with perhaps also a sharp decrease along the western map margin where the sections become dominantly conglomeratic and volcanic. Calcareous rocks, largely ignored in the present study, would show patterns of distribution related to areas of maximum marine occupation and perhaps remoteness from source areas.

In order to determine whether the detailed lithology of the Upper Cretaceous rocks supports some of the inferences derived from the more generalized maps, the junior author collected samples from several localities in the Black Mesa basin and adjoining parts of western New Mexico and southwestern Colorado. Sandstone beds were sampled at top, bottom, and at 10-foot intervals. Shales were sampled at top, bottom, and at 20-foot intervals. The samples were examined under the binocular microscope, and the lithologic properties were recorded on special check sheets. The data were then assembled into various compositional and textural end members and analyzed graphically.

The results of the study showed that the Dakota sandstone is typically quartzose, with well rounded grains. The Mesa Verde and other Upper Cretaceous sands, including the Pictured Cliffs, are composed largely of angular quartz grains, with 10 per cent or more of non-quartz detrital constituents. The angular components decrease northeasterly in the area studied. Sandstones above the Pictured Cliffs are tuffaceous, with a wide range of texture. The shales in the Dakota, Mesa Verde, and some Mancos horizons are drab, micaceous, carbonaceous, and more or less silty. Much Mancos shale is dark gray, more or less calcareous, and relatively free of silt. Shales above the Pictured Cliffs are largely tuffaceous.

In general, the sedimentary rock types comprising the Upper Cretaceous in northeastern Arizona are more closely related to unstable shelf associations (Krumbein and Sloss, 1951, p. 362) than to other lithologic associations. The sandstones, although not true subgraywackes, have certain subgraywacke characteristics (poor rounding, fair sorting, *et cetera*). Many of the sands appear to be "washed subgraywackes" (Pettijohn, 1949, p. 257), associated with beaches or stream channels.

The lithologic associations in the Black Mesa area suggest moderate contemporaneous subsidence and a moderate rate of deposition during most of Cretaceous time, perhaps interrupted occasionally by mild oscillatory movements. The writers concur with Sears, Hunt, and Hendricks (1941) and with Pike (1947) that the large-scale intertonguing observed in the deposits can be formed in a moderately subsiding area with variations in rates of subsidence and rates of detrital supply.

OIL AND GAS IN "UPPER CRETACEOUS" ROCKS

The stratigraphic units employed in this study are too thick to permit significant analysis of the relation between facies and the occurrence of oil and gas. Nevertheless, several interesting features are apparent when oil and gas occurrences are plotted on the several maps. This has been done on Figures 3, 4, 6, and 7.

Re-examination of Figures 3 and 6 shows that for the Dakotan-Coloradoan sub-unit, the majority of fields producing from that unit are clustered along the line of sand-shale ratio $\frac{1}{4}$, in areas where the stratigraphic unit is between one and two thousand feet thick. Moreover, most of the fields occur east of the zone of maximum number of sands, in the belt of two to six sands, with net sand isopachs between 100 and 400 feet. Exceptions to these statements, both for gross lithology and number of sands, occur mainly in western Wyoming and parts of the Big Horn basin.

Figures 4 and 7, for the Montanan-Laramian sub-unit, show too few oil and gas occurrences to make definite statements about facies relations. However, certain vague relations appear to be present. In general the oil and gas occur in areas with sand-shale ratio between $\frac{1}{4}$ and 1, but there is no discernible relation to total unit thickness. The occurrences lie in the band where 10–15 sands occur, and as in the Dakotan-Coloradoan, they are all on the eastern or marine side of the zone of maximum number of sands.

Despite the limitations of the maps, the following statements regarding oil and gas occurrences in Upper Cretaceous rocks of the Rocky Mountain region appear to be reasonable: (1) oil and gas occur generally in the eastern (marine) side of the zone of maximum number of sands; (2) oil and gas are present most commonly where there are relatively few comparatively thin sandstones in a dominant shale section; (3) in areas where sandstones are relatively thick and numerous, with small content of intervening shales, the chances for commercial production seem to be reduced.

Exceptions to presumed relations between facies and oil or gas occurrence are so numerous in the stratigraphic column that almost any generalizations are probably gratuitous. This is especially true of generalized maps of thick stratigraphic units. However, even such maps afford interesting suggestions, and provide a basis for more detailed study of thinner units, commonly on a more specific lithologic level.

SELECTED BIBLIOGRAPHY

BARTRAM, J. G. (1937), "Upper Cretaceous of Rocky Mountain Area," *Bull. Amer. Assoc. Petrol. Geol.*, Vol. 21, pp. 899–913.

BUSCH, D. A. (1950), "Subsurface Techniques," in *Applied Sedimentation*, ed. by P. D. Trask, pp. 559–78. Wiley and Sons, New York.

COBBAN, W. A. (1951), "Colorado Shale of Central and Northwestern Montana and Equivalent Rocks of Black Hills," *Bull. Amer. Assoc. Petrol. Geol.*, Vol. 35, pp. 2170–98.

———, AND REESIDE, J. B., JR. (1951), "Lower Cretaceous Ammonites in Colorado, Wyoming, and Montana," *ibid.*, pp. 1892–93.

CROWLEY, A. J. (1951), "Possible Lower Cretaceous Uplifting of Black Hills, Wyoming and South Dakota," *ibid.*, pp. 83–89.

ELDRIDGE, G. H. (1889), "Some Suggestions upon the Method of Grouping the Formations of the Middle Cretaceous and the Employment of an Additional Term in the Nomenclature," *Amer. Jour. Sci.*, Vol. 138, pp. 313–21.

GRIFFITHS, M. O. (1949), "Zones of Pierre Formation of Colorado," *Bull. Amer. Assoc. Petrol. Geol.*, Vol. 33, pp. 2011–28.

HALL, JAMES, AND MEEK, F. B. (1856), "Descriptions of New Species of Fossils from the Cretaceous Formations of Nebraska, with Observations upon *Baculites ovatus* and *B. compressus*, and the Progressive Development of Septa in *Baculites*, *Ammonites*, and *Scaphites*," *Amer. Acad. Arts Sci. Mem.* 5 (New Ser.).

HEATON, ROSS L. (1933), "Ancestral Rockies and Mesozoic and Late Paleozoic Stratigraphy of Rocky Mountain Region," *Bull. Amer. Assoc. Petrol. Geol.*, Vol. 17, pp. 109–68.

HUNTER, L. D. (1950), "Evidence of Uplift in the Bighorn Mountains During Upper Cretaceous Time" (abst.), *Bull. Geol. Soc. America*, Vol. 61, p. 1554.

KRUMBEIN, W. C., AND SLOSS, L. L. (1951), *Stratigraphy and Sedimentation*. W. H. Freeman and Company, San Francisco.

LEE, W. T. (1915), "Relation of the Cretaceous Formations to the Rocky Mountains in Colorado and New Mexico," *U. S. Geol. Survey Prof. Paper 95*, pp. 27–58.

MCKEE, E. D. (1949), "Facies Changes in the Colorado Plateau," *Geol. Soc. America Mem. 39*, pp. 35–48.

——— (1951), "Sedimentary Basins of Arizona and Adjacent Areas," *Bull. Geol. Soc. America*, Vol. 62, pp. 481–506.

MEEK, F. B. (1876), "A Report on the Invertebrate Cretaceous and Tertiary Fossils of the Upper Missouri Country," *U. S. Geol. Survey Terr.*, Vol. 9.

PETTIJOHN, F. J. (1949), *Sedimentary Rocks*. Harper and Brothers, New York.

PIKE, W. L., JR. (1947), "Intertonguing Marine and Non-Marine Upper Cretaceous Deposits of New Mexico, Arizona and Southwestern Colorado," *Geol. Soc. America Mem. 24*.

REESIDE, J. B., JR. (1944), "Thickness and General Character of the Cretaceous Deposits in the Western Interior of the United States," *U. S. Geol. Survey Prelim. Map 10*, Oil and Gas Inves. Ser.

SEARS, J. D., HUNT, C. B., AND HENDRICKS, T. A. (1941), "Transgressive and Regressive Cretaceous Deposits in Southern San Juan Basin, New Mexico," *ibid.*, *Prof. Paper 193*, pp. 101–21.

SPIEKER, E. M. (1946), "Late Mesozoic and Early Cenozoic History of Central Utah," *ibid.*, *Prof. Paper 205-D*.

——— (1949), "Sedimentary Facies and Associated Diastrophism in the Upper Cretaceous of Central and Eastern Utah," *Geol. Soc. America Mem. 39*, pp. 55–82.

STANTON, T. W. (1893), "The Colorado Formation and Its Invertebrate Fauna," *U. S. Geol. Survey Bull. 106*.

WALTON, P. T. (1944), "Geology of the Cretaceous of the Uinta Basin, Utah," *Bull. Geol. Soc. America*, Vol. 55, pp. 91–130.

WHITE, C. A. (1891), "Correlation Papers—Cretaceous," *U. S. Geol. Survey Bull. 82*.

ZAPP, A. D. (1949), "Geology and Coal Resources of the Durango Area, La Plata and Montezuma Counties, Colorado," *ibid.*, *Prelim. Map 109*, Oil and Gas Inves. Ser.

Reprinted from:
BULLETIN OF THE AMERICAN ASSOCIATION OF PETROLEUM GEOLOGISTS
VOL. 42, NO. 3 (MARCH, 1958), PP. 609–639, 2 TABLES

STRATIGRAPHIC FACIES DIFFERENTIATION AND NOMENCLATURE[1]

J. MARVIN WELLER[2]
Chicago, Illinois

ABSTRACT

Growing interest in stratigraphic facies relations has resulted in the proposal of numerous terms intended to distinguish different kinds of facies. Instead of promoting accuracy in thinking and expression, however, this has resulted in much confusion because little agreement is evident in the current usage of facies terms. Twenty-four terms are listed together with the varying definitions that have been presented for them and illustrations of their varied use.

The many ideas expressed in connection with stratigraphic facies show that several different types, classes, and aspects of varyingly objective facies can be distinguished. This introduces so many variables into facies considerations that no simple classification seems possible. Also if names were to be provided for all kinds of facies, nomenclature would become unreasonably burdensome. Correction of the present unsatisfactory situation demands that facies terminology be systematized and simplified.

The twenty-four facies terms are re-examined to determine their usefulness and necessity. The conclusion is reached that only two are important in application to stratigraphy. These are *facies* and *lithofacies* which are redefined in such a way that each can be used in more than one sense without confusion. A few other terms may be useful in special situations or in theoretical considerations and definitions are provided for them. More than half of the twenty-four terms are judged to be superfluous and they can be rejected without loss.

INTRODUCTION

Stratified rocks vary in character both vertically and laterally. Vertical variation is obvious in outcrops and changes are abrupt at many bedding planes. The grosser features of vertical variation provide the basis for differentiating most rock-stratigraphic units. Lateral variation generally is much more gradual and it is not so plainly apparent. Complications with respect to lateral stratigraphic relations, subdivision, nomenclature, and interpretation, that result from the variable characters of stratified rocks, introduce what commonly is recognized as the facies problem.

The common meaning of *facies*[3] is *general appearance*. A facies has no significance, however, except as it is in contrast to one or more other related facies. Therefore, it implies variation, comparison, and differentiation. As used in connection with stratified rocks, it also implies a certain degree of constancy and con-

[1] Manuscript received, December 24, 1956; revised, November 12, 1957.

[2] Department of Geology, University of Chicago.

The writer expresses his sincere appreciation for comments and criticisms regarding facies concepts and nomenclature received from the following persons. None of them, however, is to be held responsible for the opinions presented in the section on *Facies Nomenclature.* Manuel Alvarez, Jr., Sociedad Geológica Mexicana; W. C. Bell, University of Texas; E. R. Cumings, Indiana University, emeritus; J. C. Frye, Illinois State Geological Survey; James Gilluly, United States Geological Survey; H. D. Hedberg, Gulf Oil Corp.; Marshall Kay, Columbia University; S. W. Lowman, Rensselaer Polytechnic Institute; E. D. McKee, United States Geological Survey; R. C. Moore, University of Kansas; L. C. Sass, Gulf Oil Corporation; R. R. Shrock, Massachusetts Institute of Technology; L. L. Sloss, Northwestern University; J. W. Wells, Cornell University; and H. E. Wheeler, University of Washington.

[3] In this article, *facies* and other facies terms are italicized where reference to these *words* is intended but where the meaning is the strata or concept for which the word stands it is not italicized. This distinction does not apply to quotations.

tinuity, both horizontal and vertical, within whatever rock mass is being compared with or differentiated from others. Facies relations in stratified rocks have been recognized for more than a century but the word *facies* has been used in so many different ways that a general vagueness attaches to this term.

In recent years, increasing awareness of lateral lithologic variation in stratified rocks and more critical interest in facies relations and the problems that they present have resulted in efforts to clarify and systematize various useful facies concepts and the nomenclature to be applied to them. Proposals to sharpen the definitions of old terms have been made and new terms have been introduced for various kinds of facies. Significant advances have been made in both of these respects but no fully integrated consideration of this subject has yet been presented and considerable confusion regarding concepts and disagreement concerning terms persists.

FACIES CONCEPTS

Stratigraphic literature provides examples of the word *facies* used with many shades of meaning. Facies have been differentiated or identified on the basis of differences in (a) lithologic characters, (b) biologic composition, (c) stratigraphic relations, (d) structural form or condition, (e) metamorphic alteration, (f) geographic location or distribution, (g) temporal relations, (h) environmental influences, (i) tectonic situation or control, (j) genetic implications, and possibly others. All of these provide valid basis for some kind of differentiation but all are not equally objective or equally significant in stratigraphy.

Also, the word *facies*, as commonly used in stratigraphy, and other more restrictive terms have been used in such ways that they embody four related but different concepts. These are (a) the *appearance* of a rock body, (b) the *composition* or actual nature of a rock body, (c) the rock body *itself* as identified by its appearance or composition, and (d) the *environment* that is recorded by a rock body. As used by different persons or by the same person at different times, the meanings of the terms shift back and forth between these ideas or include more than one of them. This is amply demonstrated by the definitions and statements that are quoted in the following section.

Finally, facies have been differentiated at widely different scales or levels of importance. The differences recognized may be of any kind and they range from the plainly obvious to the very subtle. Facies may be generalized and, with the exception of more or less minor differences, include a varied assemblage of associated elements. In contrast, they may be highly exclusive and be identified by a complete set of specific detailed characters. Thus, facies considerations relate to sediments and rocks ranging from large bodies of more or less heterogeneous but characteristic composition to thin and restricted beds or parts of beds of uniform composition, texture, structure, biologic content, *et cetera*.

FACIES TERMS

Various terms have been introduced or adapted to designate particular kinds of facies and to differentiate kinds of facies from each other. At present there is little apparent agreement with respect to the precise meanings of many of these terms or the ways in which they should be used. The following definitions, statements, and observations have been culled from a review of many publications. References are made to all of the American articles that specifically consider the nature of facies and their relations. These are supplemented by other references to recent literature in which facies terms have been used somewhat incidentally. These last have been selected more or less at random and they are by no means exhaustive.

Bioherm.—Cumings and Shrock (1928) pointed out the many different meanings of the word *reef* and objected to use of the expression *coral reef* in stratigraphy because it is inaccurate and misleading. They proposed *bioherm* as a substitute (p. 599) but did not define it. Subsequently, Cumings (1930, p. 207) provided the definition "any dome-like, mound-like, lense-like or otherwise circumscribed mass, built exclusively or mainly by sedentary organisms such as corals, stromatoporoids, algae, brachiopods, molluscs, crinoids, etc., and enclosed in normal rock of different lithologic character." Later, he (1932, p. 333) again defined bioherms as "reeflike, moundlike, lenslike or otherwise circumscribed structures of strictly *organic origin*, embedded in rocks of different lithology." At this time he recognized that *bioherm* is a general term for several different kinds of mound-shaped bodies which he described under the headings "coral and algal reefs" ("true reefs," "typical coral reefs"), "algal banks," "shell banks," and "crinoidal bioherms." Regarding the last, he wrote "the term bioherm is preeminently applicable to those structures" (p. 344). Stockdale (1931a, p. 251) defined bioherms (crinoidal bioherms) as "Irregular calcareous masses of variable size and distribution, built in considerable part by crinoids and in less degree by bryozoans, and completely surrounded by rock of different lithology . . . ." He (1931b, p. 711) also wrote "The term 'bioherm' is particularly suitable for the calcareous, organically made masses which are so prominent in the upper Borden rocks." Shrock (1939, p. 532) defined bioherms (Silurian reefs) as "organic reefs built not dominantly by corals but rather from the contributions of many animals and of certain algae" and he described them as "reefy mounds."

Bioherm clearly was proposed as a structural term, although composition was stressed in all the definitions, but also it certainly involves facies differentiation (Moore, 1949, p. 14). It was applied by both Cumings and Shrock primarily to Silurian reefs of the Great Lakes region but it also was intended to include the Mississippian crinoidal mounds of southwestern Indiana. These two kinds of structures are quite different. The first, which are true organic reefs, grew as rigid wave-resistant limestone masses whose central non-bedded cores rose conspicuously above the sea bottom and may have reached nearly to water level. The others consist of bedded limestone masses originally formed of loose more or less fragmented organic debris that accumulated locally on the sea bottom but did not rise much above contemporaneously deposited land derived sediment.

Most recent authors have appeared to accept *bioherm* as originally defined (*e.g.*, Wilson, 1950, p. 181; Link, 1950, p. 264). Twenhofel (1950, p. 183) wrote

"The . . . term has merit and should be extensively used, but the word reef is so generally ingrained by use that complete displacement is unlikely." This prediction seems to have been justified and, if *bioherm* is mentioned at all, it generally is quoted as a subordinate synonym of *organic reef*. Newell (1953, p. 11) remarked "It does not clarify the meaning to employ 'bioherm' for organic reef . . . . A spade is not sharpened by being called a geotome."

Actually, retention of *bioherm* has perpetuated confusion because it does not differentiate true organic reefs, that were potentially wave resistant, and moundlike structures that were not. Cloud (1952, p. 2127) has suggested that the definition of *bioherm* be restricted and this term "be reserved for reef-like organic masses of uncertain potential or doubtful wave-resistant nature." MacNeil (1954, p. 390) disagreed and proposed that this term be restricted to the "growth lattice" of an organic reef by which apparently he meant its rigid framework to the exclusion of reef flank and other associated fragmental deposits.

Biostrome.—This term was proposed by Cumings (1932, p. 334), "For purely bedded structures, such as shell beds, crinoid beds, coral beds, *etcetera*, consisting of or built mainly by sedentary organisms, and not swelling into moundlike or lenslike forms" and he added, "*Biostrome* means literally an organic layer."

Biostrome was intended as a structural term, in contrast to *bioherm*, for limestone deposits some of which have been called coral or other reefs but do not have biohermal form. The definition makes no mention of enclosing sediment or rock but, as compared with other sedimentary deposits, *biostrome* has facies connotations. The last sentence quoted in the preceding paragraph probably is the most significant part of the definition. Thus, almost any highly fossiliferous limestone bed that does not thicken conspicuously might be a biostrome. According to this interpretation, a bioherm of the second type may consist of a succession of biostromes.

Several questions concerning biostromes remain unanswered, however. For example, does reference to "sedentary organisms" imply that autochthonous origin is necessary? Many "crinoid beds" consist largely of disarticulated plates that have been moved some distance from the places where the crinoids grew. What degree of fragmentation is permissible? If no limit is set, many limestone beds are biostromes because they consist of organic calcium carbonate. Also, are biostromes exclusively calcareous or are other organic beds, like coals, to be included? Regarding these questions, Shrock (letter, January 2, 1957) writes "if the remains of crinoids were spread evenly over an extensive bottom, I would not be as much inclined to call such an accumulation a bioherm or biostrome." Also, "I do not think that bioherms or biostromes need necessarily be calcareous, but I would balk a little at coals being included unless it could be shown that the vegetable material accumulated in a particular area and underwent very little transportation."

Moore (1957, p. 1789) states that *biostrome* "is a term of little value" and it should be forgotten. This is reasonable if it is to mean no more than an organic layer. Although *biostrome* might be redefined as the counterpart of a lithostrome (*q.v.*), Sloss (letter, December 24, 1956) is of the opinion that this can not be done. His view probably is influenced by the interpretation of *lithostrome* favored by the Northwestern University group.

Magnafacies and parvafacies.—Caster (1934a, p. 348) introduced these terms and wrote "A magnafacies is a 'complete lithic member,' the limits of which transcend stratigraphic boundaries under conditions such as existed in the area influenced by the Catskill delta. The bounds which separate one magnafacies from the adjoining magnafacies are *facies planes* (hypothetically planar). Those portions cut off by the respective stratigraphic boundaries (stratigraphic or *contemporaneity planes*) are termed *parvafacies*. The stratigraphic unit is therefore comprised of laterally grading contemporaneous parvafacies of different lithology and faunas, whereas a magnafacies is made up of varitemporaneously deposited parvafacies of like lithology and similar faunas." According to this definition, magnafacies are regressive lithologic units consisting of a succession of laterally limited, similar, imbricating time-rock units or parvafacies. Laterally equivalent intergrading parvafacies of different kinds compose extensive time-rock stratigraphic units or *monothems*. Upper and lower parvafacies boundaries are horizontal time planes. Lateral parvafacies boundaries correspond with magnafacies boundaries and are arbitrary cutoff planes inclined at a low angle to the bedding and to time planes.

Later authors appear to have overlooked the very definite spacial relations of magnafacies and parvafacies indicated by Caster. The "planar" boundaries of a magnafacies transgress time planes and one magnafacies does not intertongue with others. Actually, a magnafacies is a perfect rock-stratigraphic unit and no name other than a rock unit name (*formation*, *group*, or *member*) is needed for its designation. Moore (1957, p. 1786) agrees that the term *magnafacies* should be rejected. He states, however, that magnafacies "differ in no way from lithosome" and although "magnafacies has priority . . . lithosome is preferable." This is not true if the original definitions of these terms are to be accepted.

A parvafacies is very close to a lithofacies (*q.v.*) and, depending on the definition of the latter term, it might actually be a kind of lithofacies. The distinguishing features of a parvafacies are its boundaries. The upper and lower boundaries are time planes and adjacent laterally distinguishable parvafacies are parts of a time-stratigraphic unit and not of a rock-stratigraphic unit of any kind. The lateral boundaries are arbitrary planes but these do not extend at right angles to the bedding like the boundaries separating statistically differentiated lithofacies because they cut the bedding at low angles and correspond with time-transgressive rock-stratigraphic boundaries. As Moore (1957, p. 1787) remarks, parvafacies has no exact synonym, but he is mistaken in stating that within "a body of sedimentary strata defined as a lithosome, each part could be called a *parva-*

facies." Actually, little more than theoretical interest attaches to *parvafacies*. Since its introduction, this term has found no practical application and probably it never will.

McKee (1949, p. 42, Fig. 1) wrote, "vertically continuous units of the Muav limestone may be seen to be separable into facies of different age (parvafacies)." King (1949, p. 169), in referring to proposals made by Moore, remarked that "Caster's classification into 'megafacies' [*sic*] and 'parvafacies' has much more virtue in logic, but . . . why do we not merely adopt McKee's simple, workable classification into (1) 'facies of different lithology but the same age,' and (2) 'facies of similar lithology but different age'?" These quotations illustrate a common misconception of others also. Parvafacies are not simply different facies of the same age or similar facies of different ages. They are defined by particular boundaries and, if definitions were to be changed, both *parvafacies* and *magnafacies* would become synonymous with other terms.

Caster's magnafacies were defined in relation to "conditions such as exist in the area influenced by the Catskill delta" and, therefore they are regressive stratigraphic units. Possibly, transgressive units also might meet the requirements of the definition but intertonguing facies resulting from fluctuating sea-level provides a distinctly different situation. Consequently, *magnafacies* can not be equated with *lithosome*.

Megafacies.—*Magnafacies* has been mistakenly written "*megafacies*" by several persons. King's error (1949, p. 168) is plainly evident. Cooper and Cooper's intention (1946, p. 68) when they used this word is not so clear. Their statement "the Virginia Athens [shale] is a megafacies of the lower Middle Ordovician . . ." suggests no connection with *magnafacies* and their *megafacies* seems to be an especially large intertonguing lithologic body (see their use of *facies*).

Biotope.—This is a term used by ecologists and biologists meaning an area of uniform ecology and organic adaptation (Hesse *et al.*, 1939, p. 135). Paleontologists and stratigraphers commonly have used it in the sense of *environment* (Wells, 1944, p. 284; 1947, p. 119; Moore, 1949, p. 17). Sloss *et al.* (1949, pp. 95–96) considered biotopes (or "paleobiotopes") to be "faunal (or floral) units which may be interpreted environmentally" and stated "biotopes reflect the influence of the environment on the biota." Krumbein and Sloss (1951) variously define *biotope* as "An area inhabited by a uniform community adapted to its environment" (p. 227), "ancient environment(s)" (p. 227), "the fossil record of an organic environment" (p. 277), and considered it to be a part of a stratigraphic section as shown in their Figure 8-2, titled "Generalized section . . . showing vertical succession of biotopes" (p. 231). Swain (1956, p. 635) wrote a biotope is "an association of organisms. . . ." Newell (1957, p. 433) used *biotope* in the sense of an ecological zone. Dunbar and Rodgers (1957, p. 137) stated "the environment itself . . . is called a *biotope*." Moore (1957, p. 1780) accepts the ecological definition but states "Actually, all biotopes are three-dimensional" because benthonic organisms penetrate the bottom sediments to some appreciable distance.

The foregoing demonstrates the confusion of several distinct concepts. These are (a) an area, (b) a zone, (c) an environment, (d) a fossil unit, (e) a fossil record, and (f) a biostratigraphic unit. The most serious confusion, as Wells (letter, December 27, 1956) points out, results from failure to distinguish properly between an ecological situation and its record. A biotope is an area defined in term of ecology. It is more or less ephemeral and at any moment it is circumscribed by a boundary that is subject to expansion, contraction, or other shift in position. Whether or not it is considered to have a third dimension is entirely immaterial. The record of a biotope that persists for some appreciable time is a body of sediment or rock and its enclosed fossils. *Lithotope* (*q.v.*) was originally proposed as the designation for such a record.

Lithotope.—Wells (1944, p. 284) proposed this term "for the sedimentary rock record of a biotope" and wrote "Thus, a *biotope* with its special life community or *biocoenosis*, is preserved as a *lithotope*." Later he (1947, p. 119) defined it as "the rock record of the environment" and his usage shows that both the physical and biological features of the environment are included. Moore (1948, p. 310) stated that *lithotope* is used for "rock environment to designate . . . all attributes of a deposit bearing on its condition of origin." Also he indicated that lithotopes are parts of a stratigraphic section and that they can include such things as "coral biostromes" (p. 309, Fig. 2). Later Moore (1949, p. 17) stated that *lithotope* "refers to the environment—not to the record of it." Sloss *et al.* (1949, p. 95) considered it advantageous to segregate the physical or dynamic and the biologic attributes of environments and wrote "Each of these recognizable [physical] rock environments is a lithotope." Krumbein and Sloss (1951, p. 194) stated "The areas of uniform sedimentation are called *lithotopes*" but also defined this term as "a single depositional environment" (p. 269), and stated that lithotopes may be particular types of sediment such as beach sands or crinoidal limestones (p. 269). In addition, they indicated that a lithotope is a stratigraphic unit as shown by their Figure 12-4 with the title "Section . . . showing observed lithotopes" (p. 382). Dunbar and Rodgers (1957, p. 158) use *lithotope* in the title of one of the principal sections of their book. They wrote "black shale is a lithotope, for example—any or all black shale" (p. 137) and remark that their usage of this term, as the record of an environment, follows Wells but differs from that of Moore and of Krumbein and Sloss. Swain (1956, p. 635) wrote a lithotope is "the record of a definable sedimentary environment . . . " Wheeler and Mallory (1956, p. 2722) stated that "a *lithotope* is an area of uniform sedimentation." Sloss (letter, December 24, 1956) also now interprets *lithotope* as "an area of uniform physico-chemical environment" which has no thickness and, therefore, this is an ecologic rather than a stratigraphic term. Moore (1957, p. 1789), however, states that "Such a concept of lithotope is erroneous" and defines a lithotope as "a specified place or type of place distinguished by uniformity of principal conditions of sedimentation including occurrence of kinds of organisms associated with these environmental conditions" and adds "some degree of thickness must be associated with horizontal spread" (p. 1781).

The foregoing is another example of great confusion with *lithotope* variously considered to be (a) the rock record of an environment, (b) a particular kind of sediment or rock, (c) a stratigraphic unit, (d) a sedimentary environment, (e) a physical environment, (f) a depositional area, and (g) the area of a physical environment. The trouble here apparently is related to at least three sources of misunderstanding. First, clear distinction has not been made between an intangible environment and the tangible sediment or rock that represents or records that environment. This also occurs with respect to other facies concepts and terms. Second, *lithotope* was defined originally in relation to *biotope*. The construction of this word, however, is unfortunate because its etymology suggests that *lithotope* should bear the same relations to *lithofacies* that *biotope* bears to *biofacies*, but according to the definition this is not true. Third, there is disagreement whether or not *lithotope* includes or excludes biologic elements of the environment.

Physiotope.—Moore (1949, p. 17) suggested this as a possibly useful term to denote "all purely physiochemical elements of an environment." He considered that *lithotope* combines both physical and biological elements (others have disagreed) and, if biological elements characterize a biotope, the complementary term *physiotope* is required where the non-biological elements are distinctive. If this term is to be accepted as having a meaning similar to that of *biotope*, it should be defined as an area.

Tectotope.—Sloss *et al.* (1949, p. 96) originally defined *tectotope* as "a stratum or succession of strata with characteristics indicating accumulation in a common tectonic environment." Krumbein and Sloss (1951, p. 381) stated "Each tectonic subdivision [area] represents a particular tectonic environment called a *tectotope*. The tectotope bears the same relations to the tectonic environment as the lithotope and biotope bear to the sedimentary and biologic environments." Obviously, *tectotope* should be defined as an area rather than as a stratigraphic body or an environment. This is now recognized by Sloss (letter, December 24, 1956) and he adds that a *tectotope* "is an almost pure abstraction dependent upon interpretation of a tectosome" (*q.v.*).

Facies.—This word, meaning appearance, has been part of the English language for more than 300 years. Fenton and Fenton (1930, p. 147) reported that it was first used in geology (in French) by Gressly in 1838 and since then it has found application in the fields of stratigraphy, igneous and metamorphic petrology, sedimentation, paleontology, and paleoecology. Moore (1957, p. 1781) is mistaken in stating that "Strictly speaking, facies is a term belonging to stratigraphy. . . ." *Facies* also is a recognized technical term in biology and ecology (Carpenter, 1938, p. 107). Page's geological dictionary (1859, p. 163) states that *facies* is "A convenient term in Natural History, introduced to express any common resemblance or aspect among the rocks, plants, animals, or fossils of any area or epoch." Although it has been defined repeatedly and although there are those who believe "that if the term facies is to be meaningful, its definition must be restricted" (Wheeler, letter, February 8, 1957), it has continued to the present to be used both loosely and with several different technical meanings.

Fenton and Fenton (1930, p. 147) translated Gressly's statement that facies "in the horizontal extent of a terrane" are distinguished by "constant peculiarities in their petrographic constitution as well as in the paleontologic characters of their fossil assemblages. . . ." They (p. 148) also quoted Haug's definition that a facies is "the sum of the lithologic and paleontologic characters which a deposit presents at a given point." To these, the Fentons added their own opinion that the distinctive characters of a facies exclude "changes that are wholly independent of its original nature" such as brecciation and dolomitization. Williams (1895, p. 69) wrote "The term *facies* is used . . . in geology to apply to the particular composition or condition of a formation in a given region. . . ." Stockdale (1939, p. 38) stated "facies are simply the lateral lithologic variations, created within essentially the same time limits." Other references to facies in stratigraphic literature are mostly casual and unspecific. Kay (1945) used *facies* in the sense of *lithologic nature* (p. 431) and *environment* (p. 433). Cooper and Cooper (1946, p. 78) considered facies to be intertongued sedimentary deposits identified lithologically not falling "within the same stratigraphic [time] limits." Moore (1948, p. 310) stated that a facies is "a particular varient or aspect among two or more types of deposits belonging to a designated stratigraphic unit . . ." or to "a defined lithotope" (p. 309). Dapples *et al.* (1948, p. 1924) distinguished facies on the basis of physical or tectonic environments. Pettijohn (1949, p. 435) stated that sedimentary facies are assemblages of commonly related rock properties. Lowman (1949, p. 1948) wrote a facies "is the present aspect of a sedimentary rock."

The first attempt at a critical analysis of facies concepts was made by Moore (1949). He presented examples of some of the varied usages of the term *facies* (pp. 8–16) and on the basis of them framed two definitions as follows.

1. "Sedimentary facies are areally segregated parts of differing nature belonging to any genetically related body of sedimentary deposits" (p. 8).

2. " 'facies' is a term applicable to every distinguishable sedimentary record of a depositional environment" (p. 15).

Moore considered that these are significantly different concepts and emphasized the necessity for differentiating sedimentary deposits into (a) bodies or "facies" that are parts of some kind of stratigraphic unit and have "mutually exclusive space distribution" and (b) "kinds of sedimentary rocks without reference to their space relations" (p. 15) or relations to stratigraphic units. He proposed restricting the use of *facies* to the first (1 and a, as above) and in his conclusions stated " 'Sedimentary facies' is defined as any areally restricted part of a designated stratigraphic unit which exhibits characters significantly different from those of other parts of the unit" (p. 32). For the second (2 and b, as above) he used the term *lithofacies* (*q.v.*).

Several authors who have considered facies subsequently (*e.g.*, Krumbein and Sloss, 1951, pp. 253–55; Wheeler and Mallory, 1956, p. 2716) seem to have believed that they agreed substantially with Moore's restricted definition of *facies*. Actually, they have not because Moore's intentions are unclear and have been badly misunderstood. The main misinterpretations concern the form and mutual relations of so-called facies. Moore's facies are intertonguing sedimentary bodies within stratigraphic units whose lateral boundaries are irregular and whose upper and lower limits do not necessarily correspond with the boundaries of the stratigraphic unit which contains them or with the upper and lower limits of other facies. Therefore, Moore's facies are not necessarily laterally equivalent and they can not be mapped as conventional stratigraphic units are mapped. Moore made no such clear statement as the foregoing, and a very careful reading of his text is required before his intentions are perceived. His intentions are indicated, however, by the examples that he presented (pp. 8–14), by such remarks as "let us consider a formation composed of two interfingering deposits classed as facies" (p. 17), and especially by his Figure 19 (p. 29). Moore excluded from his conception the Lower Mississippian "facies" of Stockdale (1931, 1939) which are laterally equivalent divisions of stratigraphic units separated by arbitrary vertical cutoff boundaries. He objected to such facies because of "the arbitrary and seemingly artificial nature of the stratigraphic compartmentation" and "the lack of significance of most facies units from the standpoint of historical geology" (p. 21) and concluded "schemes like those of Caster, Stockdale, and McKee for facies nomenclature are judged unacceptable" (p. 33). He no longer holds exactly these same views.

Most subsequent authors (*e.g.*, Sloss *et al.*, 1949; Krumbein and Sloss, 1951; Wheeler and Mallory, 1956) have not attempted to define *facies*. It is clear from their usage of the term, however, that they consider it to be applicable to generally arbitrary, laterally equivalent and mappable parts of some kind of a stratigraphic unit bounded by vertical planes and identified by their generalized aspects that include elements related to physical, biological, and possibly other variables. Swain (1956, p. 635) defined *facies* as "lateral variations in a stratigraphic unit." Johnson and Wood (1956, pp. 714–15) labeled intertongued lithologic bodies as facies in their figures. Cloud and Barnes (1957, p. 168, Fig. 1) used the term *facies* for lithologic bodies that exhibit a variety of mutual relations. Among them are (a) laterally intertongued divisions of stratigraphic members, (b) restricted lens-like or bioherm-like bodies wholly enclosed in strata of different lithology, and (c) vertically successive and laterally extensive lithologic units that resemble the members of a formation. Dunbar and Rodgers (1957, p. 136) wrote "facies means the general aspect of the rocks, lithologic and biologic (and by extension, structural or tectonic and even metamorphic), as that aspect reflects the environmental conditions under which the rocks were formed" and "the term clearly refers to a facies or aspect of some definite stratigraphic unit . . ." (p. 137).

Moore (1957, p. 1781) now states "the word facies . . . has been used both to designate (1) particular kinds of sedimentary deposits (as red-bed facies . . .) that broadly correspond to certain types of environment, and (2) individual subdivisions of specified rock bodies that incorporate two or more kinds of deposits reflecting different original environments. . . . This divergent usage actually involves no obscurity. . . ." These two divisions correspond (in reverse order) with his two divisions of 1949 but Moore now admits them both under the term *facies.* The first division apparently is restricted, however, and it does not seem to include smaller sedimentary deposits closely corresponding with more restricted environments. These are the bodies that Moore had in mind principally when he used the term *lithofacies* in 1949. Also, the second appears to be expanded and includes both irregularly disposed and intertongued deposits and arbitrarily segregated facies like those of Stockdale.

Other geologists (*e.g.*, Lochman, 1956) continued to use *facies* loosely for undefined rock bodies differentiated on the basis of either lithologic characters or presumed ecologic implications. Cloud and Barnes (1957, p. 206) regarded facies as rock types characterized by lithology and fossils. Kummel (1957) considered them to be intertongued bodies identified lithologically. Fairbridge's expression (1957, p. 134) "nonbiohermal facies" shows that he used the term in a structural sense but he also regarded facies as environments (p. 159).

Phase.—This word is identical etymologically with *facies* but it is a much more recent addition to the English language, having appeared less than 150 years ago. Like *facies*, it means appearance but it also carries the somewhat more common meaning of a stage in a process. *Phase* has had a wide and general usage in geology but confusion has arisen as the results of attempts to restrict its meaning technically.

Fenton and Fenton (1930), in their definition of *facies* (*q.v.*), excluded rock characters such as dolomitization that are not indicative of the original nature of a sediment and defined *phase* as "a local or regional aspect or condition of a stratum or group of strata, as determined both by original nature and secondary change; the latter being the determining factor" (p. 150). McKee (1938, pp. 15–16) used the word in the sense of *facies* and quoted Barrell (1913, p. 465) as an authority. Barrell's *phase* has exactly the same meaning as *facies* as this latter term was generally and loosely used prior to 1949. McKee's *phase*, however, is a kind of *superfacies*. He recognized laterally equivalent and contrasting phases in formations and laterally equivalent contrasting facies in the members of the phases of those same formations. Thus, McKee's *phase* and *facies* are exactly comparable except that they were used at different levels in the scale of rock-stratigraphic classification. Stockdale (1931a, p. 76) wrote "Under the facies names the phases of each formation are described," suggesting the equivalence of facies and phase. Later he (1939, pp. 39–40) again equated *facies* and *phase* and stated that facies are "the contemporaneously formed lithologic phases within the bounding limits of a stratigraphic unit." He (1939, p. 139) also, however, referred to "limestone phases" within a formation that was not differentiated into facies. Allen (1948, p. 8) preferred an older concept said to have been originated by Arthur Vaughan in 1909. According to this, *phase* may be defined as characteristic strata repeated at various positions in a stratigraphic section that record the recurrence of a particular kind of environment.

The following are other examples illustrating the varied recent stratigraphic usage of the term *phase:* (a) early, middle and late parts of a depositional cycle (Fisk and McFarlan, 1955, pp. 290–97); (b) transgressive and regressive parts of a cyclothem (Weller, 1956, p. 30); (c) sediments consisting of more or less identical constituents deposited under relatively uniform conditions (Fisher, 1956, p. 618); (d) a time-stratigraphic division (Lochman, 1956, p. 1331); (e) a type of lithologic development (Newell, 1953, p. 115); (f) a sedimentary province (Newell, 1957, p. 413); and (g) a major biotope (*ibid.*, p. 433). Moore's conclusion (1957, p. 1787) is reasonable that "Because of vagueness in previous usage and because technical application of 'phase' is not needed, this word may well be left without restriction as to special meaning."

Intrafacies.—Cloud and Barnes (1957, p. 169) used this term "to denote a subordinate facies development within a differing major facies development." This definition suggests that their *facies* and *intrafacies* might be related similarly to McKee's *phase* and *facies*. The single intrafacies that they recognize (p. 168, Fig. 1), however, shows that this is not so. Actually the intrafacies resembles a tongue or submember extending horizontally into a facies that can be likened to a member of a formation.

Lithofacies.—The origin of this word can not be attributed to any individual. Sloss (letter, December 24, 1956) states that it came into informal use in oil-company reports in the 1930s. Evidently it was derived by contraction of the expression *lithologic facies* (*e.g.*, "litho-facies," Lowman, 1945, p. 1514). Before attention was directed to facies terminology in 1949, it was loosely used for a *kind of rock* (Kay, 1945, p. 433), as a synonym of *lithology* (Kay, 1947, p. 165) and for a *lithologic unit* (Lowman, 1949, p. 1949). Krumbein (1948, p. 1909) provided the first definition, "The sum total of the lithological characteristics of a sedimentary rock is its lithofacies." Up to this time no particular distinction appears to have been made between the meanings of *lithofacies* and *facies* as the latter was used with reference to rocks.

Moore (1949) distinguished two concepts to both of which the term *facies* (*q.v.*) had been applied and he proposed to restrict the usage of *lithofacies* to the second. Accordingly he wrote, a "facies" (meaning the second type of facies, or lithofacies) consists of rocks "formed under like environments without regard to age or geologic setting" (p. 14); *lithofacies* "signifies the record of a sedimentary environment as represented by the sum total of the lithological characteristics of a sedimentary rock" (p. 17) and it is "the rock record of any sedimentary environment, including both physical and organic characters" (p. 32). The last closely parallels the original definition of *lithotope* (*q.v.*), and Moore used the word *lithotope* instead of *lithofacies* in the oral presentation of his paper (p. 16).

Moore's second definition of 1949 with reference to the older use of *facies* includes two concepts that were not clearly expressed and have resulted in subsequent confusion. The first concept has to do with such generalized things as the rocks of the "red-beds facies" without regard to stratigraphic units and Moore stated "There is value in this sort of classification" (p. 14). There seems to be little difference between this and *facies* as defined by Moore at that time except that *facies* was restricted in terms of some particular stratigraphic unit. The second lithofacies concept has to do with a particular kind of rock such as "black graptolite bearing shale" or "black slabby trilobite bearing limestone" that is one among two or more constituents of a formation (p. 17), or the rock in an individual bed such as a member or part of a member in a Pennsylvanian cyclothem (p. 31, Fig. 21), and he stated that "A sedimentary facies comprises one or more lithofacies" and "Any sedimentary facies is made up of one or more lithofacies having distribution different from the lithofacies components of other facies of the sedimentary unit which is divided into facies" (p. 32). This is the concept that he emphasized particularly. It is not entirely clear from any statement in Moore's 1949 paper whether he intended *lithofacies* to refer to all rocks of some particular kind, formed under the influence of a particular environment wherever they may occur, or whether this term also can be applied to a single bed or succession of similar beds differing in character or origin from adjacent strata.

King (1949, pp. 167–68) recognized and was troubled by the inconsistency of the two concepts included by Moore in *lithofacies*. He remarked "First, he tells us that many geologists use the term facies for broad designations, such as 'red-bed facies,' 'black shale facies,' and 'geosynclinal facies.' Then he tells us that these are not facies but lithotopes [lithofacies]. Next he defines lithotope [lithofacies] as a collective reference for all features of lithology, and says that one facies may include one or many lithotopes [lithofacies]." (As already noted, Moore replaced the *lithotope* of his oral presentation by *lithofacies* in his published paper.)

Sloss *et al.* (1949) either misunderstood Moore's definition of *lithofacies* or disagreed with it. (Sloss observes that Moore's 1949 "use of lithofacies runs counter to . . . previous practice"—letter, December 24, 1956.) They stated that lithofacies "are groups of strata demonstrably different in lithologic aspect from laterally equivalent rocks, the aspect being controlled by the lithotopes or combinations of lithotopes of which the lithofacies is composed" (p. 95). As *lithotope* is used here, it has a meaning very close if not identical to Moore's principal concept of *lithofacies*, *i.e.*, a particular kind of rock. Although Sloss *et al.* made no direct statement, it is clear that their lithofacies was considered to be one of several more or less comparable but different types of specialized facies as they used this latter term. Thus their lithofacies is a facies "aspect" characterized by "physicochemical" but not by biological features. (Sloss points out that fossils and fossil materials, insofar as they are sedimentary particles, are physicochemical features—letter, December 24, 1956.) They considered lithofacies to be qualitatively distinguishable, laterally equivalent, and mappable segments of some kind of stratigraphic unit that are separated by arbitrary vertical boundaries. Similarly, Krumbein and Sloss (1951, p. 268) stated "The expressions of variation in *lithologic aspect* are *lithofacies*." Another definition was provided by Wheeler and Mallory (1956, p. 2718) who wrote, lithofacies are "laterally segregated, statistical, lithic variants of a stratigraphic interval."

McLaughlin and Gerhard (1953, p. 136) used *lithofacies* as a substitute for *formation* "since 'formation' may be misconstrued as a time unit." Greenman (1955) followed a similar practice. He reported that lithofacies are conformable (p. 27), interfinger (p. 31), occur as alternating bands (p. 32), and he subdivided them into members (p. 33). His geologic map shows many repeated alternations of lithofacies. Hedberg (1954, p. 217) stated that lithofacies is "the manifestation of lithologic characters. . . ." Swain (1956, p. 635) wrote, lithofacies are "Lateral variations in a rock unit (formation, group, member) or perhaps in a larger 'time rock' [unit] (series, system). . . ." Lochman (1956) employed *lithofacies* for lithologic character and for strata of a certain type. Bergquist and Cobban (1957, p. 872) used this word as an adjective meaning lithologic. Kummell (1957, p. 454) differentiated lithofacies in time-rock units. To Cooper (1957, p. 259) they are "sediments representing environments." Newell (1957, p. 411) used *lithofacies* for "a lithologic type." Mullens and Freeman (1957, p. 505) define *lithofacies* as "lithologic aspect" and they use it for a body of strata, stating "ore deposits . . . occur in a lithofacies . . ." Dunbar and Rodgers (1957, p. 137) wrote that lithofacies is the lithologic aspect of a facies of some definite stratigraphic unit. Moore's latest definition (1957, p. 1783) states that a lithofacies is "any part of a stratigraphic unit (or arbitrarily limited body of sedimentary deposits) that is distinguished by noteworthy lithologic characters differing from those of other parts of the unit (or arbitrarily delimited body)," and he adds, "The lithologic changes typically are gradational but may be accompanied by intertonguing of unlike kinds of strata."

Biofacies.—Like *lithofacies*, this word was used informally for several years before it appeared in print. It was derived by contraction from the expression *biologic facies* ("bio-facies," Lowman, 1945, p. 1514) and it was used in this sense by Kay (1947, p. 165). Lowman (1949, p. 1949) indicated that biofacies are "fossil zones." Moore (1949) defined *biofacies* as "the strictly organic elements of a particular sedimentary environment" (p. 16) and "the total biological characteristics of a sedimentary deposit" (p. 17). Sloss *et al.* (1949, p. 96) stated, "*Biofacies* are composed of one or more biotopes differing in biologic aspect from laterally equivalent biotic assemblages." Krumbein and Sloss (1951, p. 268) wrote, "the expression of variation in *biologic aspect* are *biofacies*."

Hedberg (1954, p. 217) defined a biofacies as "the manifestation of biologic character." Imbrie (1955, p. 450) stated, "Biofacies [are] the total recognizable organic content of a designated portion of a stratigraphic unit." To Lochman

(1956) a biofacies was a biologic environment. Cooper (1957, p. 259) defined it as "Sediments characterized by a fauna or an animal." Newell (1957, p. 411) stated that biofacies are "distinctive associations of fossils." Natland (1957, p. 555) considered them fossil assemblages. Dunbar and Rodgers (1957, p. 137) wrote that biofacies are the biologic aspects of a facies of some definite stratigraphic unit. Moore (1957, p. 1785) now holds the opinion that the recognition of biofacies is not useful unless it distinguishes "within the lithologically uniform sedimentary body, each significant spacially segregated part defined by organic content," and adds there is "no need to recognize biofacies," "unless it is desired to put primary emphasis on distribution of organisms" (p. 1784).

Physiofacies.—One of the principal disagreements that has arisen concerns the relations of *biofacies* to *facies* and *lithofacies*. Sloss *et al.* (1949) regarded biofacies and lithofacies as comparable but different expressions of facies. Some of their followers consider that *facies* equals *lithofacies* plus *biofacies* plus *tectofacies* (*q.v.*). Moore (1949) included biologic characters among the "collective characters" by which a lithofacies is recognized and he (1957, p. 1784) now states, "Biofacies is not correlative with lithofacies but a component of it." Accordingly, he (1949, p. 17) proposed the term *physiofacies* and defined it as "the total inorganic characteristics of a sedimentary rock." He considers that *biofacies* plus *physiofacies* equals *lithofacies*. Thus, *physiofacies* has a meaning close to if not identical with that of *lithofacies* as the latter term has been interpreted by some. Moore (1957, pp. 1784–85) now believes, however, that "the concept of physiofacies . . . may well be forgotten."

Tectofacies.—This is another term comparable with the foregoing. It was proposed by Sloss *et al.* (1949, p. 96) and defined as "a group of strata of different tectonic aspect from laterally equivalent strata." Krumbein and Sloss (1951, p. 383) stated that tectofacies are "laterally varying tectonic aspects of a stratigraphic unit." The Northwestern group considers that facies have three expressions that are designated *lithofacies*, *biofacies*, and *tectofacies*, respectively. Evidently a tectofacies is a special kind of lithofacies because both are identified by lithologic characters. The important difference seems to be that lithofacies are empirical units whereas tectofacies are interpretative and they are dependent on, or designed to distinguish, differences in tectonism. Moore (1957, p. 1785) states "This term . . . has dubious value. Little need . . . is seen . . . for it because lithofacies units can be readily classified as to their inferred tectonic significance. . . ."

Microfacies et cetera.—The term *micro-facies* was attributed by Fairbridge (1954, p. 685) to Cuvillier and to this he added *micro-lithofacies* and *micro-biofacies* (p. 683). None was defined but each was said to be recognized by a "certain look" or "appositional fabric" (p. 686) recognizable in thin sections of sedimentary rocks. Presumably, microlithofacies and microbiofacies are complementary expressions of microfacies. All these facies are textural and they are related to unexplained environments. They are not rock bodies and they seem to play no part in stratigraphy except that they may be useful in stratigraphic correlation.

Monothem.—Caster (1934b, p. 18) proposed the term *monothem* "For the more ordinary, and perhaps more normal major subdivisions of a stage." In this sense it is a time-stratigraphic unit and has no place in facies nomenclature. Moore (1949, p. 19), however, expressed the opinion that "A monothem is a local deposit having essentially uniform lithologic character; it corresponds to formation or member as commonly recognized by American geologists." This definition embodies two ideas, one of which carries definite facies implications, but both of them are entirely foreign to the original concept.

Lithosome.—This term was introduced by Wheeler and Mallory (in Fischer *et al.*, 1954, p. 929). They pointed out that stratigraphic bodies are separated into lateral parts, whose lithologic characters differ, in two ways. First, if lithologic change is gradual, boundaries are placed at arbitrary vertical cutoff planes. Second, if transition results from the intertonguing of different kinds of rock, boundaries are irregular lithologic contacts and a restricted sedimentary body of one kind bears both lateral and vertical relations to a similar body of a different kind. Moore's 1949 concept of *facies* (*q.v.*) admitted only intertongued deposits but Wheeler and Mallory, following Sloss *et al.*, considered facies to be bounded by vertical cutoff planes. Therefore, they believed that the other kind of facies required a name. They (1956) defined *lithosome* as "a lithostratigraphic body (or unit) which is mutually intertongued with one or more bodies of differing lithic composition" (pp. 2719, 2722), and they stated this was intended "to replace the non-stratigraphic term lithotope as a unit used in the analysis of a columnar section" (p. 2721).

Sloss (letter, December 24, 1956) defines *lithosome* as a "body of sediment deposited under uniform physico-chemical conditions" and presumably differentiates it from a lithofacies on the basis of uniform lithology rather than irregular lithologic boundary relations. Moore (1957, p. 1787) accepts *lithosome* but he states that Wheeler and Mallory placed "mistaken emphasis on intertonguing at lateral margins" although this was the fundamental basis of their concept. Moore defines *lithosome* as "(1) an independent body of genetically related sedimentary deposits of any sort, or alternatively, as (2) a spatially segregated part of any genetically related body of sedimentary deposits" (pp. 1787–8) and adds "a 'genetically related body . . .' may have almost any magnitude from . . . lenses or members . . . to formations or groups" (p. 1788). These statements are confusing and seem to suggest that a lithosome might be subdivided into other lithosomes. Moore also provides as another definition "the stratigraphic record of a more or less uniform lithotope" (p. 1788) and adds it "may . . . contain . . . two or more measurably distinct lithofacies" (p. 1789). These statements certainly are contradictory. Also Moore remarks that the "lower and upper boundaries [of a lithosome] . . . Commonly . . . coincide with bedding planes," and "borders may be of varied nature" (p. 1789), "subvertical . . . cutting across stratification planes . . . or . . . varyingly oblique to bedding planes" (p. 1788).

Other statements demonstrate that Moore accepts almost every kind of lateral boundary, transitional or abrupt, stable in location, oscillatory, or steadily shifting in position with the passage of time. Altogether, Moore's account is so unclear that the discovery of any means of distinguishing between his concept of a lithosome and either facies or lithofacies seems to be impossible.

Biosome.—This is a new term used by Wheeler (1958). Wheeler (letter, April 25, 1957) furnishes the definition, "Biosomes are mutually intertongued biostratigraphic units." This word is being used by the Northwestern group, and Sloss (letter, December 24, 1956) defines it as a "body of sediment deposited under uniform biological conditions." The relations of *biosomes* and *biofacies* are similar to those of *lithosomes* and *lithofacies*. Wheeler and Mallory emphasize intertonguing relations whereas Sloss and his associates emphasize uniform biological characters.

The similarity of *biosome* and *biostrome* is sure to cause confusion.

Tectosome.—This is an unpublished term that is in use at Northwestern University. Sloss (letter, December 24, 1956) defines it as a "body of strata indicative of uniform tectonic conditions," and it replaces *tectotope* as this word was used in 1949 and 1951.

Lithostrome.—Wheeler and Mallory (1956, p. 2721) introduced this term for "a lithostratigraphic layer consisting of one or more beds of essentially uniform or uniformly heterogeneous lithologic character." This word is comparable with *biostrome* (*q.v.*) but Wheeler and Mallory did not state whether or not they consider it to be a companion or a more inclusive term. They added, however, that a lithostrome is the "three-dimensional counterpart of a lithotope" (p. 2722). Thus it seems that *lithostrome* is nearly if not exactly the same as *lithosome* as the latter is employed at Northwestern. Sloss (letter, December 24, 1956) writes that he and his associates consider lithostromes to be "individual tongues projecting from a lithosome," and adds (letter, April 24, 1957), "I see no need for 'lithostrome' . . . the word 'tongue' conveys the information without confusion." Moore (1957, p. 1789) states that *lithostrome* "is synonymous with stratum . . . formation, member or tongue" and "may be equivalent to lithosome" and adds "lithostrome is a term serving no useful purpose."

The similarity of *lithostrome* and *lithosome* is certain to cause confusion.

SEMANTIC MAZE

Understanding and appreciation of the importance of facies relations in sedimentary rocks have grown with increasing momentum over a period of about 30 years. As one concept after another was recognized and formulated, terms to express various facies relationships have appeared. As knowledge increased, ideas changed and multiplied. Old terms were judged to be inadequate to express the more complex relations that became evident. Some old terms acquired specialized meanings and others were redefined or used with more restricted or otherwise altered meanings. Because facies problems have been approached from more than one standpoint, however, redefinitions of old terms, expressed or implied, have differed. Also careless or inexact definitions or ill considered usages have confused a variety of concepts under single terms. Consequently, at present, very few available facies terms have accepted and exact meanings that seem adequate for the identification of the concepts that they are intended to represent.

The foregoing rather tedious recapitulation of usages with respect to facies terminology may distort to some extent the authors' intended meanings because quotations are presented out of context, but they surely demonstrate the confusion that now exists. If stratigraphers are to extricate themselves from the maze of semantics that now engulfs the entire field of facies relations, facies concepts must be more clearly realized and facies terminology must in some way be rationalized and standardized and if possible simplified.

FACIES CLASSIFICATION

Much of the confusion so evident in facies nomenclature is the result of failure to distinguish clearly between different kinds of facies. This suggests that a practical classification of facies clarifying important differences is much needed. No such classification has yet been presented although the pattern in recently proposed facies terms indicates that some elements of a classification are realized by the Northwestern group and by Wheeler and Mallory. Their systems show that they recognize different kinds of facies principally in two ways. First, they differentiate facies on the basis of their form and mutual relations, and such terms as *lithofacies*, *lithosome*, *lithostrome*, and *lithotope* may be considered to identify different *facies classes*. Second, they differentiate facies on the basis of one or another elements of their nature which are considered to be important, and such terms as *lithofacies*, *biofacies*, and *tectofacies* may be considered to identify different *facies aspects*.

The facies concepts that have developed at Northwestern University are adapted to a system of regional facies analysis but they are not necessarily closely related to conventional stratigraphic units. According to the Northwestern group, a stratigraphic unit analysed with respect to facies may be a standard rock or time-rock unit, a biostratigraphic or tectonic unit, or an "operational unit" determined by lithology, electric-log characters, or any other practical means. Also it may be a "slice" consisting of some arbitrary fraction of the vertical extent of an otherwise indivisible unit (Sloss, letter, December 24, 1956). Facies as recognized at Northwestern can be arranged as shown in Table I.

Wheeler and Mallory have been more directly concerned with concepts related to conventional stratigraphic units and they view facies as specialized units of this kind. The relationships that they visualize are shown in Table II.

FACIES TYPES

Stratigraphers are concerned primarily with facies that are stratigraphic units. A review of past and present facies definitions and the concepts that are indicated reveals, however, that other factors are important in facies discrimina-

TABLE I. CLASSIFICATION OF STRATIGRAPHIC FACIES ACCORDING TO NORTHWESTERN GROUP

Facies Classes	*Facies Aspects*		
	Litho-stratigraphic	*Bio-stratigraphic*	*Tecto-stratigraphic*
Depositional *Area* of uniform environment	Lithotope	Biotope	Tectotope
Body of sediment representing uniform environment	Lithosome	Biosome	Tectosome
Vertically bounded *Lateral Parts* of stratigraphic unit (Facies)	Lithofacies	Biofacies	Tectofacies

tion. Thus facies that are definitely stratigraphic units grade off into others that are not. In one direction they pass into petrographic facies, or some other comparable kind, and in the other they merge into environmental facies. Facies, therefore, can be differentiated in another way that appears to be more fundamental than divisions into the classes and aspects recognized in the foregoing classifications. Accordingly, the following three different *types* of facies can be differentiated.

Type I.—Facies differentiated primarily on the basis of appearance or composition without respect to their form, boundaries, or mutual relations. These may be termed *petrographic facies.* Even though some may be characterized by fossils, the fossils can be regarded as rock constituents. Included in this type are redbed facies, clastic facies, limestone facies, marine facies, geosynclinal facies, and others. These are all generalized lithologic facies even though interpretation enters into the recognition of some of them. From one point of view, reef facies also might be included here in spite of its possible implications regarding form because it is a similarly generalized facies. Lithosomes if broadly interpreted according to the definition of Sloss and his associates are facies of this type.

Type II.—Facies distinguished primarily on the basis of form, nature of boundaries, and mutual relations to which appearance and composition are subordinated. These are all stratigraphic bodies of one kind or another and they may be termed *stratigraphic facies.* Many of the specialized facies terms that have

TABLE II. CLASSIFICATION OF STRATIGRAPHIC UNITS ACCORDING TO WHEELER AND MALLORY

Stratigraphic Classes	*Stratigraphic Aspects*		
	Litho-stratigraphic	*Bio-stratigraphic*	*Chrono-stratigraphic*
Vertical relations	Formation	Zone	Stage
Intertongued relations	Lithosome	Biosome	
Lateral relations	Lithofacies	Biofacies	
Layer of uniform character	Lithostrome		

been proposed refer to facies of this type. Some of these facies are differentiated among themselves on the basis of environmental interpretations. Lithofacies of Sloss *et al.* and lithosomes of Wheeler and Mallory are facies of this type.

Type III.—Facies that are concerned solely with environment. An environment is a combination of mutually interacting influences and conditions. It is not something that can be seen although some of its factors are visible. It can only be inferred from its results as these are exhibited in the form of sedimentary types and organic communities. *Environmental facies* are not sediments, rocks, or stratigraphic bodies of any kind; in fact they are not material. They are related more closely to sedimentology, biology, or tectonics than to stratigraphy although their applications to stratigraphy are important. Lithotopes, biotopes, and tectotopes are facies of this type.

FACIES CLASSES

In accordance with current concepts, facies of Types I and II can be subdivided into *classes*. The two classes of Type I or petrographic facies are the following.

Class A.—Facies consisting of actual large bodies of unspecified or indefinite form, extent and mutual relations identified by generalized features of their composition or appearance that occur in certain areas and more or less restricted parts of the stratigraphic section. Examples are redbed facies, evaporite facies, paralic facies, geosynclinal facies, and others. There are as many facies of each kind as there separate rock bodies.

Class B.—Facies consisting of all rocks of a single kind without any reference to their form or occurrence. Examples are black shale facies, oölitic limestone facies, graywacke facies, and others, and there can be only one facies of each kind. A general term for such facies is *lithosome* as defined by Sloss.

The three classes of Type II, or stratigraphic facies, are the following.

Class C.—Facies occurring in vertical succession whose boundaries are more or less horizontal stratigraphic planes. These generally do not require specialized facies nomenclature because they are conventional stratigraphic units such as systems, formations, and zones. Lithostromes are facies of this class.

Class D.—Facies that are laterally intergrading parts of some kind of a stratigraphic unit and are separated at more or less arbitrary vertical cutoff planes. These are the mappable facies that have become important in various kinds of facies analysis. Lithofacies as defined by Sloss *et al.* are the best known and most commonly employed.

Class E.—Facies that bear both lateral and vertical relations to each other, whose boundaries are irregular and record laterally shifting transgressive and regressive relations. These include the intertongued facies for which the term *lithosome* was proposed by Wheeler and Mallory.

This subdivision of stratigraphic facies into classes appears logical but it is not entirely practical because difficulties are encountered when an attempt is made to apply it to actual stratigraphic bodies. Note should be made of the fact that facies of these classes are defined on the basis of their form, but attention has been devoted almost exclusively to the mutual relations of adjacent facies and their boundaries, and consideration of the form of the facies as a whole has been neglected. Thus, two laterally intertongued facies would be termed *lithosomes* by Wheeler and Mallory who overemphasize the boundary relations at the expense of actual facies form. To meet the original definition of *lithosome* strictly, facies must intertongue in all directions. As Moore (1957, p. 1789) notes, this is not likely to occur and a facies that is intertongued on one side and arbitrarily cut off on the other might be termed either *lithosome* or *lithofacies* according to the direction of approach. Also a formation or other similar stratigraphic unit must terminate laterally somewhere. Unless it is cut out at unconformities, it must be arbitrarily cut off or intertongue; consequently, it does not really differ in form from horizontally differentiated facies.

Facies of Type III, or environmental facies, constitute a single class.

Class F.—Facies that are areas and not sedimentary or rock bodies of any kind. They include lithotopes, biotopes, and tectotopes.

FACIES ASPECTS

The aspect of a facies is determined by the kind of characters relied on for its recognition. The aspects that have been noted, as shown in Tables I and II, are (a) lithostratigraphic, (b) biostratigraphic, (c) tectostratigraphic, and (d) chronostratigraphic. With few and comparatively unimportant exceptions, only lithostratigraphic facies have found application in stratigraphic analysis. The relations of lithostratigraphic facies and biologic characters, however, are not agreed upon completely. Moore (1949, p. 17) wrote "the definition of lithofacies which I have presented . . . [embraces] both inorganic and organic features of the environment," and he continues to maintain this view at the present time (Moore, 1957, p. 1784). Krumbein and Sloss (1951, p. 268) stated, "It is profitable . . . to attack the problem of aspects by dissecting it into its lithologic and biologic components." Sloss *et al.* (1949) discussed the physical and biologic factors of facies separately (p. 93) in a manner exactly parallel with their consideration of lithofacies and biofacies (p. 95). Furthermore, they specifically cited "organically derived sediment" as a biological factor (p. 94). They do not, however, make as complete a distinction as these remarks suggest.

Fossils and biologic products, such as calcium carbonate and organic material, that affect the appearance, composition, or texture of a sedimentary deposit certainly are parts of the lithologic aspect and they are recognized as such by Sloss and his associates. In contrast, fossils considered only taxonomically or as environmental indices are not, and in these respects they are exclusively biostratigraphic characters. Such a division, however, is rather artificial and thus lithostratigraphic and biostratigraphic aspects overlap importantly.

Some confusion also attends the concept of biofacies. If this is a stratigraphic facies, it is a body of sediment or rock that is characterized biologically. *Biofacies* also have been defined, however, as "associations of fossils" (Newell, 1957, p. 411) and "fossil assemblages" (Natland, 1957, p. 555). These are perfectly logical definitions but such biofacies are biologic and not stratigraphic facies.

Tectostratigraphic facies are a special kind of lithostratigraphic facies. They are characterized lithologically in whatever way is considered to be most significant tectonically. Thus they are interpretative rather than strictly objective facies.

The foregoing observations show that these three aspects of facies are not as distinct and different as the names *lithofacies*, *biofacies*, and *tectofacies* seem to indicate.

The chronostratigraphic aspect has no direct application to facies except as a time-rock unit is divided laterally into facies. No facies concepts of this kind have been formulated or named.

It is possible to consider facies in terms of other aspects. Wheeler and Mallory (1956, p. 2713) wrote, "there are as many potential kinds of stratigraphic facies as there are kinds of data directly relatable to stratal units. . . ." Thus specialized facies might be based on exclusively (a) physical, (b) chemical, or (c) mineralogical characters, or (d) organic matter. These are, however, all restricted lithologic aspects and they have not yet been given much separate recognition.

INTERPRETATIVE FACIES

The recognition and differentiation of all facies useful in stratigraphy are dependent on observable features of sediments and rocks. In this sense, facies are all objective and real. Facies are interesting and most important, however, because they provide the basis for interpretation which is a subjective mental process. If differentiating features are selected with the purpose of effecting some particular interpretation, the resulting facies necessarily possess distinctively subjective qualities. There is significant difference, for example, between (a) a sandstone facies that is recognized simply because it differs lithologically from adjacent shale, and (b) a facies that is recognized because sandstone is believed to record a shallow-water environment. The first is an objective facies but it can be interpreted environmentally. The second is semi-subjective. It is objective because it can be recognized on the basis of real differentiating characters but it also is subjective because those characters were singled out for special notice in the belief that they are indicative of some particular environment. Redbed facies and oölitic limestone facies are objective because they are defined lithologically. Evaporite facies and geosynclinal facies are also characterized lithologically but they are semi-subjective because they are defined environmentally.

SUMMATION

The foregoing consideration of the elements of facies classification indicates that the differentiation of facies concepts as applied to stratigraphy into types and classes and with respect to aspects and objectivity is no simple matter. No simple system like that of Table I or Table II is adequate to organize all current facies concepts and show their relations, because too many variables are involved. Also, several of the concepts are not sharply distinguishable among themselves and separate classification would indicate more clear-cut differentiation than actually occurs. In the interests of accurate thinking, an analysis of the factors on which some kind of facies differentiation can be made appears to be worthwhile. This analysis, however, does not suggest any simple pattern of relationships that might be used as the basis for a complete, unified, and unambiguous practical facies nomenclature.

FACIES NOMENCLATURE

When this review of stratigraphic facies differentiation and nomenclature was first outlined, there seemed to be a possibility for developing a system of nomenclature whereby different facies could be distinguished and defined by related names consisting of two parts, the first indicating the facies aspect and the second indicating the facies class. This was suggested by the parallelism displayed by a majority of the terms that are now in use as illustrated by the arrangements in Tables I and II. Further consideration, however, that owes much to criticisms and suggestions offered by several interested and helpful correspondents, indicates that this is neither practicable nor desirable. In the first place, all facies concepts are not clear-cut and easily distinguishable and a more complex method of word formation would be required to make the meanings of all terms unmistakable. Second, if every facies concept were to receive a distinctive name, the number would be so great that the most retentive memory would be severely taxed and the possibility of the acceptance of so many names would be negligible. Third, it became apparent that such fine distinctions of facies concepts, accompanied by an elaborate terminology, are not necessary because different shades of meaning can be conveyed adequately by context and, where greater precision is required, common adjectives can be used to modify somewhat general terms and make distinction clear. Consequently, the preliminary objective, which was a system of detailed nomenclature, has been abandoned.

The most systematic usage of facies terminology is that which has been developed by the Northwestern group (Table I). This has been more or less widely adopted by others and, in order to minimize confusion, it seems necessary to accept its general features. Wheeler and Mallory's work has been particularly significant because it has directed attention to precision in some important details. This suggests certain modifications of the Northwestern system that seem to be desirable and advantageous. Moore has contributed to these matters by revealing some inconsistencies in logic and in practice that demand attention and must be considered if a simple, practical and widely acceptable terminology is to be attained.

Hedberg (1954, p. 217) observed that "the word facies has been greatly overworked" and this appears to be true also for some other facies terms. In many connections *facies* and also *lithofacies* used as nouns and adjectives mean no more than *lithology* or *lithologic*, and no advantage is gained in using these words in such a way. Recommendations are made in the following paragraphs regarding the acceptance or rejection and the use of all facies terms that have been proposed. Those that are considered useful are defined or, if necessary, redefined. Reference should be made to foregoing pages for original definitions and subsequent usages.

Facies.—This is a general term that finds application in many ways. In stratigraphy, it should be used exclusively for actual sediments or rocks and not for their appearance, composition, or environment although appearance, composition, or environment may determine the recognition of facies.

Facies may be used in stratigraphy in three ways. First, it may be a generic term including all restricted or specialized types or kinds of facies. The need for such a term is demonstrated by the way in which this word has been used throughout this article.

Second, it may be used for bodies of some specific kind or combination of kinds of sediments or rocks irrespective of their form, relations, and geographic and stratigraphic occurrence. Facies of this kind are provided with informal descriptive names such as redbed facies, shale facies or lagoonal facies. If it is desirable to limit a facies of this kind stratigraphically, use can be made of an expression like "the limestone facies of the Paint Creek formation."

Third, *facies* may be used for a formally recognized, areally restricted, and mappable part of a formation. Facies of this kind are provided with formal geographic names (Stockdale, 1931, 1939). If formations are lithologic units, they are in themselves more or less broadly conceived facies. The naming of locally restricted parts of formations as facies provides synonyms that may be useful in detailed stratigraphic work. These synonyms can be ignored, however, by anyone who is not concerned with such details.

Usage of the word facies in these three ways is not likely to be confusing because the differences in meaning are apparent in their context and they are indicated by the addition of a descriptive word or phrase, or a geographic name, or by the absence of such a modifier.

Geographic names should not be used to identify indeterminate facies with the purpose of indicating that they are similar in character to some particular formation or other conventional stratigraphic unit.

Facies may be identified by any observable feature of sediments or rocks including secondary features resulting from diagenesis or other alteration except those produced by weathering, metamorphism or structural disturbance. Thus, dolomite and limestone or cherty and non-cherty facies may be recognized.

Facies should not be restricted to lateral parts of a stratigraphic unit except as previously noted. If *facies* is so restricted, it has no important meaning different from *lithofacies* as this term is commonly used, and a more general and more elastic term is needed.

Phase.—Phase has been used in stratigraphy in many ways and no technical meaning has gained even a semblance of general acceptance. There seems to be no need to employ it in respect to facies or to restrict its usage in any way.

Intrafacies.—The manner in which Cloud and Barnes used this term is illogical and its use is likely to cause confusion. The single intrafacies that they recognized resembles a tongue extending into a more inclusive facies. At the same time, these authors applied *facies* to less extensive and apparently less important lithologic bodies wholly enclosed in a different facies. There seems to be no good reason for attempting to make a distinction between *facies* and *intrafacies*, and the latter term serves no useful purpose.

Lithofacies.—A lithofacies is a lateral subdivision of a stratigraphic unit differentiated from other adjacent subdivisions by its lithologic characters.

Lithofacies may be separated laterally in two different ways. First, their boundaries may be arbitrary vertical cutoff planes in which case they are *statistical lithofacies.* This is the kind of boundary that separates intergrading lithofacies. This is the kind of lithofacies that has been recognized most commonly. It is a mappable stratigraphic unit.

Second, lithofacies boundaries may be irregular lithologic boundaries separating intertonguing stratigraphic bodies of contrasting characters. Bodies separated in this fashion are *intertongued lithofacies.* They are not units that can be mapped in the normal way. The term *lithosome* originally was proposed for facies of this kind. Bodies of intertongued strata also can be separated at arbitrary vertical planes to produce lithofacies of the former kind.

Confusion of these two kinds of lithofacies is unlikely because the context in which the term is used ordinarily will make the meaning clear. If there is danger of confusion, the modifying adjectives can be used.

The stratigraphic unit that is subdivided into lithofacies may be of any kind. It may be a formally recognized rock, time-rock, or biostratigraphic unit. Also it may be a similar unit not recognized in conventional stratigraphic classification. The latter is known as an *operational unit.* In addition, it may be some vertical fraction of an otherwise indivisible stratigraphic unit in which case it is termed a *slice.*

Lithofacies are differentiated on the basis of lithologic characters. These include all physical and chemical characters and also biological characters insofar as these influence the appearance, composition, or texture of the sediment or rock. Lithologic characters reflect both original characters of the sediment and also secondary characters resulting from diagenesis and alteration such as compaction, cementation, dolomitization, and silicification, but not alterations produced by weathering, metamorphism, or structural disturbance such as leaching, brecciation, and shearing.

Statistical lithofacies are by far the most commonly recognized and used kind of facies. They can be differentiated on any basis such as differences in gross lithologic composition (percentages of shale, sandstone, and limestone), on various more subtle differences (white, conglomeratic, coarse, well sorted, cross-bedded, calcareous, fossiliferous, arkosic, garnetiferous, *et cetera* sandstones), or on environmental interpretations (terrestrial, fluvial, marine, geosynclinal, *et cetera*). Adjacent intertongued lithofacies commonly differ in very obvious lithologic characters such as shale *versus* sandstone, *et cetera*.

Lithofacies should not be given formal geographic names. There is no essential difference between a formally named facies and a statistical lithofacies that is part of a recognized stratigraphic unit. In the past, however, named units generally have been considered to be facies, and it seems advisable not to change this practice.

Biofacies.—A biofacies in stratigraphy is a lateral subdivision of a stratigraphic unit differentiated from other adjacent subdivisions by its biological characters without respect to the non-biologic features of its lithology. Most of the foregoing observations made with respect to lithofacies are equally applicable to biofacies.

The concept of biofacies overlaps that of lithofacies because many biologic features contribute importantly to lithologic character. Therefore, there generally seems to be no point in recognizing biofacies except on characters that do not affect lithology, such as the taxonomic identity or environmental implications of fossils. Only if they are restricted in this way do biofacies present an aspect that is different from and complementary to that of lithofacies.

If biofacies are recognized on the basis of taxonomic differences, they may be compared with conventional biostratigraphic units but there are important differences. Biostratigraphic zones have upper and lower boundaries determined by fossil ranges and lateral limits that are less significant. In contrast, the upper and lower boundaries of biofacies correspond with the boundaries of some specific stratigraphic unit and their lateral limits are emphasized. In other words, biostratigraphic units are considered principally in vertical sequence but the significance of biofacies is concerned with their lateral relations.

The adjustment of organisms to environment is so close that fossils reveal many environmental variations not indicated by differences in the sediments. If environmental interpretations are important, great care must be taken to prevent the mixing of fossils from different succeeding zones each of which may bear a fossil assemblage with somewhat different environmental implications even though the strata are practically identical lithologically. Almost every stratigraphic unit, unless it is a very thin or restricted one, is likely to consist of various zones. The biologic character of this unit is the sum of the characters of its zones; consequently, it is a mixture. Such a mixture is almost certain to present a distorted and confused picture of biological relations whose interpretation is difficult if not impossible. For this reason a biofacies is not likely to be as satisfactory and unified a facies as a comparable lithofacies.

Lithofacies can be recognized and traced in the subsurface on the basis of drilling records but generally this is not true of biofacies. Wells furnish such tiny samples of a stratigraphic section that they can not be relied on to reveal the biological characters of strata adequately. Therefore, the differentiation of biofacies commonly is restricted to zones of outcrop and very rarely is it possible to map biofacies with a reasonable degree of accuracy throughout any extensive region.

The biological characters of sediments and rocks are very important, and the concept of biofacies is interesting and useful theoretically. The differentiation, mapping, and interpretation of biofacies, however, are surrounded by many difficulties and uncertainties; therefore, their practical value is very limited. Various attempts have been made to coordinate biofacies and lithofacies without particular success. Consequently, it is doubtful if much emphasis on biofacies in a system of facies classification is warranted.

Tectofacies.—A tectofacies is a lithofacies that is interpreted tectonically.

Tectonism certainly has influenced all sedimentation because, without uplift of source areas, the production of sediments would soon cease and, without subsidence of basins, sediments would not continue to accumulate. The influence of tectonism on the quality of sediments that are deposited in any area, however, has been greatly overemphasized. Actually, the determining factors in sedimentary deposition are (a) local conditions that permit the deposition rather than the bypassing of sediments, (b) the kind and quantity of sediment delivered to an area, and (c) the organic environment that controls the local production of organic sediment. The first and third of these factors are exclusively local and are not related to tectonic activity except that tectonic movements can alter them. The second factor is related to tectonism insofar as quantity of sediment produced is concerned. Sediment quality, however, is dependent on (a) the kind of rocks attacked by erosion, (b) the kind of erosion and degree of weathering, and (c) changes in sediments accomplished during transportion and before final deposition. Thickness of accumulated sediment is a measure of tectonism but no matter what tectonic processes are in operation, no sediment can be deposited unless it is delivered to or produced within an area.

Some sediments are much more tectonically significant than others but on the whole tectofacies appear to be exceedingly theoretical and of very limited practical value. Actually, the patterns of most tectofacies maps have been influenced much more by sedimentary thickness than by lithologic character. Generally the nature of a tectofacies is noted only after the area of the tectofacies has been outlined on the basis of other considerations.

Physiofacies.—This term has little or no value either practically or theoretically. *Physiofacies* is essentially identical with a *lithofacies* as already defined. Many sedimentary rocks are mixtures of mineral and organic constituents, and the organic constituents are no less important than the mineral in determining the physical characters of a rock.

Lithosome.—This has been defined as (a) an intertongued stratigraphic rock unit, and (b) the sedimentary record of a physico-chemical environment. These definitions actually are rather similar. The first and original one is strictly objective but the second is semi-subjective because it depends on environmental interpretation. These concepts are interesting theoretically but they are not likely to find much application except in connection with some very specialized stratigraphic or facies considerations. For all ordinary purposes, the words *lithofacies* and *facies* seem adequate to designate these concepts. For the same reasons that are presented in connection with *lithofacies* and *physiofacies*, most organic features can not be eliminated.

Biosome.—This is an intertongued biostratigraphic unit or the record of a biologic environment. These concepts are interesting theoretically but they are likely to have very little practical usefulness.

Tectosome.—A tectosome is the sedimentary record of a tectonic environment. This may be a useful concept but it can be designated adequately as a *facies*, for example, geosynclinal facies.

Lithotope.—This should be defined as an area of uniform sedimentary environment. This is a useful term for a simple concept. Organisms contribute so importantly to many environments that their influences can not be ruled out.

Biotope.—This is an area of uniform ecology and organic adaptation. It is a very important concept in paleoecology but it is not likely to find much actual application in stratigraphic facies considerations.

Physiotope.—This word and the concept for which it stands have little value and *physiotope* may be eliminated from facies terminology without loss.

Tectotope.—A tectotope is an area of uniform tectonic environment. This is a useful concept but the term is superfluous because *area* is adequate. Tectonic areas are generalized and extensive. They are not subject to the differentiation possible in the consideration or description of sedimentary environments.

Lithostrome.—A lithostrome is nothing more than a rock stratigraphic unit and the words *layer*, *stratum*, *bed*, *member*, and *tongue* are all available and adequate to express this idea. *Lithostrome* should be rejected.

Biostrome.—This term also is superfluous because a biostrome as originally defined is only a particular kind of lithostrome. If it were to be redefined in a biologic sense, it would be a kind of paleobiologic zone.

Magnafacies.—A magnafacies is a time-transgressive rock stratigraphic unit which properly may be designated a *formation* or a *group*. There is no need for this term.

Megafacies.—This term generally seems to have been written by mistake where *magnafacies* was intended. It should not be used.

Parvafacies.—Parvafacies might be considered a special kind of lithofacies. As originally defined, it is interesting theoretically but its limitations are such that it is not likely to find practical application.

Monothem.—This is a time-stratigraphic term and it has nothing to do with facies. Also *monothem* is superfluous because its meaning is the same as *substage*.

Bioherm.—This should be used as a structural term as originally intended. Insofar as it means *organic reef*, no very good reason can be offered for its retention. As an organic reef, it has only a generalized facies significance. This is because reefs commonly consist of structureless core rock that grades into steeply dipping, fragmental, flanking strata and possibly also into nearly horizontal back-reef deposits. These differ greatly in lithologic characters and biologic contents and represent quite different environmental conditions. On the other hand, if *bioherm* is accepted as meaning a *non-reef limestone mound*, it acquires definite facies significance as the record of an integrated organic community that perpetuated itself in the midst of a generally hostile environment. Such structures need a name and restricting the use of *bioherm* to them (they were included in the original definition) requires less in the way of redefinition than applying it to only reef core structures.

Microfacies et cetera.—*Microfacies*, *microlithofacies*, and *microbiofacies* are in no sense stratigraphic facies terms. There seems to be no good reason for retaining them for any kind of facies because the word *texture* can be used to convey meanings much more clearly.

Appearance and composition.—Facies terms should not be used if reference is made to the appearance or composition of sediments or rocks rather than to the actual sediments or rocks themselves. Other words, such as *lithology*, are available and more accurately express intended meanings.

Environment.—Stratigraphic facies terms should not be used with reference to environments rather than to actual sediments or rocks. The intended meaning can be indicated more accurately by *environment* or some other applicable word.

CONCLUSIONS

The foregoing review of facies concepts as applied to stratigraphy shows that the development of an increasingly complex nomenclature, designed to promote precision in thought and expression, has had the opposite effect. At the present time, disagreement and misunderstanding regarding the exact meanings of most facies terms are general, and result in much confusion. This situation may have progressed to a point where it is beyond the control of individuals operating independantly, and group action may be necessary to bring about the compromise and agreement obviously required. Therefore, this subject is recommended to the attention of the American Commission on Stratigraphic Nomenclature.

Evaluation of facies concepts suggests that facies nomenclature has become unnecessarily complex. At least half of the facies terms that have been considered here appear to be wholly superfluous. Some of them only serve to conceal, under the covering of technical expressions, meanings that can be conveyed clearly by common words. Others are intended to make distinctions that have very little practical usefulness and do not require recognition except in theoretical or highly specialized considerations.

The conclusion is reached that only two facies terms applied to sedimentary rocks are important. These are *facies* and *lithofacies*. If they are accepted in the senses here recommended, all ordinary facies relations can be described. In addition, two environmental facies terms are important. These are *lithotope* and *biotope*. Finally, *bioherm* would be a useful structural term if its meaning were restricted as suggested.

Six other terms are not likely to be of much practical value but they may be useful in theoretical discussions or in the consideration of special problems or situations. These are *lithosome*, *biofacies*, *biosome*, *tectofacies*, *tectosome*, and *tectotope*. The recommendation is made, however, that these words be avoided unless their use is really necessary.

All of the other terms considered in this article can be rejected wthout loss.

REFERENCES

ALLEN, R. S., 1948, "Geological Correlation and Paleontology," *Bull. Geol. Soc. America*, Vol. 59, pp. 1–10.

BARRELL, JOSEPH, 1913, "The Upper Devonian Delta of the Appalachian Geosyncline," *Amer. Jour. Sci.*, Ser. 4, Vol. 36, pp. 429–72.

BERGQUIST, H. R., AND COBBAN, W. A., 1957, "Mollusks of the Cretaceous," *Geol. Soc. America Mem. 67*, Vol. 2, pp. 871–84.

CARPENTER, J. R., 1938, *An Ecological Glossary*, reprint 1956. Hafner, New York.

CASTER, K. E., 1934a, "Facies Nomenclature and the Upper Devonian" (abst.), *Geol. Soc. America Proc. 1933*, pp. 348, 349.

———, 1934b, "The Stratigraphy and Paleontology of Northwestern Pennsylvania, Pt. 1: Stratigraphy," *Bull. Amer. Paleon.*, Vol. 21, pp. 1–185.

CLOUD, P. E., 1952, "Facies Relationships of Organic Reefs," *Bull. Amer. Assoc. Petrol. Geol.*, Vol. 36, pp. 2125–49.

———, AND BARNES, V. E., 1957, "Early Ordovician Seas in Central Texas," *Geol. Soc. America Mem. 67*, Vol. 2, pp. 163–214.

COOPER, B. N., AND COOPER, G. A., 1946, "Lower Middle Ordovician Stratigraphy of the Shenandoah Valley, Virginia," *Bull. Geol. Soc. America*, Vol. 57, pp. 35–114.

COOPER, G. A., 1957, "Paleoecology of Middle Devonian of Eastern and Central United States," *Geol. Soc. America Mem. 67*, Vol. 2, pp. 249–78.

CUMINGS, E. R., 1930, "Lists of Species from the New Corydon, Kokomo, and Kenneth Formations of Indiana, and from Reefs in the Mississinewa and Liston Creek Formations," *Proc. Indiana Acad. Sci.*, Vol. 39, pp. 204–11.

———, 1932, "Reefs or Bioherms?," *Bull. Geol. Soc. America*, Vol. 43, pp. 331–52.

———, AND SHROCK, P. R., 1928, "Niagaran Coral Reefs of Indiana and Adjacent States and Their Stratigraphic Relations," *ibid.*, Vol. 39, pp. 579–619.

DAPPLES, E. C., KRUMBEIN, W. C., AND SLOSS, L. L., 1948, "Tectonic Control of Lithologic Associations," *Bull. Amer. Assoc. Petrol. Geol.*, Vol. 32, pp. 1924–47.

DUNBAR, C. O., AND RODGERS, JOHN, 1957, *Principles of Stratigraphy*. John Wiley and Sons, New York.

FAIRBRIDGE, R. W., 1954, "Stratigraphic Correlation by Micro-Facies," *Amer. Jour. Sci.*, Vol. 252, pp. 683–94.

———, 1957, "The Dolomite Question," *Soc. Econ. Paleon. Min. Spec. Pub. 5*, pp. 125–78.

FENTON, C. L., AND FENTON, M. A., 1930, "Ecologic Interpretations of Some Biostratigraphic Terms, II, Zone, Subzone, Facies, Phase," *Amer. Midland Natur.*, Vol. 12, pp. 145–53.

FISCHER, A. G., WHEELER, H. E., AND MALLORY, V. S., 1954, "Arbitrary Cut-Off in Stratigraphy," *Bull. Amer. Assoc. Petrol. Geol.*, Vol. 38, pp. 926–31.

FISHER, D. W., 1956, "Intricacy of Applied Stratigraphic Nomenclature," *Jour. Geol.*, Vol. 64, pp. 617–27.

FISK, H. N., and MCFARLAN, E., 1955, "Late Quaternary Deltaic Deposits of the Mississippi River," *Geol. Soc. America Spec. Pap. 62*, pp. 279–302.

GREENMAN, D. W., 1955, "Ground Water Resources of Bucks County, Pennsylvania," *Pennsylvania Top. Geol. Survey Bull. W 11*. 66 pp., map.

HEDBERG, H. D., 1954, "Procedure and Terminology in Stratigraphic Classification," *19th Int. Geol. Cong., C.R.*, Sec. 13, pp. 205–33.

HESSE, R., ALLEE, W. C., AND SCHMIDT, K. P., 1937, *Ecological Animal Geography*. 597 pp. John Wiley and Sons, New York.

IMBRIE, JOHN, 1955, "Biofacies Analysis," *Geol. Soc. America Spec. Pap. 62*, pp. 449–63.

JOHNSON, R. B., AND WOOD, G. H., 1956, "Stratigraphy of Upper Cretaceous and Tertiary Rocks of Raton Basin, Colorado and New Mexico," *Bull. Amer. Assoc. Petrol. Geol.*, Vol. 40, pp. 707–21.

KAY, MARSHALL, 1945, "Paleogeographic and Palinspastic maps," *ibid.*, Vol. 29, pp. 426–50.

———, 1947, "Analysis of Stratigraphy," *ibid.*, Vol. 31, pp. 162–68.

KING, P. B., 1949, "Discussion *in* Sedimentary Facies in Geologic History," *Geol. Soc. America Mem. 39*, pp. 165–71.

KRUMBEIN, W. C., 1948, "Lithofacies Maps and Regional Sedimentary-Stratigraphic Analysis," *Bull. Amer. Assoc. Petrol. Geol.*, Vol. 32, pp. 1909–23.

———, AND SLOSS, L. L., 1951, *Stratigraphy and Sedimentation*. 497 pp. W. H. Freeman and Company, San Francisco.

KUMMEL, BERNHARD, 1957, "Paleoecology of Lower Triassic Formations of Southeastern Idaho and Adjacent Areas," *Geol. Soc. America Mem. 67*, Vol. 2, "Paleoecology," pp. 437–67.

LINK, T. A., 1950, "Theory of Transgressive and Regressive Reef (Bioherm) Development and Origin of Oil," *Bull. Amer. Assoc. Petrol. Geol.*, Vol. 34, pp. 263–94.

LOCHMAN, CHRISTINA, 1956, "Stratigraphy, Paleontology, and Paleogeography of the *Elliptocephala asaphoides* Strata in the Cambridge and Hoosick Quadrangles, New York," *Bull. Geol. Soc. America*, Vol. 67, pp. 1331–96.

LOWMAN, S. W., 1945, "Research Committee Program," *Bull. Amer. Assoc. Petrol. Geol.*, Vol. 29, pp. 1512–15.

———, 1949, "Sedimentary Facies in Gulf Coast," *ibid.*, Vol. 33, pp. 1939–97.

MCKEE, E. D., 1938, "Environment and History of the Toroweap and Kaibab Formations of Northern Arizona and Southern Utah," *Carnegie Inst. Washington Pub. 492*. 268 pp.

———, 1949, "Facies Changes in the Colorado Plateau," *Geol. Soc. America Mem. 39*, pp. 35–48.

MCLAUGHLIN, D. B., AND GERHARD, R. C., 1953, "Stratigraphy and Origin of Triassic Fluviatile Sediments, Lebanon and Lancaster Counties," *Proc. Pennsylvania Acad. Sci.*, Vol. 27, pp. 136–42.

MACNEIL, F. S., 1954, "Organic Reefs and Banks and Associated Detrital Sediments," *Amer. Jour. Sci.*, Vol. 252, pp. 385–401.

MOORE, R. C., 1948, "Stratigraphical Paleontology," *Bull. Geol. Soc. America*, Vol. 59, pp. 301–25.

———, 1949, "Meaning of Facies," *Geol. Soc. America Mem. 39*, pp. 1–34.

———, 1957, "Modern Methods of Paleoecology," *Bull. Amer. Assoc. Petrol. Geol.*, Vol. 41, pp. 1775–1801.

MULLENS, T. E., AND FREEMAN, V. L., 1957, "Lithofacies of the Salt Wash Member of the Morrison Formation, Colorado Plateau," *Bull. Geol. Soc. America*, Vol. 68, pp. 505–26.

NATLAND, M. L., 1957, "Paleoecology of West Coast Tertiary Sediments," *Geol. Soc. America Mem. 67*, Vol. 2, "Paleocology," pp. 543–72.

NEWELL, N. D., 1957, "Paleoecology of Permian Reefs in the Guadalupe Mountains Area, *ibid.*, pp. 407–36.

———, RIGBY, J. K., FISCHER, A. G., WHITEMAN, A. J., HICKOX, J. E., AND BRADLEY, J. S., 1953, *The Permian Reef Complex of the Guadalupe Mountain Region, Texas and New Mexico*. 236 pp. W. H. Freeman and Company, San Francisco.

PAGE, DAVID, 1859, *Handbook of Geological Terms and Geology*. 416 pp. Blackwood, London.

PETTIJOHN, F. J., 1949, *Sedimentary Rocks*. 526 pp. Harper and Brothers, New York.

SHROCK, R. R., 1939, "Wisconsin Silurian Bioherms (Organic Reefs)," *Bull. Geol. Soc. America*, Vol. 50, pp. 525–62.

SLOSS, L. L., KRUMBEIN, W. C., AND DAPPLES, E. C., 1949, "Integrated Facies Analysis," *Geol. Soc. America Mem. 39*, pp. 91–123.

STOCKDALE, P. B., 1931a, "The Borden (Knobstone) Rocks of Southern Indiana," *Indiana Dept. Conserv. Pub. 98*. 330 pp.

———, 1931b, "Bioherms in the Borden Group of Indiana," *Bull. Geol. Soc. America*, Vol. 42, pp. 707–18.

———, 1939, "Lower Mississippian Rocks of the East-Central Interior," *Geol. Soc. America Spec. Pap. 22*. 248 pp.

SWAIN, F. M., 1956, "Stratigraphy of Lake Deposits in Central and Northern Minnesota," *Bull. Amer. Assoc. Petrol. Geol.*, Vol. 40, pp. 600–53.

TWENHOFEL, W. H., 1950, "Coral and Other Organic Reefs in Geologic Column," *ibid.*, Vol. 34, pp. 182–202.

WELLER, J. M., 1956, "Argument for Diastrophic Control of Late Paleozoic Cyclothems," *ibid.*, Vol. 40, pp. 17–50.

WELLS, J. W., 1944, "Middle Devonian Bone Beds of Ohio," *Bull. Geol. Soc. America*, Vol. 55, pp. 273–302.

———, 1947, "Provisional Paleoecological Analysis of the Devonian Rocks of the Columbus Region," *Ohio Jour. Sci.*, Vol. 47, pp. 119–26.

WHEELER, H. E., 1958, "Primary Factors in Biostratigraphy," *Bull. Amer. Assoc. Petrol. Geol.*, Vol. 42, pp. 640–55.

———, AND MALLORY, V. S., 1954, "Analysis and Classification of Stratigraphic Units" (abst.), *Bull. Geol. Soc. America*, Vol. 65, p. 1324.

———, AND ———, 1956, "Factors in Lithostratigraphy," *Bull. Amer. Assoc. Petrol. Geol.*, Vol. 40, pp. 2711–23.

WILLIAMS, H. S., 1895, *Geological Biology*. 395 pp. Henry Holt, New York.

WILSON, W. B., 1950, "Reef Definition," *Bull. Amer. Assoc. Petrol. Geol.*, Vol. 34, p. 181.

Reprinted from:
BULLETIN OF THE AMERICAN ASSOCIATION OF PETROLEUM GEOLOGISTS
VOL. 42, NO. 11 (NOVEMBER, 1958), PP. 2718–2744, 1 FIG.

CONCEPTS OF FACIES[1]

CURT TEICHERT[2]
Denver, Colorado

ABSTRACT

Stratigraphic facies was recognized, defined, and named by Gressly in 1838. Before the end of the nineteenth century the concept of facies became firmly established, through the works of Mojsisovics, Renevier, and Walther, as referring to the sum of lithologic and paleontologic characteristics of a sedimentary rock from which its origin and the environment of its formation may be inferred. Similar or identical rock types have isopic facies, different rock types heteropic facies. Facies changes must be studied in horizontal as well as vertical direction, with the aim of reconstructing changes of environment in space and time. Genetically interconnected isochronous facies form facies tracts; genetically interconnected facies tracts form facies families. Heteropic facies in vertical succession form facies sequences.

When used to designate major stratigraphic sequences occurring in certain geographic, oceanographic, or inferred tectonic environments, the term facies loses its descriptive objectivity. Nonstratigraphic uses of facies are discussed. Recent American facies terminology is discussed in terms of earlier established European terminology. Facies and biofacies as used in ecology differ importantly from stratigraphic facies and biofacies. Rock characteristics now referred to as "biofacies" in stratigraphy should be called paleontologic facies. Finally, some modern Russian facies literature is reviewed.

INTRODUCTION

Probably no geological term has been used more indiscriminately and for a wider range of concepts than *facies*. This word, not so long ago a generally understood and well defined item of geological terminology, is now bereft of meaning, unless used with qualifying adjectives or accompanied by lengthy definitions. At international meetings geologists no longer speak in mutually comprehensible terms, if facies concepts are being discussed (see discussion by Frijlinck, Cuvillier, and Sloss, 1955). At a recently held symposium on problems of correlating and interpreting sedimentary facies two of six speakers avoided use of the term facies, and the other four used it for three rather different concepts: (1) for physical or chemical characteristics of a rock; (2) for the environment in which a rock was formed; (3) for associations of living or fossil Foraminifera.

The following discussion is offered in support of a plea by Wheeler and Mallory (1956) "to avoid a stratigraphic classificatory and nomenclatural system in which terms have different meanings to different persons." However, to achieve this aim, little benefit will result from yet another definition of the term facies and related stratigraphic terms, or from expression of personal preferences. Rather, to determine the meaning of a scientific term two guides are available: (1) the original definition by the author who proposed it, (2) subsequent usage by a majority of those who have found the term useful.

It is proposed to show that the facies concept was established and defined distinctly and unambiguously in 1838, and that it was consolidated and firmly

[1] Publication authorized by the director of the United States Geological Survey. To P. E. Cloud, Jr., R. W. Fairbridge, E. D. McKee, S. S. Oriel, and V. E. Swanson the writer is indebted for reading and criticizing the manuscript. Manuscript received, May 26, 1958.

[2] United States Geological Survey.

rooted in geological literature well before the end of the nineteenth century. It is further suggested that during the last 20–30 years the term facies has been applied to numerous scientific concepts, unrelated to the original concept, and for which introduction and use of other terms might therefore be preferable.

FACIES CONCEPTS IN WESTERN EUROPE

ORIGINS OF FACIES CONCEPT

Facies is a Latin word meaning face, figure, appearance, aspect, look, condition. It signifies not so much a concrete thing, as an abstract idea. The word was introduced into geological literature by Nicolaus Steno (1669, pp. 68–75) for the entire aspect of a part of the earth's surface during a certain interval of geologic time. As thus used facies specifically included any sediments deposited during such intervals, but periods and areas of denudation were also included in the concept.

Modern scientific usage of the term facies dates from the year 1838 when the Swiss geologist Gressly published the first part of his geological investigations in the Jura Mountains near Solothurn, Switzerland.[3] This classic work has often been quoted, but apparently little read in recent years. Gressly arrived at the facies concept inductively, through comparative stratigraphic studies in an area characterized by rapid lateral and vertical changes in the characteristics of rock units. Important parts of his definition of facies have recently been translated by Dunbar and Rodgers (1957, p. 136), but in order to understand fully the original meaning of this concept as well as its subsequent history some more extensive reading of Gressly's paper is essential.

A translation of Gressly's definition (with slight modifications from the version given by Dunbar and Rodgers) reads as follows.[4]

"Instead of being satisfied with a number of vertical sections . . . I have followed each stratigraphic unit[5] horizontally as far as possible, in order to study all its changes.

"I have thus succeeded in recognizing, in the horizontal dimension of each stratigraphic unit, diverse, well defined modifications which present constant characteristics in respect to their petrographical composition as well as to the paleontological features of their fossil assemblages, and which are governed by special, only slightly variable laws.

"To begin with, two principal facts characterize the sum total of the modifications which I call *facies* or *aspects of a stratigraphic unit:* one is that *a certain lithological aspect of a stratigraphic unit is*

[3] Markevich (1957, p. 3) has recently claimed that the Russian naturalist and poet Lomonosov in a book entitled *O sloyakh zemnykh* (On Earth Strata) and published in 1757–1759, was the first to recognize facies in sedimentary rocks. Unfortunately, I have not seen Lomonosov's book, but from quotations cited by Markevich it appears that Lomonosov recognized and defined physical differences in successive rather than in contemporaneous strata and thus failed to recognize an important aspect of facies as later understood.—C. T.

[4] Throughout this article quotations from foreign language publications are given in English translation. All translations are by the writer. Translations from older publications are generally rendered in a free, modernized style which conveys the original meaning more accurately than a stilted literal translation. All translated quotations are preceded by the indication "(transl.)." I. Mitten, U. S. Geological Survey, assisted in the translation of some Russian papers.

[5] Dunbar and Rodgers translated the French *terrain* as "formation," but gave "stratigraphic unit" as an alternative. Actually, a *terrain* in French may be composed of several formations in the American sense.

linked everywhere with the same paleontological assemblage; the other is that *from such an assemblage fossil genera and species common in other facies are invariably excluded.*

"If it happens occasionally that certain genera and species, characteristic of one facies, also occur in another facies, it is the rule that they are much less abundant, less well developed and less characteristic than in the facies or general assemblage to which they really belong."

Gressly then points out that these facies changes are caused by, as we would now say, changes in the depositional environment, such as may be observed in the oceans of the present day.

Gressly proceeds to summarize his conclusions on the nature of facies changes in five facies laws (1838, pp. 20–23) which, freely translated from the cumbersome and somewhat archaic original, read as follows.

First law: Each facies of a stratigraphic unit presents very distinct lithological or paleontological characteristics, quite different from the characteristics of other facies of the same geological horizon.

Second law: In different stratigraphical units the same lithological facies affects the paleontological characteristics in much the same way, not only in the horizontal direction, but also if they follow each other vertically in stratigraphical sequence.

This law is further elaborated in the following passage (transl.).

"The organic forms of fossils resemble each other very much in analogous facies even though they may appear in different stratigraphical units. . . . This has led to the opinion, now generally accepted, that identical fossils may occur not only in different subdivisions and groups of a stratigraphical unit, but also in units which, in a vertical geological section, are separated by other very thick units."

Third law: Sometimes different facies have sharp boundaries in a horizontal direction, sometimes they pass into each other through transitional intermediate varieties.

Fourth law: The diversity of facies increases from bottom to top in the stratigraphical section.

This "law" evidently refers to conditions prevalent in the area investigated by Gressly and has limited validity.

Fifth law: The diversity of facies may be greater or less in different areas. Gressly discusses at some length that the degree of facies variability is determined by distance from the shore and that there is greater variability in littoral than in off-shore areas.

Finally Gressly (1838, pp. 24–25) sums up the advantages of the facies method as follows.

1. The many paleontological details which may at first seem to be devoid of meaning can be reduced to a few simple laws by demonstrating the close relationships between them and the lithological and geological aspects of the rocks.

2. The lithological and geological properties of sedimentary rocks may be interpreted more successfully by advancing from routine mineralogical descriptions to geologic interpretation, demonstrating their relationships with the evolution of life as manifested in the different epochs of the history of our planet.

3. The submarine relief of various periods and the submarine processes, which

have affected the stratigraphic units and their facies, can be determined with greater accuracy.

4. Periods of mountain uplift can be determined with greater precision by recognition of the littoral character of the rocks which have subsequently been deposited.

Thus, Gressly fully realized the implications of the results of his studies in the Jura Mountains of Solothurn and he did not hesitate to point out some very broad generalizations. It is also evident that Gressly restricted the facies concept to sedimentary rocks, that for him the core of his facies concept was the interdependence of lithological and paleontological facies, and that he recognized vertical as well as horizontal facies changes or identity of facies.

The French geologist Prevost arrived at very similar ideas, apparently independently from Gressly and at about the same time. He may, therefore, rightly be regarded as the co-founder of the facies concept. He used, however, a terminology which was different from that of Gressly, and it was not his but Gressly's term that was subsequently accepted. Prevost's approach was essentially deductive. At a meeting of the Geological Society of France in December, 1837, he (Prevost, 1838) pointed out that just as today many different kinds of sediments are being formed at the same time, so it is possible to distinguish different "formations" in the same "epochs" of the past as well as identical "formations" formed in different "epochs." In a later paper Prevost (1845) made important contributions to early thinking about time-stratigraphical and rock-stratigraphical concepts. He defined the French word *terrain* as a term for rocks of any kind and origin which had been formed during the same time interval and *formation* as a term for rocks which had been formed by the same causes (or in the same environment). Prevost's *formation* is thus identical with Gressly's *facies*. When Prevost says that a *terrain* consists of diverse *formations* he means, in modern parlance, that a time-stratigraphic unit may be composed of rocks of diverse facies.

It is interesting to observe that whereas Gressly arrived at his generalizations by painstaking detailed stratigraphical observations, for Prevost the *synchronisme des formations* (=contemporaneity of facies) was a "principle deduced by reasoning and observation." Thus, although much has been written about these relationships in recent years, the concepts and the approaches to their treatment are anything but modern.

FURTHER DEVELOPMENT OF FACIES CONCEPT

The development of the facies concept in geological literature during the nineteenth century for the most part followed closely the principles laid down by Gressly and Prevost. First, however, the pendulum had to swing back from recognition of and emphasis on facies fossils to re-emphasis on the chronological value of fossils, with consequent development of the zone or guide fossil concept by Oppel in 1856–1858 (see Teichert, 1958). Oppel recognized the fact that in most fossil assemblages only restricted numbers of species are useful for stratigraphic correlation. These he used to characterize paleontological zones. Oppel thus was the first, just 100 years ago, to establish, in effect, the distinction between facies fossils and zone or guide fossils, even though he did not spell out these concepts in clear-cut definitions.

The facies concept was found useful by German, Austrian, Swiss, and French geologists to the extent that since at least the late 1880s every geological textbook published in German and French included a thorough treatment of this subject: Neumayr, 1887; Reyer, 1888; Walther, 1893, 1894; Credner, 1902 (and earlier editions); Haug, 1907; Kayser, 1912; Dacqué, 1915; Tornquist, 1916; Keilhack, 1921; Diener, 1925; Salomon, 1926—to mention only some of the better known and most widely used books of several decades ago.

Fundamental to most European discussions were contributions made by the Austro-Hungarian Mojsisovics von Mojsvár, the Swiss Renevier, and the German Walther.

By 1879, when Mojsisovics published his classic work on the Triassic reefs of the eastern Alps, he was able to regard the facies concept as firmly established in contemporary geological thinking (transl.).

> "Following Gressly and Oppel one now customarily applies the term facies to deposits formed under the influence of different environmental conditions."

He clarifies the concept as follows (transl.).

> "It is important to remember that the facies concept expresses the general mutual relationships between environmental conditions on the one hand and the sediments and the dwelling places of organisms [=biotopes] on the other" (Mojsisovics von Mojsvár, 1879, p. 5).

Mojsisovics objects to application of the term facies to rocks formed in different geographical provinces. He then proceeds to set out the principles of what appears to be the first environmental classification of sedimentary rocks (1879, pp. 5–10). Sedimentary rocks may be classified according to (1) the surrounding medium in which they were deposited: marine, terrestrial, lacustrine, etc., (2) the geographical province of deposition, (3) facies. Rocks formed in the same medium are called *isomesic*, rocks formed in different media *heteromesic;* rocks formed in the same basin or province *isotopic*, in different basins or provinces *heterotopic;* rocks of the same facies are called *isopic*, no matter what their relative age, rocks of different facies *heteropic.*

The important proposal here is the distinction between isopic ("same-looking") and heteropic ("different-looking") facies (transl.): "Just as different facies are formed simultaneously and in juxtaposition in the same sedimentation area, identical facies appear in different sedimentation areas and at different times. The former are called heteropic, the latter isopic formations" (pp. 6–7). This distinction was subsequently adopted, maintained, and elaborated by most European geologists. We shall return to it later.

Another important contribution was made by Renevier (1884) who called facies the "physiognomy" of a rock and established a classification of facies, essentially based on petrographic and paleontologic criteria. He emphasized that (transl.) "facies are the different kinds of formations, sedimentary or other, which can be formed simultaneously at a certain moment of geologic time, just as it happens at present." Among conditions determining facies he lists: environment (*milieu*), geographic position, chemical and mineralogical composition, animal and plant communities, climate, altitude, and depth below sea-level. Thus, littoral formations may appear as coralliferous, marly, muddy, sandy, or conglomeratic facies, and so on.

Finally, Johannes Walther made important theoretical contributions to the facies concept in this period. Walther (1893, 1894) defined facies as the (transl.) "sum of all primary characteristics of a sedimentary rock" (1894, p. 989) and this definition was adopted by Haug (1907) and the majority of European geologists up to the present day. Walther also developed the important concept of the *Faciesbezirk* (facies tract) as a system of different (heteropic), but genetically interconnected facies, and including the areas of erosion from which the sediments of these facies are derived. In this definition an erosional interval represents part of a facies tract. A facies tract is thus somewhat related to the original facies concept of Steno (1669).

Necessity to recognize and to study vertical facies changes, in addition to lateral ones, was strongly emphasized by Arbenz (1919). In German literature successions of vertical facies became known as *Faziesreihen* (see Frank, 1936; Spengler, 1951, and many others). This concept will be known in this paper as *facies sequence* (Fig. 1). It is an old established concept and not at all a matter of recent argument, as, for example, Wheeler and Mallory (1956) seem to believe. The concept of recurrent facies is as familiar to European thinking as that of recurrent faunas to geologists everywhere. *Facies evolution* is a gradual change of facies in time, indicating gradually changing sedimentational conditions (see also Lombard, 1952).

In continental European literature the flow of discussion on the facies con-

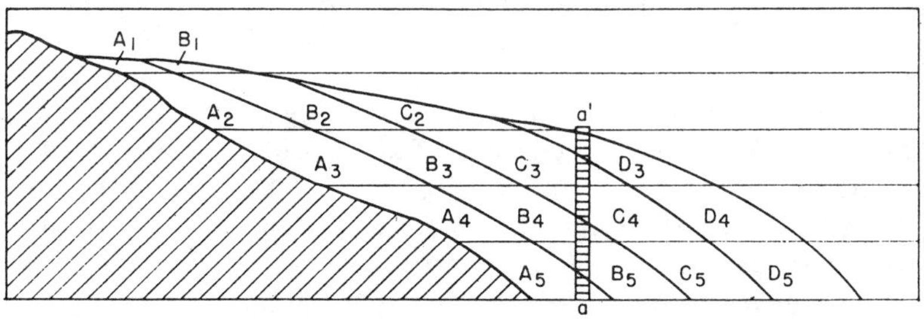

THREE WAYS OF CONSIDERING FACIES RELATIONS

I. vertico-laterally — II. laterally — III. vertically

I. $A_1+A_2+A_3+A_4+A_5$ = isopic heterochronous facies ("magnafacies," "lithosome")

II. $A_3+B_3+C_3+D_3$ = facies tract of isochronous heteropic facies (each element = "parvafacies")

III. Columnar section *a-a'* $[A_5+(B_5, B_4)+(C_4, C_3)+D_3]$ = facies sequence of heterochronous heteropic facies

FIG. 1.—Generalized rock sequence formed under conditions of rising sea-level. Oblique lines are facies boundaries, horizontal lines time-planes. Equal facies indicated by Roman letters, time-equivalent deposits by numerals.

cept continues uninterruptedly to the newest treatises by Simon (1948), Bubnoff (1949a, 1949b), Gignoux (1950, 1955), Kuenen (1950), Schindewolf (1950), Fourmarier (1950), Termier and Termier (1952), Pomerol and Fouet (1953), Brinkmann (1956), Lombard (1956), and others. In all of these, facies is defined in one way or another as the expression of the collective lithological and paleontological characteristics of a sedimentary rock; isopic and heteropic facies are being distinguished (see, for example, Fourmarier's isopic maps); and horizontal facies changes as well as vertical "facies evolution" are being discussed.

As defined and used by these authors a facies is not a rock body, but the expression of all the primary properties of a rock body. Facies, therefore, is an abstraction, as indeed implied by the original Latin meaning of the word: look, aspect, and so on.

While the ideas of Gressly, Mojsisovics, Renevier, and Walther were thus being preserved and developed in Europe, there appeared in European literature another trend which emphazied application of the term facies to paleogeographically, tectonically, or otherwise geologically defined stratigraphic or other rock sequences and complexes. As early as 1888 Reyer had spoken of eolian, deep sea, flysch, crystalline, volcanic, and many other kinds of facies, indicating sedimentary and non-sedimentary environments rather than objectively defined rock characteristics. Deecke (1913) was among those who maintained strongly that the facies concept was a strictly stratigraphical one. However, in the terminology of many authors, facies came to be used for the designation of generalized and sometimes ill defined environmental situations, forgetting that according to definition, facies indicates rock properties from which such environments should be inducible.

This situation had been aggravated by Haug's influential *Traitê de Gêologie* (1907). Here Haug defined facies correctly as (transl.) "the sum of lithologic and paleontologic characters which represent a deposit at a given point." He also adopted Mojsisovics' divisions of heteropic and isopic facies, but he then proceeded to present an entirely environmental and genetic classification of rock units into volcanic, eolian, marine, glacial, neritic, bathyal, and other facies. This kind of classification, of course, attempts to define facies by assumed environmental conditions instead of reconstructing environment by means of study of facies. Strausz, in an important facies study (1928), was among those who called attention to (transl.) "the error frequently committed to confuse facies studies with paleogeography," and Tercier (1939) gathered forceful arguments against Haug's over-generalizations. However, the application of facies to environment rather than to rock characteristics became deep-rooted in much of European literature. Thus, as late as 1955 Moret calls facies (transl.) "the totality of local geographical and biological conditions which determine the lithologic nature of a deposit and its contained animal and plant associations." This is indeed putting the cart before the horse!

Even the briefest review of the history of the facies concept in European literature would be incomplete without reference to the role it has played in Alpine geology. Here, facies was early defined in a broad sense to refer to the characteristics of large belts or basins of sedimentation (Bertrand, 1897). Correlation of such broad facies belts became important in the unravelling of complex nappe structures when it was discovered that rock sequences which were originally laterally contiguous had been brought into vertical superposition by horizontal thrusts. These conditions were analyzed in detail by Heim (1921), Bubnoff (1921), and many others (for a brief résumé see Gilluly *et al.*, 1952, p. 517). Soon every nappe was identified with a distinct facies belt, a trend to which objections were raised by Heritsch (1927) and others. In more recent years these nappe-facies relationships have been less strongly emphasized (see Cadisch, 1953).

Further broad generalizations developed. The entire sedimentary contents of paleogeographical provinces were designated as helvetic, penninic, and other facies. Sedimentary columns built up in areas of differing tectonic history were differentiated as representing geosynclinal, orogenic, foredeep, foreland facies, and others (Arbenz, 1919; Stille, 1924; and many others). Kobayashi (1956) and others have named such "facies" after orogenic cycles and phases. All these concepts proved very useful for analysis and synthesis of highly complex paleogeographic and tectonic situations, but they had little to do with the original facies concept and it would have been better had a different terminology been created for them.

That such indiscriminate use of the term facies in stratigraphy has tended to blur the concept to the point of impairing its usefulness has been forcefully pointed out by Moore (1949, 1953).

The term microfacies, proposed by Cuvillier and Sacals in 1951 (see also Cuvillier and Sacals, 1956; Fairbridge, 1954), refers to only microscopically recognizable lithological and paleontological rock properties. It is thus not a new concept, but indicates refinement of the classical facies concept. It is a useful term which is gradually gaining ground.

Sonder (1956) has recently made the interesting suggestion to group facies characteristics according to whether they supply information on the source of the sediments or on the conditions in the place of sedimentation. The former features he proposed to classify as "alimentation facies," the latter as "precipitation facies." Alimentation facies is expressed by rock composition (sandstone, clay, chert, etc.), precipitation facies by rock textures (cross-bedding, ripple marks, etc.) and by primary (autochthonous) constituents, especially fossils. However, the distinction between these two groups of facies characteristics seems to be anything but clear-cut.

NON-STRATIGRAPHICAL USE OF FACIES

At this point a brief diversion is necessary to explain uses of the term facies in branches of geology other than stratigraphy.

In 1915, the Finnish geologist Eskola introduced the term "metamorphic facies" for rocks which have been metamorphosed under identical conditions. He emphasized that he selected this term in analogy with stratigraphical facies and that the metamorphic facies gave no information on the pre-metamorphic conditions of the rocks. Some years later Eskola (1922) broadened his facies concept by contrasting metamorphic with igneous facies and uniting both in the category of "mineral facies." These major facies groups were subdivided by mineral content and geochemical characters into individual facies such as green schist, amphibolite, diabase, eclogite facies, and many others. In spite of early warnings by Becke (1921), Tilley (1924), and Grubenmann and Niggli (1924) against duplication of a scientific term already well established in another branch of geology the concept of metamorphic facies has become widely accepted by petrographers (Eskola, 1939; Mason, 1952; Williams, Turner, and Gilbert, 1954). Nevertheless, it is unsatisfactory that in the description and interpretation of the history of a metasediment we should have to operate with two entirely unrelated facies concepts.

In 1912, Sander introduced the concept of "tectonic facies" for rocks which owe their present characteristics mainly to tectonic movement, such as mylonites, some phyllites, and others. This concept should not be confused with the tectofacies of Sloss *et al.* (1949) which is discussed elsewhere in this paper.

Rivière (1952) introduced "granulometric facies" as an interpretative term for semilogarithmic cumulative curves representing grain size analyses. Subdivisions are "linear facies" for unsorted sediments and parabolic, logarithmic and hyperbolic facies according to the shape of the granulometric curves. The use of the term facies to describe such statistical representations of one single property of a sedimentary rock seems inadmissible.

Some other not strictly stratigraphical definitions of facies appear in some recent Russian literature to be discussed later.

FACIES CONCEPTS IN ENGLAND AND NORTH AMERICA

For reasons not easily understood the stratigraphic facies concept failed to gain acceptance in English-speaking countries until comparatively recent years. A search of such well known textbooks as Dana (1880), Chamberlin and Salisbury (1904–1906), and Pirsson and Schuchert (1915) reveals absence of any mention of the word facies. Geikie (1903) spoke of a "facies of organic remains," but had in mind the animal ecological facies to be discussed below and made, in fact, no distinction between facies fossils and index fossils. H. S. Williams stands out as a lone figure who in 1895 discussed facies as "the particular composition or condition of a formation in a given region." On this basis he distinguished an argillaceous or calcareous facies from an arenaceous facies of the Hamilton group in New York state.

For all practical purposes, however, as Dunbar and Rodgers (1957) have stated, introduction of the facies concept into American literature dates from Grabau (1913, 1920, and other publications). Even Grabau was inconsistent in his application of the facies concept and never really applied it to the solution of stratigraphical problems. In 1913 (quoted from Grabau, 1931) he used facies for different kinds of shore, or littoral, deposits but did not distinguish clearly between sedimentary and faunal facies. Thus, in the "rocky cliff facies" he included submarine talus formed by erosion of cliffs, but did not mention faunas, and under "organic facies" fauna and flora are mentioned, but little said about the sediments. Peculiarly, Grabau did not use the facies concept in the discussion of bathyal, abyssal, or any other deposits. In 1920 Grabau contrasted index fossils and facies fossils, but again did not discuss or define facies as such.

Yet, the facies concept was slowly gaining ground. Fenton and Fenton (1930) stated that "facies is regularly employed in the writings of European geologists" and gave a brief digest of Gressly's and Haug's ideas on the subject. In 1931, Stockdale recognized the importance of study of lateral facies changes in the Mississippian of southern Indiana for the correct stratigraphic interpretation of these rocks, but did not come up with an integrated picture of facies relationships and development.

In an address given in 1933 Kindle (1934) discussed the role of facies in stratigraphic paleontology. He gave an excellent review of the ecologic facies concept, but ignored the European literature and did not distinguish clearly between ecologic facies, stratigraphic facies, time-stratigraphical, and biostratigraphical concepts.

The first major stratigraphical paper which applied facies analysis and dealt with facies concepts was by Caster in 1934. It was followed by an equally important paper by McKee (1938) and some years later discussion of facies was beginning to be included in textbooks (Schuchert and Dunbar, 1941). On the other hand in two of the most penetrating and instructive facies studies of that period (Elias, 1937; Payne, 1942) the term itself was not used, although its application would have been highly appropriate. Moore (1941) was among the first to put the facies concept into focus in relation to major issues of North American stratigraphy.

In England, more general recognition of the facies concept dates only from Jones (1938) and Hatch, Rastall, and Black (1938). Jones recognized the importance of facies studies in his analysis of the Paleozoic geosyncline of Great Britain and followed essentially Haug's environmental classification.

In Australia the facies concept was successfully applied to the unravelling of complex lithological and biostratigraphic patterns in the Upper Devonian (Teichert, 1943, 1949; see also Termier and Termier, 1952, pp. 162–163). David and Browne (1950) in their monumental *Geology of the Commonwealth of Australia* referred extensively to facies in comparative studies of isochronous formations and sequences on the Australian continent.

Weller (1958) has recently published a thorough review, accompanied by an exhaustive bibliography, of United States facies literature since about 1930. To this the reader may turn for additional references.

In the following brief discussion of recent American thinking general familiarity with the literature is, therefore, assumed and emphasis will be placed on interpretation of American terminology in terms of earlier established European concepts.

Caster (1934) recognized that belts of identical or similar facies may extend obliquely through several defined time intervals and may be divisible into, or assignable to, several time-stratigraphic divisions. For a continuous and homogeneous facies belt he proposed the name "magnafacies," for its time-stratigraphically defined parts the name "parvafacies." It will be clear from the earlier discussion of European facies concepts that a magnafacies is a sedimentary rock unit of isopic, but heterochronous facies, and that parvafacies are isochronous, heteropic elements of a facies tract (Fig. 1).

In 1938, Kleinpell used the facies concept in foraminiferal analysis for geographic-climatical provinces (tropical, temperate, boreal facies) as well as to indicate depth zonation (littoral, neritic, bathyal, abyssal). In general, Kleinpell equated this type of environmental facies with ecologic facies, a concept which will be discussed later.

In the development of the stratigraphic facies concept in America two papers by McKee (1938, 1945) were important steppingstones. In his study of Permian stratigraphy of the Grand Canyon McKee (1938) placed the main emphasis on lateral facies variations and he reconstructed environments by careful facies analyses of individual rock-stratigraphic units. In his study of the Cambrian in the Grand Canyon the same author (1945) distinguished most consistently and carefully between rock-stratigraphic units (formations and members) and their facies, using the latter term to designate different lithological aspects of the rocks, regardless of whether these were found in the same or in different rock units. For example, a *Girvanella* limestone facies is recognizable in different members of the Muav limestone.

In 1949 the publication of three works gave a stimulus to the facies discussion that is still being felt today: (1) Pettijohn's *Sedimentary Rocks*, the first American textbook in which considerable space was given to facies, (2) the Geological Society of America Memoir on "Sedimentary Facies in Geologic History" (Longwell, 1949), with contributions by Moore, McKee, Sloss, Krumbein, and Dapples, and others, and (3) an important paper by Lowman on facies in Gulf Coast sediments. In the tradition of Haug, Lowman at first defines facies as a "rock-descriptive term including both the biological and mineralogical components," and then uses the term throughout his paper exclusively for environmentally defined concepts: continental, littoral, neritic, bathyal.

Pettijohn (1949) reviewed some of the European facies literature and used the term in the broadest, essentially environmental sense, introducing and discussing geosynclinal, foreland, orogenic, post-orogenic, euxinic, saline, and other kinds of facies.

An innovation was initiated by Moore (1949, 1953) who defined sedimentary facies as "any areally segregated part of a designated rock division in which physical-organic characters differ significantly from those of another part or parts." This definition was adopted as a working basis by Sloss, Krumbein, and Dapples (1949), Krumbein and Sloss (1951), and in further publications by these and other authors, too numerous to mention here. "Facies" in this garb are heteropic facies elements of a stratigraphic unit or of a facies tract. In some instances such a facies will represent a parvafacies in the sense of Caster.

While Moore, Sloss, Krumbein, and others restricted the use of facies to lateral rock variants, McKee (1949) found it useful to study "facies of different lithology but same age" as well as "facies of similar lithology but different age." The former are referred to more simply as heteropic, the latter as isopic facies. In 1957, Moore conceded admissibility of use of facies for "particular kinds of sedimentary deposits . . . that broadly correspond to certain types of environment," that is for isopic facies as defined by Mojsisovics.

Rich's (1951b) undathem, clinothem, and fondothem are clearly facies tracts, as defined above, because these terms indicate assemblages of rocks of genetically interconnected facies. Thus, an undathem is formed from sediments accumulated in the "unda environment" (=environment characterized by agitated waters of waves and currents); characteristic rock types are: conglomerate, sandstone, coarse siltstone, fragmental limestone, oölite, and coquinite. Rich (1951a) himself seemed to restrict use of the term facies by implication to biologic facies characterizing different parts of his unda, clino, and fondo environments.

"Tectofacies" was defined by Sloss *et al.* (1949) as "a group of strata of different tectonic aspect from laterally equivalent strata." Objections to this concept were raised by Nabholz (1951) and by Moore (1953). Such groups of strata, if recognizable, should not be grouped as a facies. The term itself is somewhat close to Sander's (1912) "tectonic facies" which has already been discussed.

Two terms which have been much in favor in recent years are "lithofacies" and "biofacies." As used and understood by many authors stratigraphic lithofacies comprises the mineralogical and petrographical aspects as well as certain mass properties of a sedimentary rock, stratigraphic biofacies its paleontological characteristics. The latter term requires more critical discussion which is given in the next chapter.

The term lithofacies seems to have been introduced by the Russian geologist Eberzin (1940: *fide* Markevich, 1957). When and by whom it was launched in America is not clear from available references. Soon after 1945 it appeared in publications and was used in the sense given above. It was, of course, only a new word for an old concept, already defined by Gressly.

Moore (1949) has pointed out with considerable justification that, etymologically viewed, "lithofacies," meaning rock facies, is a strict synonym of facies, because "the root 'litho' as applied to sedimentary rock necessarily embraces the organic constituent of such rock, wherever present, along with the inorganic." In other words, fossil remains form part of the rock (see also Teichert, 1958).

This is in harmony with Gressly's ideas who regarded the petrographic and paleontologic properties of a rock as inseparable parts of its facies. Nevertheless, for analytical purposes separation of treatment of the mineralogic-petrographical properties and of paleontological content is almost always advisable as one method of establishing relationship between sediment and biota at the time of the formation of the sediment. Moore's proposed term "physiofacies" for the "purely inorganic elements in the record of a sedimentary environment" has considerable merit.

The "lithotope" of Wells (1947) is by original definition a "rock record of environment." Lithotopes are, therefore, rock units of identical, or isopic, lithofacies (or "physiofacies" of Moore). Wells subdivided lithotopes into "facies" on the basis of fossil assemblages. His "facies" are, therefore, comparable with "biofacies" of other authors. Krumbein and Sloss (1951) and Wheeler and Mallory (1956) have, without apparent justification, redefined lithotope as an "area of uniform sedimentation," thus reducing the concept to two dimensions. For the purpose of vertical differentiation of a stratigraphic unit Krumbein and Sloss suggested recognition of successive lithotopes and biotopes rather than facies. Since fossil biotopes are rarely recognizable and lithotopes must be defined in terms of facies, recognition of vertical facies changes follows logically.

The terms lithostrome and lithosome, proposed by Wheeler and Mallory (1954, 1956), apply to rock units which can be defined in terms of facies. "Lithostrome" has been defined as a "stratigraphic layer consisting of one or more beds of essentially uniform or uniformly heterogeneous lithologic character." It is difficult to see how such a layer differs from a formation or member in the rock-stratigraphic sense, or from a lithotope as defined by Wells (1947). Its facies is isopic and isochronous.

A lithosome, according to Wheeler and Mallory is a "vertico-laterally" delineated stratigraphic unit. I can not agree with these authors that such units differ in concept from either the vertically or the laterally segregated units. Lithosome is essentially identical with Caster's magnafacies. Both are rock bodies of isopic and heterochronous facies (Fig. 1).

The "consanguineous association" of Pettijohn (1957) was defined as a "natural group of sedimentary rocks related to one another by origin." It is a useful concept which, however, as already pointed out by Fairbridge (1958), should not be directly equated with "facies" as such. Rather, a consanguineous association might be defined as a group of rocks whose facies indicate that they are related to one another by origin. Consanguineous associations thus combine features of facies sequences and facies tracts or facies families.

More recently, Adams and Weaver (1958) have introduced the concept of "geochemical or elemental facies" with particular reference to the Th/U ratio in sedimentary rocks. Thus, they distinguish three major facies as high, low, and intermediate Th/U facies. This concept of geochemical facies refers to precisely determinable rock characteristics which serve as aids in environmental analysis. It is concerned with one aspect of "lithofacies" as commonly understood and is, therefore, logical and precise in its meaning. Adams and Weaver, however, do not refer to the concept of "geochemical facies" as developed by Pustalov and Teodorovich. Because of previous use of the term "geochemical facies" in a somewhat different meaning and in order to avoid misunderstanding, it seems preferable to use Adams and Weaver's alternate term "elemental facies" for elemental constitution and ratios of sedimentary rocks.

Cloud and Barnes (1957) have recently suggested the term intrafacies "to denote a subordinate facies development within a differing major facies expression." This term should be useful in specialized and detailed studies of facies sequences.

To Weller (1958) we owe an up-to-date, comprehensive review of recent American facies terminology. He did not, however, refer to the huge facies literature published before 1930 and none of the three definitions of facies which seem acceptable to him is in harmony with the original facies concept. Weller, however, rightly deplored the unnecessary complexity of contemporary American facies literature and advocated rejection of many recently created terms. In this I wholeheartedly concur, although fundamental differences in our concepts and approaches to the facies problem will be obvious to every reader.

ECOLOGIC VS. STRATIGRAPHIC FACIES AND BIOFACIES CONCEPTS

Facies is used in both plant and animal ecology. In plant ecology facies has been defined as "the developmental unit of associes characterized . . . by the grouping of dominants" (Weaver and Clements, 1938, p. 101). Since the stratigrapher is almost never concerned with plants in their original habitat (floral biocoenoses), chances for misunderstandings or conflicts to arise through use of facies by plant ecologists are small.

The cases of the animal ecological and the marine ecological facies, however, are different. Both are closely related, but refer to something quite different from the stratigraphic facies. For example, the designation "sandy facies" in stratigraphy indicates a rock unit which is either a sandstone or some other sedimentary rock with a noticeable sandy component. A "sandy facies" in marine ecology, on the other hand, is an association of animals or plants or both, which live on a sandy sea floor. The following discussion is restricted to the facies concept in marine ecology, with emphasis on animal ecology.

Introduction of the facies concept into marine ecology is due to Pruvot (1895, 1897) who proposed the term to designate different types of bottom condition: sandy, rocky, and muddy. In his own words (Pruvot, 1895, p. 635) (transl.):

> "A natural classification of sea-bottoms from the faunistic point of view should consist of *zones* of wide extent whose upper and lower limits should be kept quite elastic; each zone should be subdivided into distinct *facies*, the differences being almost complete at the same level between animal associations that inhabit rocky and sandy parts of the same coast . . ."

Thus, each depth zone is subdivided into facies according to the same criteria: sand, mud, rock. Naturally, the animal communities inhabiting identical facies at different depths are not alike. Pruvot's facies, thus, indicates bottom conditions and nothing else. In French marine-ecological papers facies continues to be used in the sense of Pruvot and Walther (see Feldmann, 1940; Prat, 1940). We shall see shortly that this concept of facies is overlapping with the modern "biotope" as defined by Hesse (1924) and now generally used in German and American ecological literature.

The geologist Walther, in his classic facies study of sediments and faunas in the Gulf of Naples, referred to the sediments themselves as facies, so that in his terminology different sedimentary facies were characterized by different animal communities, the latter not being part of the recent facies (Walther, 1910). From the standpoint of the stratigrapher, this makes good sense.

The ecologist Doflein (1914), however, used facies in an entirely different sense for (transl.) "the biologic association occurring on a certain type of bottom." Some kinds of facies (rocky, sandy, pebbly, muddy, silty) are conditioned by the mineralogic composition of the sea floor; others, the "biologic facies," are mainly characterized by the organisms themselves (coral reefs, *Serpula* banks, shell banks, algal banks).

In limnology, facies had come to be used in a similar sense at an even earlier date for biological associations restricted to certain bottom conditions (Forel, 1901), thus: rock facies, pebble facies, sand, mud, silt facies, and so on. These terms were used not to indicate varieties of lake bottoms, but the faunas and floras that live on them. In modern American limnological literature, the facies concept does not seem to be used.

In subsequent marine-ecological literature the meaning of facies underwent further changes. Thus, in 1924, Hesse used facies not for biologic associations, but for variants of biotopes, and the term continues to be applied in this sense. But although Hesse, Allee, and Schmidt (1937) believed that facies in the geological sense has "similar applications," the intrinsic differences are obvious. A biotope is the habitat, the physical basis, of a biologic association. A marine-ecological facies, therefore, implies presence of floras and faunas; many stratigraphical facies are unfossiliferous, hence not associated with floras or faunas. A marine-ecological facies may be devoid of sediment ("rocky facies"), and yet may indicate a rich fauna; the stratigrapher generally has to think of facies in connection with sedimentary rocks, and if he includes, with Walther, stratigraphic breaks in the facies concept such "facies" are devoid of both sediments and organic remains. Finally, there are biotopes, such as the pelagic biotope, which do not form part of the sea-bottom. When applying results of marine-ecological investigations to stratigraphy, it is of great importance to take heed of the essential differences between the facies concepts as used in these two sciences.

Very similar considerations must be applied to the term "biofacies," where differences in marine-ecological and stratigraphic concepts are even more fundamental.

Outstanding recent examples of "biofacies" analyses in the two sciences are the works of Imbrie in stratigraphy and of Phleger in marine ecology. According to Imbrie (1955) biofacies is "the total recognizable organic content of a designated portion of a stratigraphic unit." By "total recognizable organic content" Imbrie means only the fossils, because other matter that could be included under "organic content" such as coal, hydrocarbons, proteins, amino acids, were not listed in his table of methods of biofacies analyses (p. 451).

Phleger (1954), at about the same time, studied the microfauna in the Mississippi Sound area which he divided into four geographic and ecologic biofacies: open gulf, sound, estuary, and marsh, these terms being applied to distinctive assemblages of Foraminifera. Obviously, Phleger's biofacies is something quite different from Imbrie's. In discussing the geological applications of his ecologic studies Phleger recognized that changes in the original biofacies due to geological influences must be expected. From this we may conclude that the stratigraphic "biofacies" which will ultimately occupy the place of Mississippi Sound, may be expected to differ from the original ecologic "biofacies" and that no certainty exists that we shall be able to reconstruct the latter through analytical study of the former.

The difference between the two kinds of "biofacies" is in fact so obvious that it requires little explanation. A stratigraphic "biofacies" is determined by fossils which under the most ideal conditions may represent part of the original ecological "biofacies," but more often are quite unrepresentative of the latter. An ecological "biofacies," on the other hand, comprises many elements which can not be fossilized and which, therefore, can never become part of a stratigraphic "biofacies." Also, organic remains from two or more ecologic "biofacies" may be buried in the same place and thus form a stratigraphic "biofacies" which bears little or no resemblance to any of the original ecologic "biofacies."

The duplicity in usage of the term biofacies could be documented by references to many additional publications in stratigraphy and in marine ecology. Abrard (1927) seems to have been the first to call attention to this undesirable situation, although the term biofacies itself was apparently not yet known to him. Sometimes the expression "faunal facies" is used, as in Ladd *et al.* (1957), for an ecological biofacies. In this paper such terms as reef and inter-reef facies have a meaning which bears no resemblance to the meaning these terms have in stratigraphy, because the authors use "reef facies" for animal communities inhabiting oyster banks, "inter-reef facies" for animal communities inhabiting stretches of sea floor between oyster banks.

Just when, by whom, in what language, and for which science biofacies was first proposed, I have not been able to ascertain. The word appears in German geologic literature not later than the mid-nineteen-thirties (e.g., Frank, 1936). As here used, biofacies indicates the nature of the paleontologic record of a stratigraphic unit. The statement by Nabholz (1951), that Hesse, Allee, and Schmidt (1937) proposed biofacies for "one or more biotopes" is in error. For this concept Hesse (1924) and his American translators (Hesse, Allee, and Schmidt 1937,

1951) used the term biochore. They distinguished carefully between the physically defined biotope and the biocoenose, which is the plant and animal association inhabiting the biotope. Since Hesse and his collaborators defined variants of biotopes as facies, it would seem logical that variants of the corresponding biocoenoses should have been distinguished as biofacies. When and where this was first done, remains to be discovered. It is interesting to note that in an earlier publication Phleger (1951) did not use biofacies but stressed the concepts of fauna and assemblage.

Since the ecological concept of biofacies is closer to the literal meaning of the word than the corresponding stratigraphic concept, use of that term should be restricted to marine ecology, irrespective of possible priority conflicts. Biofacies are represented by, or are subdivisions of, biocoenoses. Since there can be no true fossil biocoenosis, it would seem advisable in stratigraphical studies to revert to the use of *paleontologic facies*, a well defined unambiguous concept of the nineteenth and early twentieth century.

Large "biofacies realms" such as those defined by Lochman-Balk and Wilson (1958) for Cambrian trilobite faunas of North America are perhaps more appropriately called paleozoogeographic provinces.

In analyzing paleontologic facies it is convenient and logical to distinguish, with Craig (1953), fossil communities and fossil assemblages. The former are associations of fossils ecologically related among themselves; the latter may consist of several fossil communities, they may be wholly composed of ecologically unrelated fossils, or they may be of mixed composition. It should also be remembered that not all fossils which make up a paleontologic facies are facies fossils, the latter term being reserved for fossils which are restricted to defined facies and, thus, may be indicative of certain environments.

To speak, in the fashion of some recent writers, of the "biologic" or "organic," instead of the paleontologic, content of sedimentary rocks can only serve to divert attention from the fact that fossils are not organisms and that fossil assemblages and biologic communities are two rather different things.

FACIES CONCEPTS IN SOVIET UNION

Lively discussion of the facies concept among Soviet geologists during the last 25 years has produced many new concepts, only some of which can be reviewed in this paper. Few of these have found their way into the geological literature of western Europe and North America. The following résumé is brief and incomplete. For a fuller discussion the reader is referred to Markevich (1957).

Up to the nineteen-thirties Russian authors had generally adhered to the facies concept as used by the majority of European geologists. The following quotation from a textbook by the prominent geologist Borisyak may be taken as representative (Borisyak, 1934, p. 16) (transl.): "By facies one understands usually the physical attributes of a given area or a given layer on the earth's surface as found in conjunction with a definite association of animals or plants."

Facies is, thus, defined by given physical conditions, fauna, and flora.

Even more clearly, Mazarovich (1937, p. 47) defines facies as (transl.) "the sum of petrographical and organic features which determine individual layers of the earth's surface by their distinctive characters of sediments and populations [=paleontologic assemblages]." Mazarovich gives considerable space to facies; his classification follows essentially that of Haug. Of particular interest are his discussion of facies changes in basin sedimentation and of recognition and interpretation of vertical facies changes (superposition of facies). He also discusses cyclic facies changes, with examples from the Paleogene of the lower Volga district.

An early protest against this use of the term facies was raised by Kasakov *et al.* (1934) who maintained strongly that facies is a paleogeographic concept, not a lithologic or paleontologic one. For such an opinion there is, of course, little justification.

Belousov (1948) devoted a chapter of a comprehensive treatise on geotectonics to discussion of the influence of tectonic movements on facies. He gave no special definition of facies, but used the term conservatively for the general expression of lithological and paleontological properties of rocks. Of particular interest are his studies of the relationships between facies and thickness, illustrated by examples from the Russian platform. Belousov's Figure 30, illustrating displacement of facies zones in conjunction with transgressive movements of the shore line, is somewhat similar in concept and intent to Figure 1 of this paper.

Simultaneously with the afore-mentioned publications very different ideas on facies were being developed elsewhere.

In 1933, Pustalov proposed the concept of "geochemical facies" which he defined as (transl.) "a part of the earth's surface throughout which there exist the same physico-chemical and geochemical conditions of accumulation and formation of sedimentary rocks." The concept was further developed by Teodorovich (1947) and by Pustalov (1954) himself who distinguished the following types of geochemical facies.

1. Sulphide facies (strongly reducing)
2. Siderite facies (reducing)
3. Leptochloritic facies (weakly reducing)
4. Glauconite facies (neutral)
5. Oxydizing facies (excess O_2)
6. Ultra-oxydizing facies (pyrolusite sediments)

Obviously, in the language of most geologists these "facies" represent distinct environments characterized by presence of chemical reactions in the sedimentation medium.

Vassoevich (1948, *fide* Markevich, 1957, p. 16) has proposed the following facies classification.

Origofacies = facies of the primary sedimentary milieu
Lapidofacies = facies of the diagenetic milieu
Densofacies = facies of the metamorphic milieu
Eksedofacies = facies of the weathering milieu

Vassoevich's densofacies is identical with Eskola's metamorphic facies, lapidofacies and eksedofacies are excluded from the facies concept by original definition, and origofacies is the same as facies of most western authors.

Strakhov (1951) published an important study of limestone and dolomite deposition in ancient and recent marine and fresh-water environments and compiled maps of the distribution of carbonate facies in different geologic periods. This is an excellent study of the origin and distribution of some specialized rock facies.

Khain (1954; *fide* Markevich, 1957) recognized the abstract nature of the facies concept and defined facies as the "reflection" of the conditions of sedimentation.

Markevich (1957, p. 50) speaks of "geological facies" which he defines as (transl.) "a certain volume of sedimentary layers characterized by similar complexes of paleontological, petrographical and physico-chemical characteristics, indicating the tectonic, physico-chemical, biotic, and geographical conditions of formation of the sediment." The geological facies turns out to be a whole group of facies.

Geochemical facies
Lithologic facies
Carbonate facies
Faunistic facies
Floristic facies
Petroliferous facies
Coal-bearing facies

Of these categories lithologic, faunistic, and floristic facies seem to come under the true stratigraphic facies concept. The other categories express broadly defined environments that must be induced from lithologic and paleontologic facies observations, or, if taken literally, carbonate, petroliferous, and coal-bearing facies are special categories of the lithologic facies.

Rukhin (1954, *fide* Lombard, 1956, p. 416) defines facies as (transl.) "the complex of primary characteristics of a sedimentary deposit . . . characterizing distinct sedimentary environment, physico-geographical, tectonic, and geochemical. Each type of sedimentary rock is, therefore, represented by several facies." This definition, too, is far removed from the original facies concept. By original definition each rock in any one locality represents one, and only one, facies. From this a multiplicity of environmental factors may be induced, but it is inadmissible to say with Rukhin that each of these environmental factors constitutes, or gives rise to, a special facies of the rock.

Rukhin (*fide* Markevich, 1957) furthermore introduced the term "macrofacies" for genetically connected similar facies. Thus, he speaks of near-shore, shallow-water, deep-water macrofacies, and so on. This concept seems to be close to Walther's *Faciesbezirk* (facies tract of this paper) and to Nalivkin's *serviya*.

Botvinkina *et al.* (1956) in an important lithological study of sediments of Carboniferous age in the Donetz Basin establish the concept of subfacies (*subfatsiya*). The facies and subfacies classification of these authors is in reality a mixed environmental and genetic rock classification. Thus under facies we find offshore carbonate sediments distinguished from near-shore carbonates. The former are further subdivided into (1) homogeneous and biomorphic sediments, (2) diverse marine carbonate sediments, and so on. These divisions would not generally be accepted as part of a facies classification.

It seems then that the term facies in the Soviet Union as in America has come to mean many different things and that in the definition of many authors the term stands for concepts which differ widely from the classical facies concept of Gressly, Mojsisovics, and Walther.

A complete reversal of this trend, however, is found in Nalivkin's (1955–1956) recent two-volume *Treatise on Facies* which must be regarded as a classic in its field. Nalivkin returns to the original concept of facies as an expression of lithologic and paleontologic characteristics of sediments and sedimentary rocks and applies it in a most instructive and logical way to analysis of sedimentary environments in space and time. This author compares facies with the species concept in biology. As species are grouped into higher categories: genera, families, orders and so on, facies may be so grouped, in order to arrive at broader generalizations.

A number of different, but genetically connected and intergrading facies form a facies group which Nalivkin calls *serviya* (plur. *servii*) and a number of genetically connected, adjacent and intergrading *servii* form a *nimiya* (plur. *nimii*). Several *nimii* form together a formation (*formatsiya*). Since no English equivalents to *serviya* and *nimiya* have been discovered, and formation is used in a sense quite different from that in America, these concepts must be discussed in greater detail.

Nalivkin defines *serviya* as a complex facies and *nimiya* as a complex of complex facies. Thus in a coral atoll a breccia deposit formed in the surf zone of the outer slope represents a *facies*. All deposits and formations of the outer slope (coral formations, coral sands, coral breccias, and so on) together represent the *serviya*. Other *servii* of the atoll are the atoll rim and the lagoon. All three together form the *nimiya* which is the atoll itself. The atoll in its turn forms part of the great marine "formation." Thus, Nalivkin's *serviya* and *nimiya* are closely related to, and in part identical with, the facies tract of the present paper (= Walther's *Faciesbezirk*). For the sake of simplicity, it is suggested to equate the *serviya* with facies tract, and to regard the *nimiya* as a facies family. The sequence facies→facies tract→facies family thus indicates environments of increasingly larger size and complexity. Following a proposal by S. S. Oriel (personal communication) the term facies suite may be suggested for the concept of Nalivkin's *formatsiya*.

The foregoing example of facies analysis based on modern deposits can now be translated in the following terms.

Facies suite: marine deposits
Facies family: coral atoll deposits
Facies tract: outer slope deposits
Facies: breccia formed in surf zone

Another example is adapted from Nalivkin.

Facies suite: continental deposits
Facies family: desert deposits
Facies tract: dry channel deposits
Facies: angular pebble deposit

Facies analysis has to proceed from bottom to top of these lists. Correct environmental interpretation as a rule presupposes comparative studies of facies tracts and facies families. Thus, in addition to the angular pebble facies, other facies, such as cross-bedded sandstone and mudcracked clay, have to be recognized in order to define a facies tract of dry channel deposits. The latter must be associated with pebble pavements, certain types of sand deposits, fanglomerates, and others, to form the facies family "desert deposits."

The advantages in this way of proceeding from detail to generalization are obvious, if we study Nalivkin's breakdown of the major facies groups. Thus the marine facies suite (*formatsiya*) may be divided into facies families (*nimii*), representing sedimentary associations of the following.

Open shelves	Continental seas	Reef areas
Enclosed shelves	Inland seas	Bathyal areas
Lagoonal areas	Archipelagoes	Abyssal areas

Each facies family can be broken down into facies tracts, for example, the open-shelf family into *servii* representing sedimentary associations of the following.

Flat coasts	Islands
Mountainous coasts	Areas of glacial seas and glacio-marine deposits
Submarine canyons	Areas of eolio-marine deposits
Submarine elevations	Areas of pseudabyssal deposits

Each of these is composed of facies. Thus a facies tract of flat coasts may consist of rocky, pebbly, sandy, silty, clayey facies, various types of carbonate facies, and others.

Nalivkin's system demonstrates the great flexibility of the facies method as a tool for environmental analysis, if the facies concept is preserved in its original form as referring to lithologic and paleontologic characteristics of a single rock unit or parts of a unit.

In the nomenclature of many authors all of the afore-mentioned categories of deposits would be called facies. No logical analytical procedure can be developed on the basis of a terminology where one and the same term is applied to categories of four different orders. This is not a sound approach to the study of natural phenomena.

CONCLUSIONS

The foregoing discussion may be summarized as follows.

1. The facies concept in stratigraphy, resting on foundations laid by Gressly, Prevost, Oppel, was firmly established in continental European thinking well before the end of the nineteenth century.

2. As thus developed, the term facies should preferably be restricted to sedimentary rocks.[6] It signifies a stratigraphic concept.

3. Facies is the sum total of all primary characteristics of a sedimentary rock from which the environment of its deposition may be induced. Facies is thus, in the last analysis, an abstraction. It is, therefore, not a particular kind of rock, but something a rock has.[7]

4. Secondary changes, including those introduced by weathering and metamorphism, are not part of the stratigraphic facies. For example, the term dolomitic facies should, if possible, be reserved for dolomites recognized as primary or penecontemporaneous. The metamorphic, igneous, and mineral facies of Eskola and others and the lapidofacies and eksedofacies of Vassoevich are conceptually different from and unrelated to stratigraphic facies.

5. Unlike facies are called heteropic, identical or closely similar facies isopic.

6. Facies tracts (=Walther's *Faciesbezirke*) are genetically interconnected systems of isochronous heteropic facies. Several facies tracts may be grouped into a facies family and several facies families into a facies suite, each of these categories characterizing facies associations indicative of increasingly larger and more varied environments.

7. Isopic facies may be recognized in vertically segregated stratigraphic succession in one sedimentation area or basin. For example, recurrent isopic facies are recognizable in the Pennsylvanian cyclothems of the Mid-Continent region.

8. Recognition of vertical facies changes, or facies sequences, is as important as that of lateral facies. Just as lateral facies changes give information on varieties of coeval environmental conditions in the same sedimentation area, observation of vertical facies changes is fundamental to study of changing environmental conditions in time. Inasmuch as stratigraphy is concerned with the study of sequences of geologic events, the concept of facies sequences is valid and important. Rejection of the concept of vertical facies changes is illogical because every stratigraphic unit has a vertical dimension and represents a time interval. Where a facies intertongues laterally with another facies, vertical superposition of facies exists. It would be impracticable to set up arbitrary standards to determine when a facies change ceases to be a lateral one and therefore would no longer be a facies change.

9. Facies should be used as a strictly descriptive term, referring to primary lithologic and (or) paleontologic characteristics of a sedimentary rock. Environmental interpretation is the ultimate aim of facies studies, but environment is not

[6] Stainforth (1958) has recently suggested that we treat sedimentary rocks as "four-dimensional continua." This is an obvious misapplication of a term borrowed from theoretical physics. Sedimentary rocks are objects whose properties may be recorded in a system of Cartesian coordinates. They are finite bodies, not continua—either four- or three-dimensional.

[7] Facies are, therefore, not to be formally named like formations, as has been done, for example, by Stockdale (1931) who divided formations into laterally segregated facies and the latter into members. Informal numbering of facies in the manner introduced by McKee (1938) and Botvinkina *et al.* (1956) is admissible and often helpful.

facies. It produces facies. Interpretations may be subject to change; and facies, which indicates littoral environment to one observer, may suggest continental (eolian) environment to another. Hence, terms like littoral and eolian facies are meaningless, or at least ambiguous and subject to misinterpretation.

10. The use of the term facies for generalized stratigraphic sequences laid down in paleogeographically or tectonically defined belts, such as transgressive, regressive, geosynclinal, foreland, neritic, bathyal, abyssal, Ouachita, Appalachian, and many other kinds of facies, is inexact. While it would be unrealistic to expect that such usages can be abolished or eradicated overnight, it is necessary to realize that, when thus used, the facies concept loses its usefulness as an analytical tool. Facies, in this sense, becomes a somewhat nebulous term for all kinds of vaguely defined paleogeographical and geotectonic situations, where authors can fit the concept to theoretical preferences.

11. The ecological facies is conceptually different from the stratigraphic facies. The former may be devoid of sediments, the latter without organic remains. Such "facies" are not comparable.

12. "Biofacies" as presently used in stratigraphy is conceptually different from the biofacies in ecology, especially marine ecology. The ecologic biofacies contains elements which can not enter into the stratigraphic biofacies which may be derived from it. The stratigraphic "biofacies" may be utterly unrepresentative of, and unrelated to, any particular preexisting ecologic biofacies. The use of the term biofacies should be restricted to ecology and the features now so designated in stratigraphy should be known as paleontologic facies.

REFERENCES

ABRARD, R., 1927, "Faciès et associations paléontologiques," *Arch. Mus. Nation. Histoire Naturelle*, Sér. 6, Vol. 2, pp. 81–109.

ADAMS, J. A. S., AND WEAVER, C. E., 1958, "Thorium-to-Uranium Ratios as Indicators of Sedimentary Processes: Examples of Concept of Geochemical Facies," *Bull. Amer. Assoc. Petrol. Geol.*, Vol. 42, pp. 387–430.

ARBENZ, P., 1919, "Probleme der Sedimentation und ihre Beziehungen zur Gebirgsbildung in den Alpen," *Naturforsch. Gesellsch. Zürich Vierteljahrsschrift*, Vol. 64, pp. 246–75.

BECKE, F., 1921, "Zur Facies-Klassifikation der metamorphen Gesteine," *Tschermaks Mineral. Petrogr. Mitt.*, Vol. 35, pp. 215–30.

BELOUSOV, V. V., 1948, *Obshchaya Geotektonika*. 599 pp., 284 figs. Gosudarstvennoe izdatel'stvo geologicheskoi literatury ministerstva geologii SSSR, Moscow-Leningrad.

BERTRAND, MARCEL, 1897, "Structure des Alpes françaises et recurrence de certain faciès sédimentaires," *Congr. Géol. Intern., C. R., VI Sess. 1894*, pp. 163–77.

BORISYAK, A. A., 1934, *Kurs istoricheskoi geologii*. 3d edit., 424 pp. Gosudarstvennoe Nauchnotekhnicheskoe gorno-geol.-neftyanse isdatel'stvo, Moscow-Leningrad.

BOTVINKINA, L. N., ZHEMCHUZHNIKOV, Y. A., TIMOFEEV, P. P., FEOFILOVA, A. P., AND YABLOKOV, V. S., 1956, *Atlas litogeneticheskikh tipov uglenosnykh otlozhenii srednego Karbona Donetskogo Basseina*. 368 pp., 106 pls. Akad. Nauk SSSR, Inst. Geol. Nauk, Moscow.

BRINKMANN, R., 1956, *Abriss der Geologie*, Bd. 1, viii+286 pp. Ferd. Enke, Stuttgart.

BUBNOFF, S. VON, 1921, *Die Grundlagen der Deckentheorie in den Alpen*. 149 pp. E. Schweizerbart'sche Verlagsbuchh., Stuttgart.

———, 1949a, *Einführung in die Erdgeschichte*, Bd. 1. viii+343 pp. Mitteldeutsche Druckerei & Verlagsanst., Halle (Saale).

———, 1949b, *Grundprobleme der Geologie*, 2d ed.. vii+246 pp. *Ibid.*

CADISCH, J., 1953, *Geologie der Schweizer Alpen*. xi+480 pp. Wepf & Co., Basel.

CASTER, K. E., 1934, "The Stratigraphy and Paleontology of Northwestern Pennsylvania, Part I: Stratigraphy," *Bull. Amer. Paleontology*, Vol. 21, No. 71. 185 pp.

CHAMBERLIN, T. C., AND SALISBURY, R. D., 1904–1906, *Geology*, Vol. 1 (1904). xix+654 pp. Vol. 2 (1906). xxvi+692 pp. Henry Holt & Co., New York.
CLOUD, P. E., JR., AND BARNES, V. E., 1957, "Early Ordovician Sea in Central Texas," *Geol. Soc. America Mem. 67*, pp. 163–214.
CRAIG, GORDON, 1953, "Fossil Communities and Assemblages," *Amer. Jour. Sci.*, Vol. 251, pp. 547–48.
CREDNER, H., 1902, *Elemente der Geologie*, 9th ed. xviii+802 pp. W. Engelmann, Leipzig.
CUVILLIER, J., AND SACAL, V., 1951, *Corrélations stratigraphiques par microfaciès en Aquitaine occidentale*. 23 pp., 90 pls. E. J. Brill, Leiden.
———, AND ———, 1956, *Stratigraphic Correlation by Microfacies in Western Aquitaine*. 33 pp., 90 pls. *Ibid.*
DACQUÉ, E., 1915, *Grundlagen und Methoden der Paläogeographie*. vii+499 pp. G. Fischer, Jena.
DANA, J. D., 1880, *Manual of Geology*, 3d ed. xiv+911 pp., 13 pls. Ivison, Blakeman & Co., New York and Chicago.
DAVID, T. W. E., AND BROWNE, W. R., 1950, *The Geology of the Commonwealth of Australia*, Vol. I. xvii+747 pp. Edward Arnold & Co., London.
DEECKE, W., 1913, "Faciesstudien über europäische Sedimente," *Naturforsch. Gesellsch. Freiburg i. Br., Berichte*, Vol. 20, pp. 1–40.
DIENER, C., 1925, *Grundzüge der Biostratigraphie*. viii+304 pp. Franz Deuticke, Leipzig and Vienna.
DOFLEIN, FRANZ, 1914, *Das Tier als Glied des Naturganzen*, Vol. 2 of *Tierbau und Tierleben*, by R. Hesse and F. Doflein. xv+960 pp. B. G. Teubner, Leipzig and Berlin.
DUNBAR, D. O., AND RODGERS, J., 1957, *Principles of Stratigraphy*. xii+356 pp. John Wiley & Sons, New York.
ELIAS, M. K., 1937, "Depth of Deposition of the Big Blue (Late Paleozoic) Sediments in Kansas," *Bull. Geol. Soc. America*, Vol. 48, pp. 403–32, 1 pl.
ESKOLA, PENTTI, 1915, "Om sambandet mellan kemisk och mineralogisk sammansättning hos Orijärvitraktens metamorfa bergarter," *Bull. Commission Géol. Finlande*, No. 44, pp. 1–145.
———, 1922, "The Mineral Facies of Rocks," *Norsk Geol. Tidsskrift*, Bd. 6, pp. 142–94.
———, 1939, "Die metamorphen Gesteine," pp. 263–467 in *Die Entstehung der Gesteine*, by T. F. W. BARTH, C. W. CORRENS, and P. ESKOLA. viii+422 pp. Jul. Springer, Berlin.
FAIRBRIDGE, R. W., 1954, "Stratigraphic Correlation by Microfacies," *Amer. Jour. Sci.*, Vol. 252, pp. 683–94.
———, 1958, "What Is a Consanguineous Association?," *Jour. Geology*, Vol. 66, No. 3, pp. 319–24.
FELDMANN, JEAN, 1940, "La végétation benthique de la Méditerranée," *Soc. Biogéographie, Mém.*, Vol. 7, pp. 181–95.
FENTON, C. L., AND FENTON, M. A., 1930, "Ecologic Interpretation of Some Biostratigraphic Terms," *Amer. Midland Naturalist*, Vol. 12, pp. 145–53.
FOREL, F. A., 1901, *Handbuch der Seenkunde. Allgemeine Limnologie*. x+249 pp. J. Engelhorn, Stuttgart.
FOURMARIER, P., 1950, *Principes de géologie*, 2 Vols. 1523 pp. Masson & Cie., Paris.
FRANK, MANFRED, 1936, "Der Faziescharakter der Schichtgrenzen der süddeutschen und kalkalpinen Trias," *Zentralbl. Mineralogie, Geologie, Paläontologie*, Abt. B, pp. 475–502.
FRIJLINCK, C. P. M., CUVILLIER, J., AND SLOSS, L. L., 1955, "Discussion" of L. L. SLOSS, "Location of Petroleum Accumulation by Facies Studies," *Proc. Fourth World Petroleum Congr., Geology, Geophysics*, Sec. 1, pp. 333–34.
GEIKIE, A., 1903, *Text-Book of Geology*, Vol. 2, 4th ed., pp. 705–1472. Macmillan & Co., London.
GIGNOUX, M., 1950, *Géologie Stratigraphique*. viii+735 pp. Masson & Cie., Paris.
———, 1955, *Stratigraphic Geology*. xvi+682 pp. W. H. Freeman & Co., San Francisco.
GILLULY, JAMES, WATERS, A. C., AND WOODFORD, A. O., 1952, *Principles of Geology*. viii+631 pp. W. H. Freeman & Co., San Francisco.
GRABAU, A. W., 1920, *A Textbook of Geology, Part I, General Geology*. xvi+864 pp. D. C. Heath & Co., New York.
———, 1931, *Principles of Stratigraphy*, 3d ed. xxxii+1185 pp. A. G. Seiler, New York.
GRESSLY, A., 1838, "Observations géologiques sur le Jura Soleurois," *Neue Denkschr. Allg. Schweizerische Gesellsch. ges. Naturw.*, Vol. 2, pp. 1–112.
GRUBENMANN, U., AND NIGGLI, P., 1924, *Die Gesteinsmetamorphose. I. Allgemeiner Teil*. xii+539 pp. Gebr. Borntraeger, Berlin.
HATCH, F. H., RASTALL, R. H., AND BLACK, MAURICE, 1938, *The Petrology of Sedimentary Rocks*. 583 pp. Thomas Murby & Co., London.
HAUG, E., 1907, *Traité de Géologie*, Vol. 1. 536 pp. Libr. Armand Cohn, Paris.
HEIM, ALBERT, 1921, *Geologie der Schweiz*, Bd. II, Erste Hälfte. xi+476 pp. C. H. Tauchnitz, Leipzig.
HERITSCH, FRANZ, 1927, "Die Deckentheorie in den Alpen (Alpine Tektonik 1905–1925)," *Fortschr. Geologie Palaeontologie*, Bd. 6, Heft 17, pp. i–iv, 75–210.
HESSE, RICHARD, 1924, *Tiergeographie auf ökologischer Grundlage*. xii+613 pp. Gustav Fischer, Jena.

———, ALLEE, W. C., AND SCHMIDT, K. P., 1937, *Ecological Animal Geography.* xiv+597 pp. J. Wiley & Sons, New York.
———, ———, AND ———, 1951, *Ecological Animal Geography*, 2d ed. xii+715 pp. *Ibid.*
IMBRIE, JOHN, 1955, "Biofacies Analysis," *Geol. Soc. America Spec. Paper 62*, pp. 449–64.
JONES, O. T., 1938, "On the Evolution of a Geosyncline," *Quar. Jour. Geol. Soc. London*, Vol. 94, pp. lx–cx, 4 pls.
KASAKOV, M., MIRCHNIK, G., STRAKHOV, N., AND SHANTSER, E., 1934, "Istoricheskaya geologiya kak ychebnyy predmet," *Byull. Moscovsk. Isp. Prirody, Otd. Geol.*, Vol. 12, pp. 441–61.
KAYSER, E., 1923, *Lehrbuch der Geologie*, Bd. 1, 6th and 7th ed. xii+740 pp. Ferd. Enke, Stuttgart.
KEILHACK, K., 1921, *Lehrbuch der Praktischen Geologie*, Bd. 1, 4th ed. xii+548 pp. *Ibid.*
KINDLE, E. M., 1934, "Rôle of Facies in Stratigraphic Paleontology," *Proc. Geol. Soc. America, 1933*, pp. 409–30.
KLEINPELL, R. M., 1938, *Miocene Stratigraphy of California*, Amer. Assoc. Petrol. Geol. x+450 pp.
KOBAYASHI, TEIICHI, 1956, "The Triassic Akiyoshi Orogeny," *Geotekton. Sympos. zu Ehren von Hans Stille*, pp. 85–101. Ferd. Enke, Stuttgart.
KRUMBEIN, W. C., AND SLOSS, L. L., 1951, *Stratigraphy and Sedimentation.* viii+497 pp. W. H. Freeman & Co., San Francisco.
KUENEN, PH. H., 1950, *Marine Geology.* x+568 pp. John Wiley & Sons, New York.
LADD, H. S., HEDGPETH, J. W., AND POST, R., 1957, "Environment and Facies of Existing Bays on the Central Texas Coast," *Geol. Soc. America Mem. 67*, pp. 599–640; 2 pls.
LOCHMAN-BALK, C., AND WILSON, J. L., 1958, "Cambrian Biostratigraphy in North America," *Jour. Paleon.*, Vol. 32, pp. 312–50.
LOMBARD, A., 1952, "Sédimentologie et évolution des lithofacies dévoniennes du bord Nord du synclinal de Namur," *Soc. Belge Géol. Paléont. Hydrol.*, Vol. 61, pp. 44–82.
———, 1956, *Géologie sédimentaire. Les séries marines.* 722 pp. Masson & Cie., Paris.
LONGWELL, C. R., editor, 1949, "Sedimentary Facies in Geologic History," *Geol. Soc. America Mem. 39.* 171 pp.
LOWMAN, S. W., 1949, "Sedimentary Facies in Gulf Coast," *Bull. Amer. Assoc. Petrol. Geol.*, Vol. 33, pp. 1939–97.
MCKEE, E. D., 1938, "The Environment and History of the Toroweap and Kaibab Formations of Northern Arizona and Southern Utah," *Carnegie Inst. Washington Pub. 492.* viii+268 pp., 48 pls.
———, 1945, "Cambrian History of the Grand Canyon Region. Part I. Stratigraphy and Ecology of the Grand Canyon Cambrian," *ibid., Pub. 563.* viii+232 pp.
———, 1949, "Facies Changes in the Colorado Plateau," in C. R. LONGWELL, 1949, pp. 35–48.
MARKEVICH, V. P., 1957, *Ponyatiye "fatsiya."* 88 pp. Akad. Nauk SSSR Inst. Nefti, Moscow.
MASON, BRIAN, 1952, *Principles of Geochemistry.* vii+276 pp. John Wiley & Sons, New York.
MAZAROVICH, A. N., 1938, *Istoricheskaya Geologiya*, 2d ed. 463 pp., 22 pls., Obedinenoe nauchno-tekhnicheskoe Isdatel'stvo IKTP SSSR, Moscow-Leningrad.
MOJSISOVICS VON MOJSVÁR, E., 1879, *Die Dolomit-Riffe von Südtirol und Venetien.* xiv+552 pp. Alfred Hölder, Vienna.
MOORE, R. C., 1941, "Stratigraphy," *Geol. Soc. America, 50th Anniversary Vol.*, pp. 177–220.
———, 1949, "Meaning of Facies," in C. R. LONGWELL, 1949, pp. 1–34.
———, 1953, "Les faciès sédimentaires," *Inst. Géol. Univ. Louvain Mém.*, Vol. 18, pp. 5–30.
———, 1957, "Modern Methods of Paleoecology," *Bull. Amer. Assoc. Petrol. Geol.*, Vol. 41, pp. 1775–1801.
MORET, LÉON, 1955, *Précis de Géologie.* ix+669 pp. Masson et Cie., Paris.
NABHOLZ, W. K., 1951, "Beziehungen zwischen Fazies und Zeit," *Eclogae Geologiae Helvetiae*, Vol. 44, pp. 131–58.
NALIVKIN, D. V., 1955–1956, *Uchenie o Fatsiyakh*, Vol. 1 (534 pp.), 1955, Vol. 2 (393 pp.), 1956. Isdatel'stvo Akad. Nauk SSSR, Moscow-Leningrad.
NEUMAYR, M., 1887, *Erdgeschichte*, Bd. 2. xii+879 pp. Bibliogr. Inst., Leipzig.
OPPEL, A., 1856–1858, "Die Juraformation Englands, Frankreichs und des südwestlichen Deutschlands, nach ihren einzelnen Gliedern eingetheilt und verglichen," *Württembergische Naturw. Jahreshefte*, Jahrg. 12, Heft 2 (May, 1856), pp. 121–312; Heft 3 (Sept., 1856), pp. 313–556; Jahrg. 13, Heft 2 (1857), pp. 141–288; Heft 3 (March, 1858), pp. 289–396; Jahrg. 14, Heft 3 (1858), pp. 129–291.
PAYNE, T. G., 1942, "Stratigraphical Analysis and Environmental Interpretation," *Bull. Amer. Assoc. Petrol. Geol.*, Vol. 26, pp. 1697–1770.
PETTIJOHN, F. J., 1949, *Sedimentary Rocks.* xv+526 pp. Harper & Brothers, New York.
———, 1957, *Sedimentary Rocks*, 2d ed. xvi+718 pp. *Ibid.*
PHLEGER, F. B, 1951, "Ecology of Foraminifera, Northwest Gulf of Mexico. Part I. Foraminifera Distribution," *Geol. Soc. America, Mem. 46*, pp. 1–88.
———, 1954, "Ecology of Foraminifera and Associated Micro-Organisms from Mississippi Sound and Environs," *Bull. Amer. Assoc. Petrol. Geol.*, Vol. 38, pp. 584–647; 3 pls.

PIRSSON, L. V., AND SCHUCHERT, CH., 1915, *A Text-Book of Geology.* x+1051 pp. John Wiley & Sons, New York.

POMEROL, CH., AND FOUET, R., 1953, *Les roches sédimentaires,* "Que sais-je?," No. 595, pp. 1–126. Press Univ. France, Paris.

PRAT, H., 1940, "Observations bionomiques sur les rivages atlantiques de l'Amérique du Nord et des îles voisines," *Soc. Biogéographie, Mém.*, Vol. 7, pp. 253–77.

PREVOST, C., 1838, "[Report of session of December 18, 1837]" *Bull. Soc. Géol. France,* Vol. 9, pp. 90–95.

———, 1845, "De la chronologie des terrains et du synchronisme des formations," *Acad. Sci. Paris, C. R.*, Vol. 20, pp. 1062–71.

PRUVOT, G., 1895, "Coup d'oeil sur la distribution générale des invertébrés dans la région de Banuls (Golfe du Lion), *Arch. Zoologie Epérim. Générale,* Sér. 3, Vol. 3, pp. 629–58.

———, 1897, "Essai sur les fonds et la faune de la Manche occidentale (côtes de Bretagne), comparés à ceux du golfe du Lion," *ibid.*, Sér. 3, Vol. 5, pp. 511–660; 5 pls.

PUSTOVALOV, L. V., 1933, "Geokhimicheskie fatsii i ikh znachenie v obshchei i prikladnoi geologii," *Problemy Sovietskoi Geologii (Problems of Geology of U.S.S.R.)*, Vol. 1, pp. 57–80.

———, 1954, "Sovremennoe sostoyaniye voprosa ob osadkonakoplenii," *Congr. Géol. Intern., C. R. XIX sess. Alger, 1952,* Sec. XIII, Fasc. XIII, pp. 171–92.

RENEVIER, E., 1884, "Les facies géologiques," *Archives Sci. Physiques et Naturelles,* Period 3, Vol. 12, pp. 297–333.

REYER, E., 1888, *Theoretische Geologie.* xiii+867 pp. E. Schweizerbartsche Verlagsbuchhandl., Stuttgart.

RICH, J. L., 1951a, "Three Critical Environments of Deposition, and Criteria for Recognition of Rocks Deposited in Each of Them," *Bull. Geol. Soc. America,* Vol. 62, p. 1–20.

———, 1951b, "Probable Fondo Origin of Marcellus-Ohio-New Albany-Chattanooga Bituminous Shales," *Bull. Amer. Assoc. Petrol. Geol.*, Vol. 35, pp. 2017–40.

RIVIÈRE, A., 1952, "Expression analytique générale de la granulométrie des sédiments meubles," *Bull. Soc. Géol. France,* Sér. 6, Vol. 2, pp. 155–67.

SALOMON, W., 1926, "Grundbegriffe der Erdgeschichte," in *Grundzüge der Geologie,* Bd. 2, W. SALOMON, editor, pp. 1–20. Schweizerbartsche Verlagsbuchhandlung, Stuttgart.

SANDER, BRUNO, 1912, "Über tektonische Gesteinsfazies," *K. K. Geol. Reichsanstalt, Verhand.*, pp. 249–57.

SCHINDEWOLF, O. H., 1950, *Grundfragen der Paläontologie.* 506 pp., 32 pls. Schweizerbartsche Verlagsbuchh., Stuttgart.

SCHUCHERT, CH., AND DUNBAR, C. O., 1941, *A Textbook of Geology, Part II, Historical Geology.* xii+544 pp. J. Wiley & Sons, New York.

SIMON, W., 1948, *Zeitmarken der Erde,* Die Wissenschaft, Vol. 98. 232 pp. Vieweg und Sohn, Braunschweig.

SLOSS, L. L., KRUMBEIN, W. C., AND DAPPLES, E. C., 1949, "Integrated Facies Analysis," in C. R. LONGWELL, 1949, pp. 9–123.

SONDER, R. A., 1956, *Mechanik der Erde.* vii+291 pp. Schweizerbartsche Verlagsbuchhandlung, Stuttgart.

SPENGLER, ERICH, 1951, "Die nördlichen Kalkalpen. Die Flyschzone und die Helvetische Zone," pp. 302–413, in *Geologie von Österreich,* F. X. SCHAFFER, editor. xv+810 pp. Franz Deuticke, Vienna.

STAINFORTH, R. M., 1958, "Stratigraphic Concepts," *Bull. Amer. Assoc. Petrol. Geol.*, Vol. 42, pp. 192–93.

STENO, NICOLAUS, 1669, *De Solido intra Solidum naturaliter contento dissertationis prodromus.* 78 pp. Florence.

STILLE, H., 1924, *Grundfragen der vergleichenden Tektonik.* viii+443 pp. Gebr. Borntraeger, Berlin.

STOCKDALE, P. B., 1931, "The Borden (Knobstone) Rocks of Southern Indiana," *Dept. Conservation Indiana Pub. 98.* xi+330 pp.

STRAKHOV, N. M., 1951, "Izvestkovo-dolomitovye fatsii sovremennykh i drevnikh bodoyemob (Opyt strabnitel'no-litologicheskogo issledovaniya)," *Trudy Inst. Geol. Nauk, Byull. 124,* Geol. Ser. (No. 45). 371 pp., 158 figs.

STRAUSZ, LASLO, 1928, "Geologische Fazieskunde," *Magyar Kiralyi Földtani Intezet Evkonye,* Vol. 28, pp. 73–272.

TEICHERT, CURT, 1943, "The Devonian of Western Australia. A Preliminary Review," *Amer. Jour. Sci.*, Vol. 241, pp. 69–94, 167–84.

———, 1949, "Observations on Stratigraphy and Paleontology of Devonian, Western Portion of Kimberley Division, Western Australia," *Australia Bur. Mineral Resources, Geology, Geophysics, Rept. 2.* 55 pp., 6 pls.

———, 1958, "Some Biostratigraphical Concepts," *Bull. Geol. Soc. America,* Vol. 69, pp. 99–120.

TEODOROVICH, G. I., 1947, "Osadochnyie geokhimicheskie fatsii," *Moskovskoe obshchestvo ispytadelei prirody,* Tome 42, *Otdel Geol.*, Tome 22, Biull. 1, pp. 3–24.

TERCIER, JEAN, 1939, "Depôts marins actuels et séries géologiques," *Eclogae Geologiae Helvetiae*, Vol. 32, pp. 47–100.

TERMIER, H., AND TERMIER, G., 1952, *Histoire géologique de la biosphère.* 721 pp. Masson & Cie., Paris.

TILLEY, C. E., 1924, "The Facies Classification of Metamorphic Rocks," *Geol. Mag.*, Vol. 61, pp. 167–71.

TORNQUIST, A., 1916, *Geologie*, Teil I, *Allgemeine Geologic.* xii+564 pp. W. Engelmann, Leipzig.

WALTHER, JOHANNES, 1893, *Einleitung in die Geologie als historische Wissenschaft*, Bd. 1, *Beobachtungen über die Bildung der Gesteine und ihrer organischen Einschlüsse.* xxx+196 pp. G. Fischer, Jena.

———, 1894, *Einleitung in die Geologie als historische Wissenschaft*, Bd. 3, *Lithogenesis der Gegenwart*, pp. 535–1055. G. Fischer, Jena.

———, 1910, "Die Sedimente der Taubenbank im Golfe von Neapel," *Königl. Preussische Akad. Wiss., 1910, Phys.-Math. Classe, Anhang*, Abh. III. 49 pp., 2 pls.

WEAVER, J. E., AND CLEMENTS, F. E., 1938, *Plant Ecology*, 2d ed. xxii+601 pp. McGraw-Hill Book Co., New York and London.

WELLER, J. M., 1958, "Stratigraphic Facies Differentiation and Nomenclature," *Bull. Amer. Assoc. Petrol. Geol.*, Vol. 42, pp. 609–39.

WELLS, J. W., 1947, "Provisional Paleoecological Analysis of the Devonian Rocks of the Columbus Region," *Ohio Jour. Sci.*, Vol. 47, pp. 119–26.

WHEELER, H. E., AND MALLORY, V. S., 1954, "Reply to Discussion [on arbitrary cut-off in stratigraphy]," *Bull. Amer. Assoc. Petrol. Geol.*, Vol. 38, pp. 928–31.

———, AND ———, 1956, "Factors in Lithostratigraphy," *ibid.*, Vol. 40, pp. 2711–23.

WILLIAMS, H., TURNER, F. J., AND GILBERT, C. M., 1954, *Petrography.* x+416 pp. W. H. Freeman & Co., San Francisco.

WILLIAMS, H. S., 1895, *Geological Biology*, xix+395 pp. Henry Holt & Co., New York.

Reprinted from:
BULLETIN OF THE AMERICAN ASSOCIATION OF PETROLEUM GEOLOGISTS
VOL. 44, NO. 10 (OCTOBER, 1960), PP. 1651–1681, 15 FIGS., 1 PLATE

FACIES ANALYSIS OF UPPER DEVONIAN WABAMUN GROUP IN WEST-CENTRAL ALBERTA, CANADA[1]

JOHN M. ANDRICHUK[2]
Calgary, Alberta

ABSTRACT

The Wabamun group, consisting mainly of carbonate rocks and anhydrite, increases in thickness westward from less than 500 feet in the Leduc and Stettler areas to approximately 1,800 feet in part of the Rocky Mountains of Alberta. East of Leduc and north of Stettler, the Wabamun has been irregularly truncated by pre-Cretaceous erosion.

For purposes of facies analysis the Wabamun is subdivided into four units: (1) basal unit including the lowermost evaporitic interval and lateral marine carbonate equivalents; (2) lower middle Wabamun representing the overlying interval to the top of the Crossfield dolomite in the Calgary area; (3) upper middle Wabamun including post-Crossfield and pre-Big Valley beds; and (4) upper Wabamun representing the Big Valley formation and approximate lateral equivalents.

The evaporites and dolomites are dominant in the southeastern part of the map area, limestones in the northwestern part (and locally at the southwest), and fine to coarse crystalline dolomite in the intervening area.

Evaporites and associated extremely fine-textured dolomites exhibit most widespread distribution in the basal Wabamun. The main evaporites extend northwest through Stettler to Leduc whereas the extremely fine-textured dolomites also occur farther northwest. In the middle Wabamun, the evaporites terminate a relatively short distance northwest and north of Stettler because of probable removal by solution and pre-Cretaceous erosion. Presumably middle Wabamun evaporites originally extended north beyond their present limit. During deposition of the basal and upper Wabamun units, a subsidiary evaporite province occurred in the foothills belt west of Calgary and may have extended north as far as the Clearwater River.

Limestones range from calcilutites to calcarenites and exhibit incipient dolomitization that commonly imparts a mottled appearance to the rock. The calcarenites are generally of the pseudo-oölitic or pelletoid type, although bioclastic limestone is characteristic of the uppermost, and locally, of lower beds.

The pelletoid limestones and calcilutites were deposited on a widespread bank on which slight restriction of the shallow waters promoted precipitation of calcium carbonate. Anhydrites and cryptocrystalline dolomite were precipitated in the Stettler-Drumheller area near the bank margin at the east, and the associated calcareous muds were diagenetically dolomitized to microsucrosic dolomite. Coarser (very fine to medium crystalline) dolomites are characteristic of a belt between the limestone province at the northwest and evaporites at the southeast. Local vuggy dolomite sections in this belt are interpreted to be possible reefs.

Petroleum has accumulated in two main types of stratigraphically controlled traps within the Wabamun group. Facies change from porous secondary dolomites to relatively non-permeable equivalents approximately up regional dip (mainly eastward) is responsible for gas entrapment in the Okotoks-East Calgary-Olds trend of fields. In the Edmonton area, oil or gas has been localized in erosional hills made up of porous Wabamun dolomites at the pre-Cretaceous unconformity.

INTRODUCTION

The Wabamun group of west-central Alberta is Late Devonian in age and consists dominantly of carbonate rocks and anhydrite. The Wabamun is conformably underlain by the Winterburn group and where not affected by pre-Cretaceous erosion, the Exshaw shale Banff formation (Mississippian) occurs above the Wabamun group. In the Rocky Mountains on the west, the subsurface Wabamun is represented by the exposed Palliser formation that is underlain by the silty Alexo formation and overlain by the Exshaw shale and Banff formation. The Palliser forms a conspicuous massive carbonate rock unit that in gross aspect is uniform and readily recognizable along the entire Rocky Mountain front in southwestern Alberta.

In latest Devonian (Wabamun) time, west-central Alberta (Fig. 1) comprised a marine shelf that was bounded on the southwest by a relatively more rapidly subsiding province along the present foothills and mountain belt. Southeastward, the Wabamun marine shelf carbonates and anhydrites grade into red and green shales, siltstones, dolomites, and anhydrites in Saskatchewan that are indicative of shoreline conditions (Andrichuk, 1951; Baillie, 1955; Belyea, 1955). Northwestward, the widespread limestone bank extended

[1] Presented before the Alberta Society of Petroleum Geologists during the Ninth Annual Field Conference, September 10, 1959.

[2] Andrichuk and Edie, consulting geologists. Appreciation is extended to Imperial Oil Limited for sponsorship of part of the studies on which this paper is based. The writer is grateful to H. R. Belyea, P. F. Brennan, R. W. Edie, W. D. MacDonald, A. R. Nielsen, and P. G. Sutterlin for critically reading the original manuscript. A. W. Kemp drafted the illustrations.

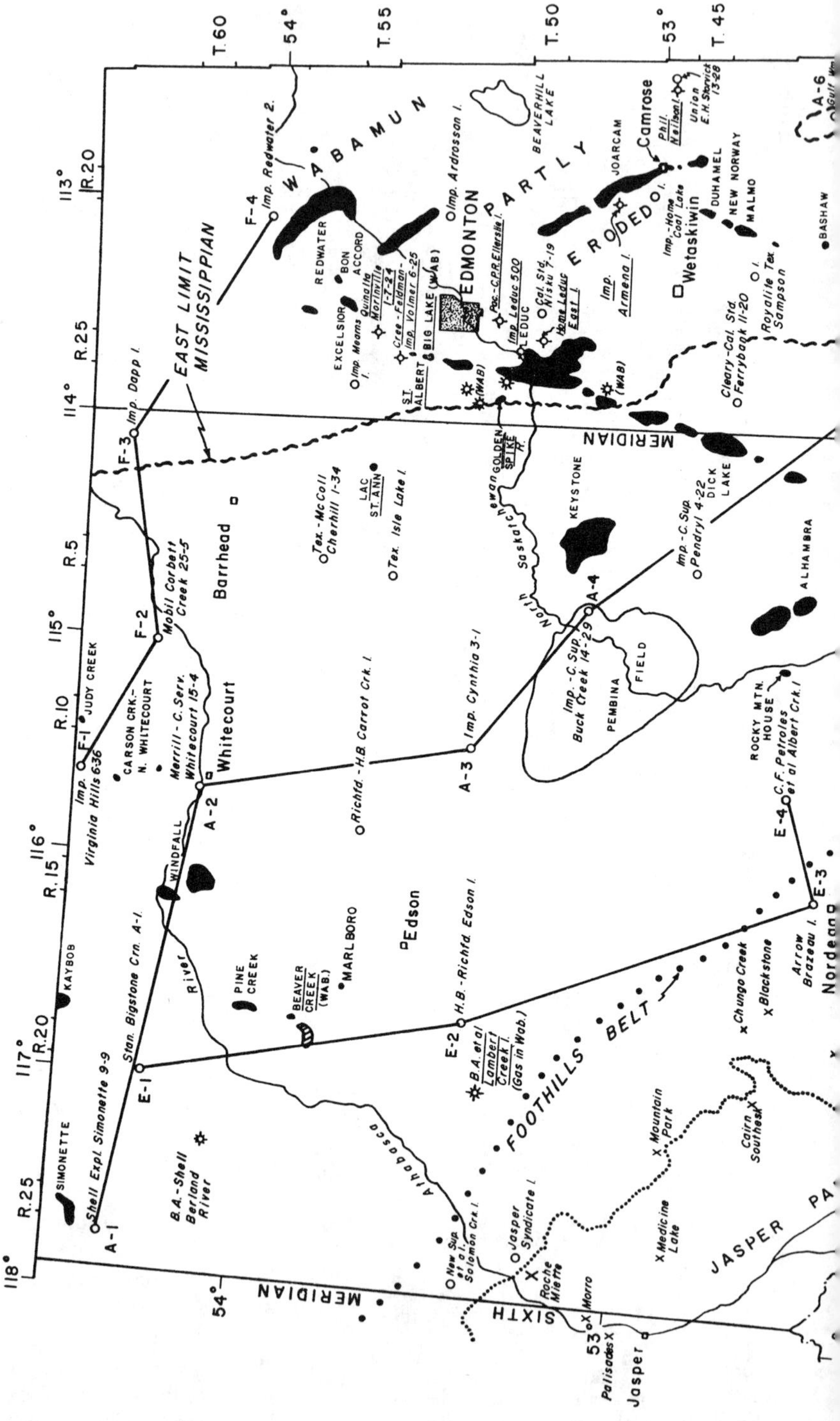
EAST LIMIT MISSISSIPPIAN
WABAMUN
EDMONTON
PARTLY
ERODED
FOOTHILLS BELT
Barrhead
Whitecourt
Edson
Camrose
Wetaskiwin
Jasper
SIXTH MERIDIAN

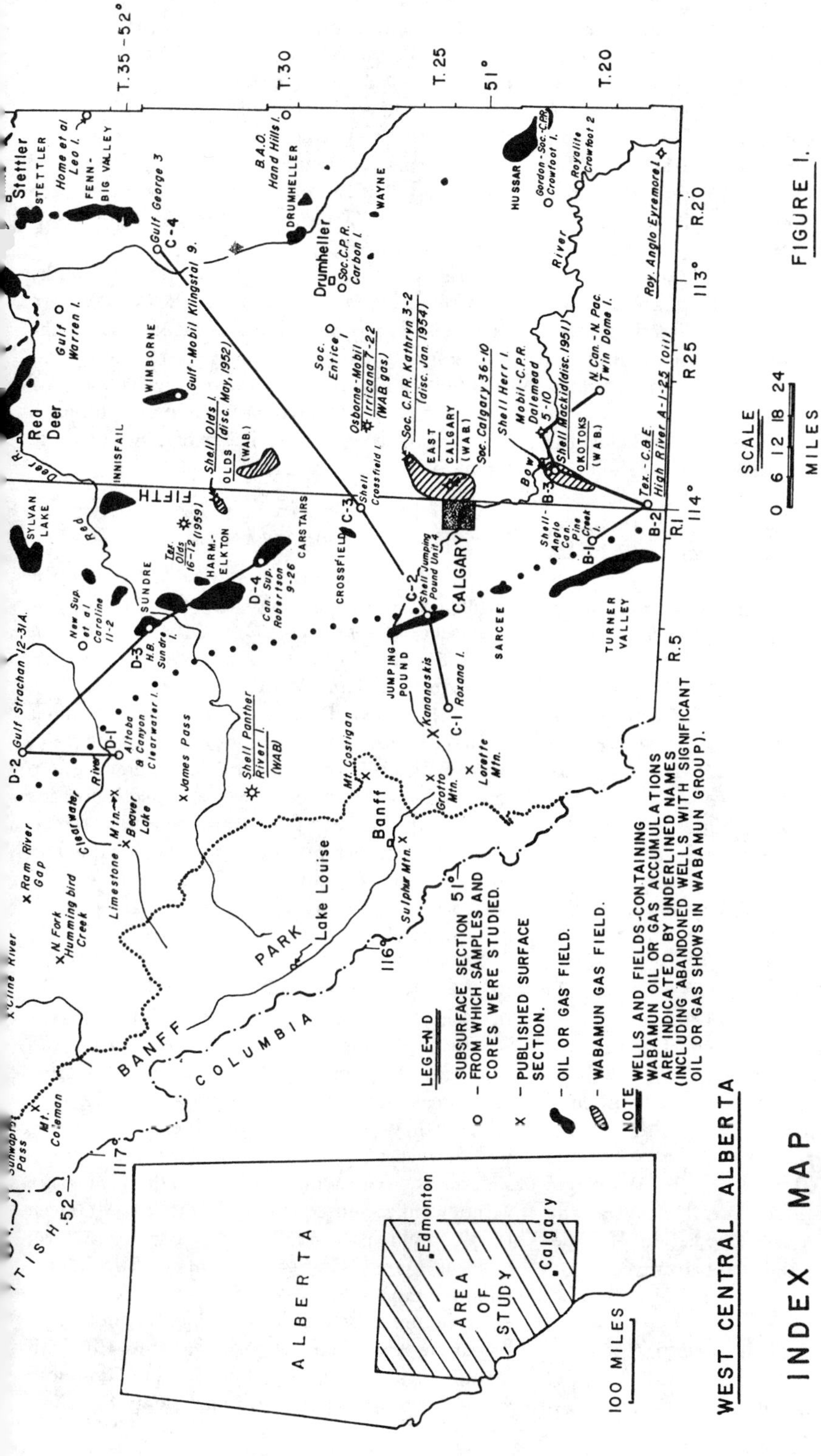

Fig. 1.—Index map of west-central Alberta showing locations of sections studied and lines of stratigraphic cross sections.

beyond the Peace River area in Alberta. According to Patton (1958), farther north along Nahanni River in the Northwest Territories, the uppermost Devonian rocks are dark gray shales and sandstones. Northeast of the map area, the Wabamun has been removed by pre-Cretaceous erosion. In general, the area under consideration contains Wabamun rocks that represent widespread bank carbonate rocks that change facies to evaporites toward the east and southeast (Beales, 1956; Sutterlin, 1958).

The Wabamun gas accumulations at Okotoks, East Calgary, and Olds fields are directly related to facies change from reservoir-type dolomite to relatively non-permeable very fine-textured dolomite and anhydrite. Oil accumulations also occur in the porous Wabamun dolomite facies below the pre-Cretaceous unconformity in the Edmonton area. Thus the main purpose of this paper is to demonstrate a practical application of the facies approach to petroleum exploration, and in the light of facies relationships to interpret the probable origin and environment of deposition of various rock types in the Wabamun group.

PREVIOUS WORK

Rocks of the Wabamun group and Palliser formation have been described by several previous workers, including Warren (1927), Beach (1943), deWit and McLaren (1950), Severson (1950), Fox (1951), Beales (1953, 1956), and Belyea (1955, 1957). Beales (*ibid.*) described and interpreted the genesis of the dolomitic mottling and limestone types in the Palliser of the Rocky Mountains of southwestern Alberta. Andrichuk and Wonfor (1954) interpreted the Wabamun sedimentation in the Stettler area where it was characteristically of the evaporitic type. The northern limit of anhydrite in the Wabamun group of central Alberta was outlined by Workman (1955). Following Beales' (1956) interpretation of a bank environment during deposition of the Palliser in the mountain area of southwestern Alberta, Sutterlin (1958) made a regional study of the Wabamun throughout southern and central Alberta and described the mechanics of deposition of widespread bank limestones and evaporites.

AREA AND SCOPE OF STUDY

The area of study (Fig. 1) includes part of west-central Alberta between T. 18 and 63, R. 17, W. 4 M., to the sixth meridian or Rocky Mountains at the west.

The writer made a detailed microscopic examination of samples and cores of the Wabamun from 55 drill holes. Three exposed sections, near Kananaskis (Bow Valley), Nordegg, and along Athabasca River (near Roche Miette), were also studied. In addition, a number of published surface sections, particularly those by deWit and McLaren (1950), were used to extend the isopach and lithofacies patterns from the plains area westward into the Rocky Mountains.

Because of substantial topographic relief on the pre-Cretaceous erosion surface in the northeastern part of the map area, the truncated edge of the Wabamun is irregular. The present study was directed mainly to outline the regional sedimentation pattern and no attempt was made to determine the detailed erosional edge of the Wabamun group or of its subunits.

STRATIGRAPHY OF WABAMUN GROUP

NOMENCLATURE

The Wabamun formation was named by the Geological Staff, Imperial Oil Limited (1950). The type section comprises 562 feet (between depths of 5,735 and 6,297 feet) of limestone and dolomite in Anglo-Canadian's Wabamun Lake 1 well (Lsd. 5, Sec. 10, T. 51, R. 4, W. 5 M.), about 40 miles west-southwest of Edmonton. Wonfor and Andrichuk (1956) raised the Wabamun to group rank and in the Stettler area subdivided it into the lower evaporitic Stettler formation and the relatively thin upper marine limestone included in the Big Valley formation. These units can be traced over wide areas in the plains region of Alberta (Belyea, 1955, 1957).

The Palliser formation was introduced by Beach (1943) for 800–950 feet of carbonate rocks (formerly included in the upper part of the Minnewanka limestone by Shimer (1926) and Warren (1927)) in the Moose Mountain-Morley map areas and Fairholme Mountains of the Bow River valley of southwestern Alberta. The name was derived from the Palliser Range, the extension of the Fairholme Range north from Lake Minnewanka. Subsequently, deWit and McLaren (1950) divided the Palliser into a lower, Morro, and an upper, Costigan, member. The Morro member ranges up to 950 feet in thickness and consists of massive to thick-bedded dolomitic limestone and dolomite, whereas the Costigan member is thinner-bedded limestone and dolomite (60–400 feet thick) and is in part richly fossiliferous.

STRATIGRAPHIC RELATIONS, CORRELATION, AND AGE

The base of the Wabamun group and equivalent Palliser formation is placed at the top of a widespread silty or arenaceous zone in the upper part of the Graminia formation of the Winterburn group (Fig. 2) and in the upper part of the Alexo formation[3] (Pl. I). The base of the Palliser formation has been chosen at slightly different stratigraphic levels by some authors and thus the variation in thickness of the Palliser locally may be attributed to inconsistent lower boundaries rather than depositional tectonics. In this study, an attempt was made to avoid this possible error by placing the base of the Palliser directly above silty and arenaceous beds in the Alexo formation. Locally, as in the Clearwater River area exposures, non-silty and non-arenaceous solution-brecciated beds that are included in the uppermost Alexo by deWit and McLaren (1950) are here considered to represent former basal anhydrites of the Wabamun. Belyea (1958, p. 98) noted that breccias in the upper part of the Alexo at James Pass are probably equivalent to evaporites in the basal Wabamun. However, contorted and brecciated beds, in association with silty and arenaceous rocks, also occur in the Alexo formation as delimited in the present study (deWit and McLaren, 1950; McLaren, 1955, p. 21).

According to Taylor (1957, p. 185) the Alexo-Palliser boundary is closely parallel with boundaries of faunizones directly above and below and thus approximates a time horizon. McLaren (1955, p. 22) notes that the Palliser-Alexo contact is commonly transitional over several tens of feet and he considers the base of the Alexo to be the most easily recognizable lithologic and faunal break, the latter defining the boundary between the Frasnian and Famennian Stages (McLaren, 1959). Crickmay (1952) states that a striking paleontological hiatus occurs in the middle of the "silt zone" between the ranges of *Cyrtospirifer* and *Cyrtiopsis*.

West and southwest of the Mississippian erosional limit that extends northwestward from Stettler, the Wabamun group is overlain by the black shale of the Exshaw formation. In a few places, such as near Medicine Lake and the Athabasca River northeast of Jasper, the Exshaw shale is absent and the overlying Banff formation rests directly on the Palliser (Harker and McLaren, 1958).

The Wabamun group and equivalent Palliser formation are generally considered to be Late Devonian in age. Crickmay (1952, p. 594) reports that at two places in the Rocky Mountains of Alberta (near Cadomin, southeast of Jasper, and on Cline River) the spiriferoid fauna in the uppermost part of the Boule (i.e., Palliser) has Mississippian affinities (see also Fox, 1951, p. 843). McLaren (1954) considers that the upper, Costigan member of the Palliser is Late Devonian in age and is typified by *Nudirostra utahensis ventricosa* (Haynes), the highest rhynchonellid zone recognized in the Devonian of the Rocky Mountains of Alberta. The *Nudirostra gibbosa seversoni* zone is indicated to occur in the Morro member of the Palliser and the uppermost Alexo (McLaren, 1954, p. 160). In a subsurface section near Stettler, Alberta, Raasch (1956) states that the uppermost 3 feet of the Big Valley formation is characterized by *Strophopleura* cf. *notabilis* Kindle, *Cyrtiopsis* cf. *kindlei* (Stainbrook) and *Cleiothyridina coloradoensis* (Girty), species that also occur in the top of the Costigan member as well as in the Ouray and Percha of Colorado and New Mexico. Raasch's zone D, containing a brachiopod faunule with *Athyris angelica* that represents the Late Chautauquan, underlies the aforementioned beds. To the northwest, in Imperial-Canadian Superior Buck Creek 14–29 well (Lsd. 14, Sec. 29, T. 48, R. 6, W. 5 M.), D. J. McLaren (in Belyea, 1958, p. 98) identified a brachiopod fauna in the Wabamun group at a depth of 7,000 feet which corresponds with zone D of Raasch.

LITHOLOGY

In west-central Alberta, the Wabamun group exhibits three major facies, namely, limestone at the northwest, anhydrite (locally halite) and dolomite at the south and southeast, and dolomite in the intermediate areas (Fig. 2). The areal distribution of these major facies is superimposed on a base of generalized total Wabamun isopachs in Figure 3. On the basis of color, the Wabamun carbonates may also be referred to two provinces. In general, the Wabamun rocks exhibit lighter

[3] This detrital silt or sand unit marks an abrupt lithologic change inasmuch as the directly overlying Palliser rocks generally contain negligible detrital quartz residue. As noted by Beales (1956) and Fox (1951), the minor quartz, feldspar, and micaceous silicates in the Palliser appear to be in considerable part of authigenic origin. Fox also indicated that locally gypsum formed up to 4 per cent of the rock in the Morro member of the Palliser formation.

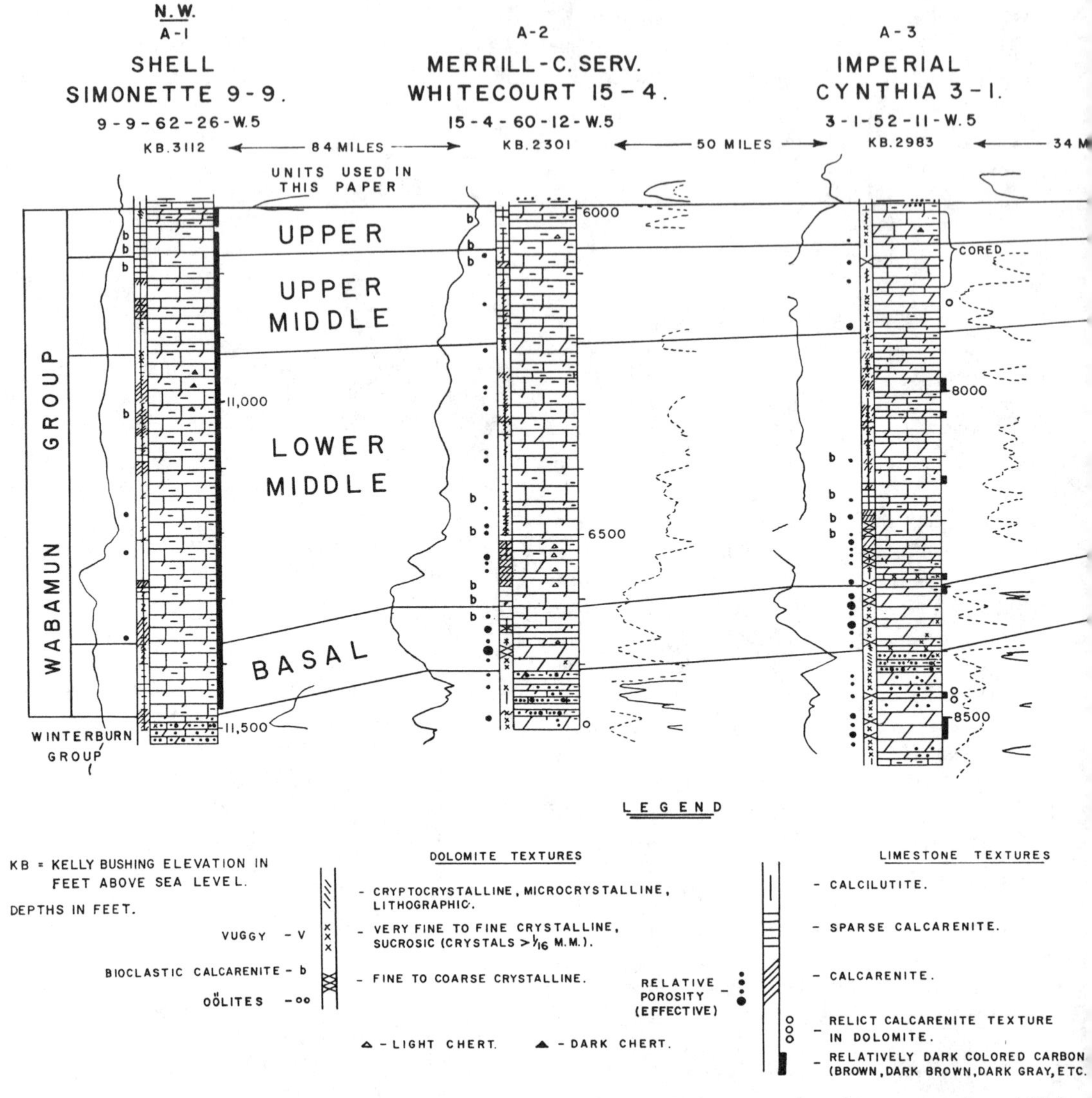

FIG. 2.—Northwest-southeast stratigraphic cross section of Waba

shades of color (cream, tan, light brown, and light gray) in the eastern and northernmost parts of the map area, whereas darker shades (brown, medium, and dark gray) are characteristic on the west. The approximate dividing line, indicated by wedge-shaped symbols in Figure 3, trends north-northwest from Calgary to Whitecourt, approximately parallel with the isopachs. Presumably, the eastern evaporitic and non-evaporitic sediments accumulated relatively slowly in shallow water, permitting the organic carbonaceous material to be oxidized. In the extreme northern part of the map area, the lighter colors are related to depositional limestone shoals.

In the vertical sequence, evaporites are concentrated in two main zones, namely, in the basal and upper parts of the Stettler formation. The uppermost beds of the Wabamun generally consist of partly fossiliferous limestones. The stratigraphic cross sections in Figures 2, 4–8 illustrate the vertical and lateral lithologic variations within the Wabamun group.

Evaporites.—The evaporitic rocks in the Wabamun group consist of anhydrite (locally

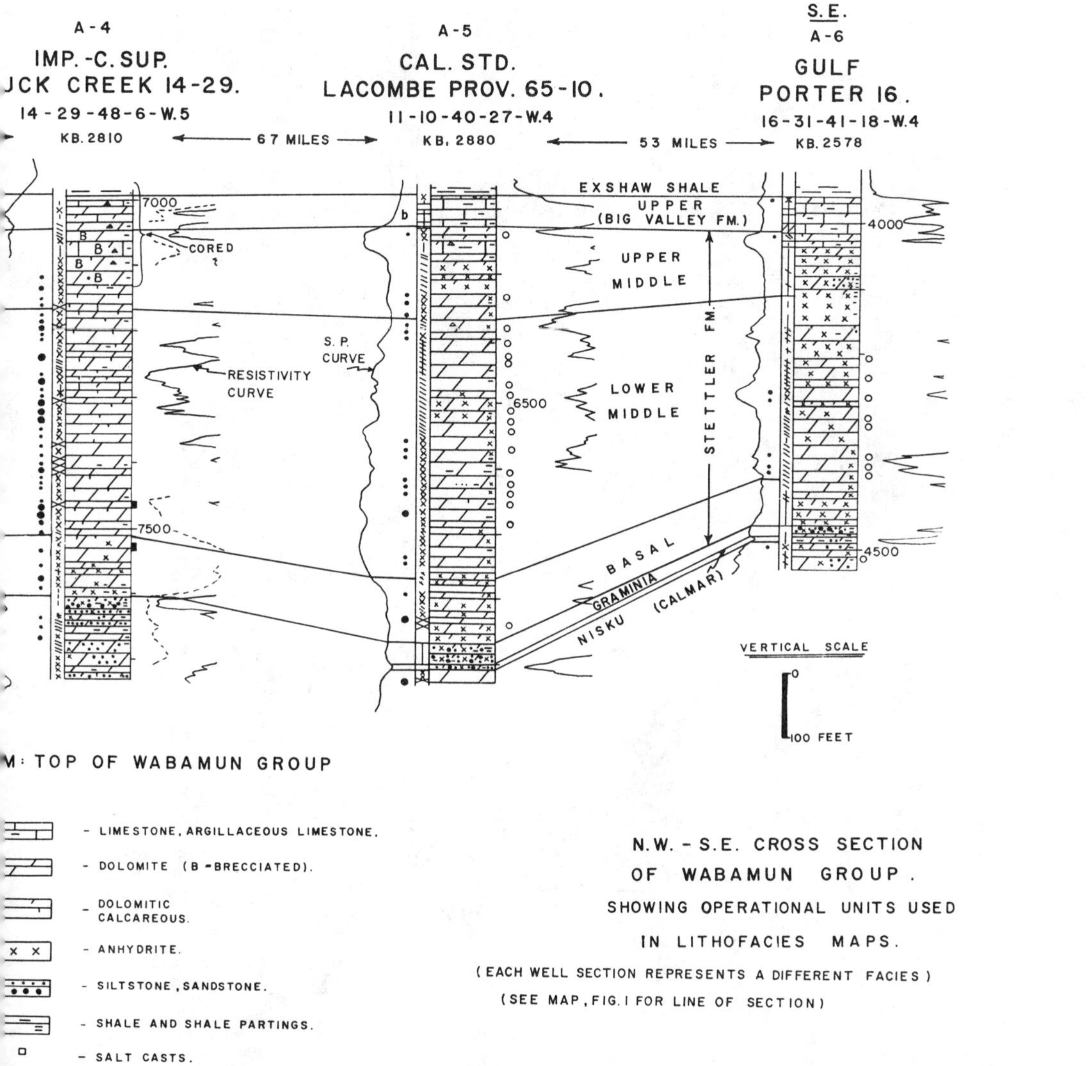

showing the four units analyzed in isopach and lithofacies maps.

halite) and primary (?) dolomite and intergradations. The primary (?) dolomite is buff-gray, tan, or light brown and cryptocrystalline. The rock is closely associated with anhydrite both laterally and as interbeds, and commonly appears finely laminated, non-fossiliferous, non-porous and locally argillaceous. These primary (?) dolomites are also associated with other extremely fine-textured, earthy, microsucrosic dolomites (crystals less than 1/20 mm. in size) that are considered of secondary origin.

Dense limestones of primary origin were not observed in close association with other evaporitic rocks; it is probable that these precipitated calcareous muds were dolomitized diagenetically to form very fine-textured dolomites. However, calcilutites are commonly associated with pelletoid limestones.

Limestones.—In terms of grain size the limestones in the Wabamun group include gradations from calcarenites to calcilutites. The calcilutites (including calcisiltites) are made up of particles generally less than 1/16 mm. in diameter. These limestones are considered to have formed in rela-

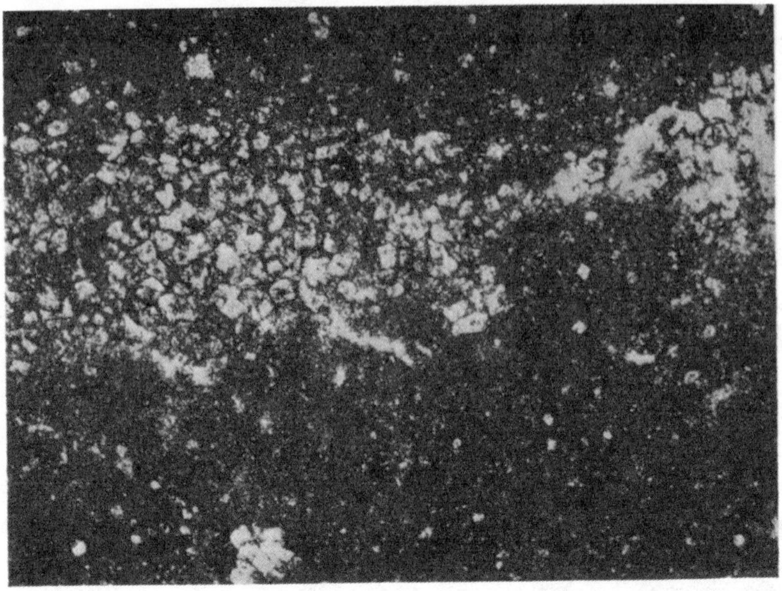

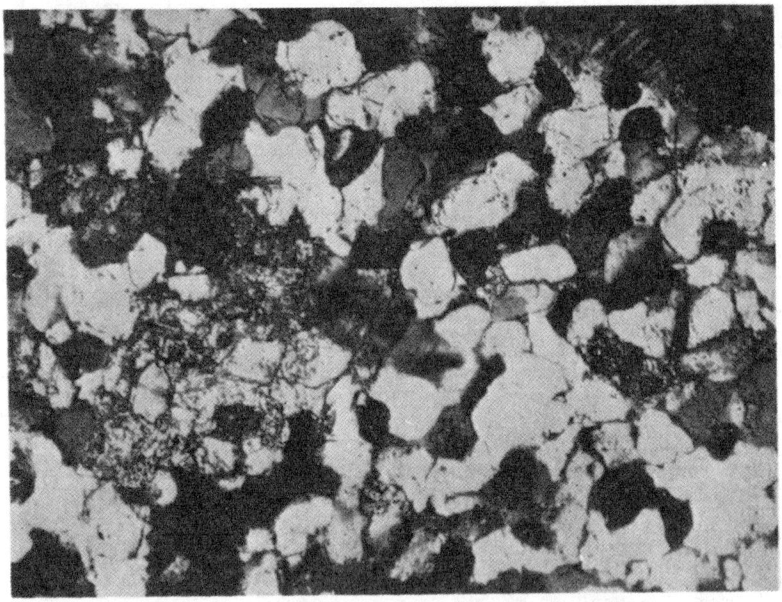

PLATE I

TOP. Ordinary light ×32. Photomicrograph of dolomitic limestone from 120 feet below top of Palliser formation exposed in front range of Rocky Mountains near Kananaskis (Sec. 35, T. 24, R. 9, W. 5 M.), Bow River Valley, Alberta. Typical dolomite mottle is shown in center. Photomicrographs of other characteristic limestone textural types in Palliser formation have been published by Beales (1956, pp. 853–55).

BOTTOM. Crossed nicols ×100. Photomicrograph of very fine-grained quartzose sandstone from Alexo formation near Kananaskis, location as above. Quartz grains show sutured contacts and secondary overgrowth. Twinned feldspar grain at upper right. Irregular patch of dolomite at left of center. Photograph illustrates marked lithologic change from relatively pure carbonate rocks of basal Palliser to silty and arenaceous beds of underlying Alexo formation.

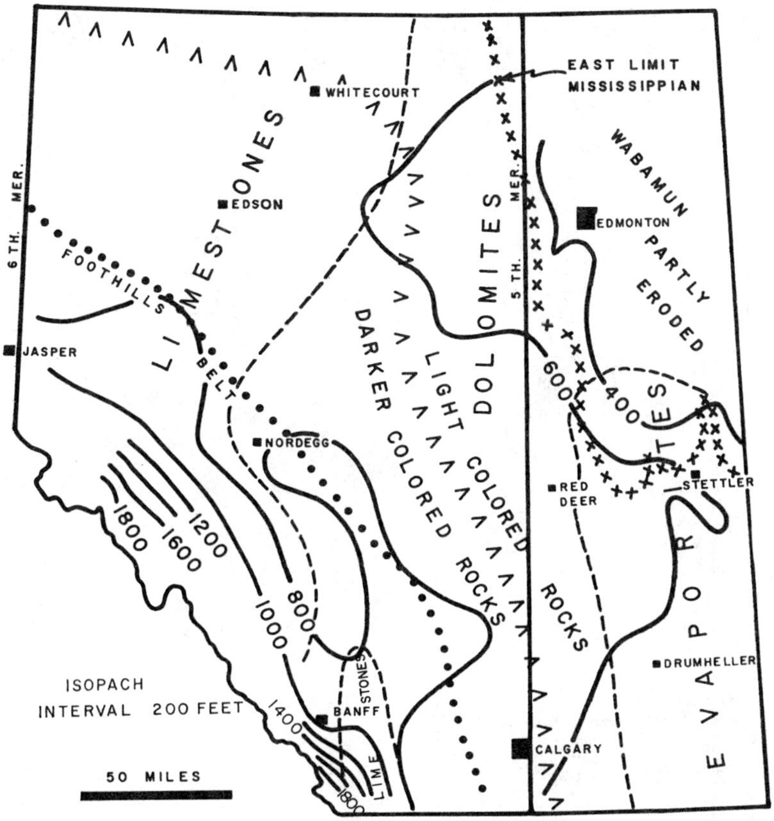

FIG. 3.—Generalized isopachs of Wabamun group in west-central Alberta. Major lithofacies provinces are shown.

tively quiet water of varying depths. The calcilutites consist generally of very fine bioclastic material or lithified calcium carbonate (drewite) muds. The latter may represent chemical or biochemical precipitates. The calcarenites are made up of rounded calcareous pellets (pseudo-oölites), locally oölites, or fossil fragments (crinoid and brachiopod fragments, *etc.*) with little or no interstitial calcilutite or calcisiltite matrix. The intermediate limestones that contain varying proportions of fossil fragments or pellets within a substantial calcilutite matrix are termed "sparse calcarenites" and correspond with biomicrites and pelmicrites of Folk (1959). Pelmicrites are characteristic of the Palliser and Wabamun limestone sections, except for the uppermost part (Costigan member and Big Valley formation) and locally lower beds. The "fossil fragment limestones" of Beales (1956), formed of fossil fragments that have collected essentially *in situ* without being subject to transport, occur in certain beds.

Beales (1956) described the limestones of the Palliser formation in the Rocky Mountains of southwestern Alberta as predominantly nonskeletal or pelletoid and suggested that they were sediments deposited in the interior of an extensive bank. However, the vuggy biostromal dolomites on Sulphur Mountain near Banff and at the southern end of the Sawback Range are believed by Beales to represent possible marginal bank deposits on the west rather than typical sediments of the bank interior. Pelletoid or pseudo-oölitic calcarenites and calcilutites may have formed by precipitation in a slightly supersaline environment in the interior of a bank as compared with the more normal salinity of waters in which bioclastic limestones were deposited (cf. Illing, 1954; Beales, 1956).

Limestones composed of crystalline calcite up to $\frac{1}{2}$ mm. in size or coarser are present in minor quantities and are believed to have formed in part by recrystallization of calcareous muds or lime-

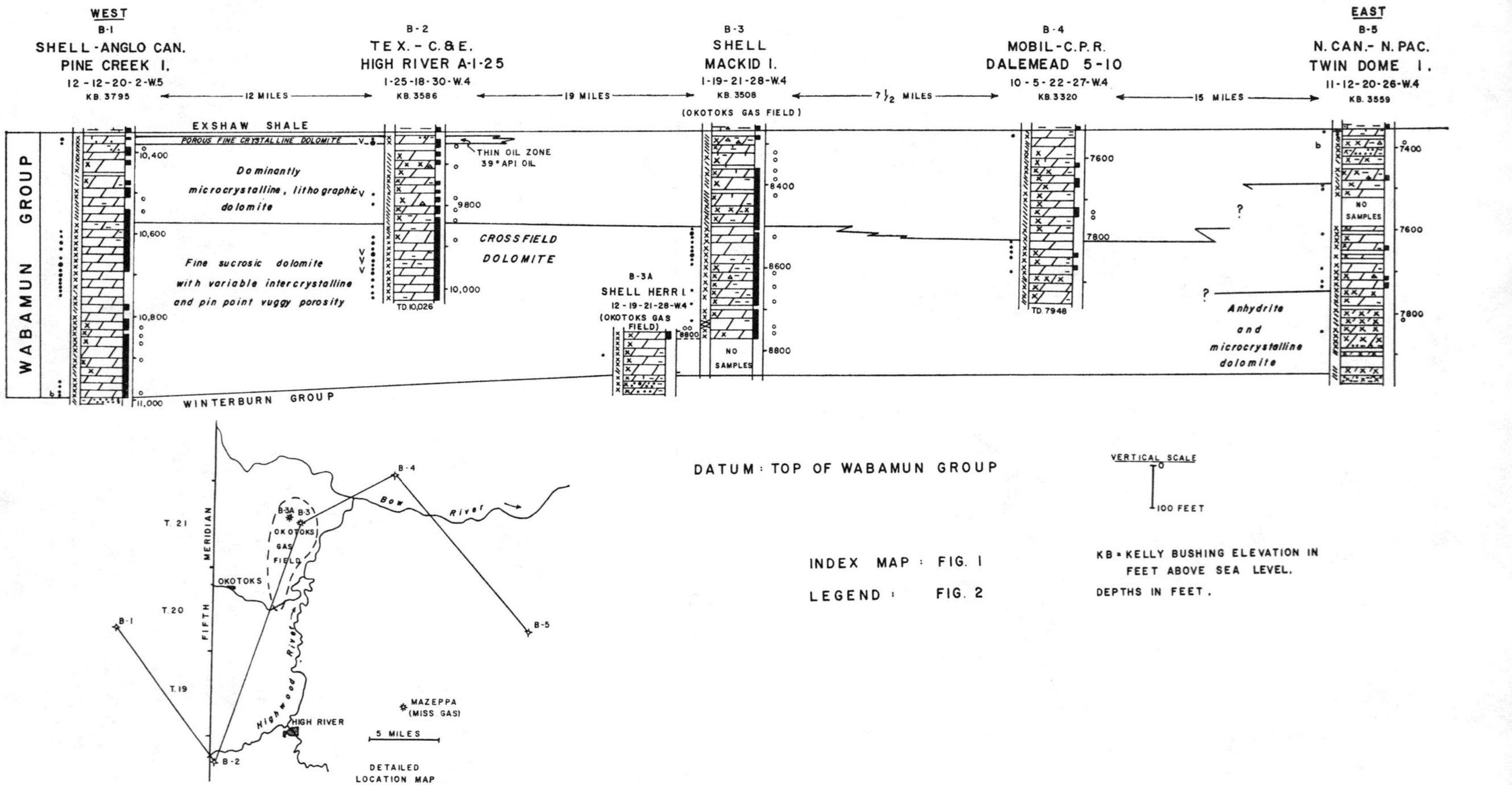

FIG. 4.—West-east stratigraphic cross section of Wabamun group, from east of Turner Valley oil field through Okotoks gas field.

stones at any time between deposition and the present.

Dolomitic limestones.—The Palliser and Wabamun limestones contain varying proportions of disseminated crystals of dolomite that impart a characteristic mottled appearance to the rock (Pl. I). Beales (1953) interpreted the dolomitic mottling to have formed by diagenetic replacement of part of the limestone by dolomite; the dolomite spread irregularly through the limestone (or matrix of calcarenites), or in part selectively replaced such structures as "worm burrows," branching sub-cylindrical "algal" structures, and certain fossils (gastropods and thin-shelled brachiopods). Generally, the dolomitic mottles are slightly coarser in crystallinity, and differential weathering produces fine dolomitic ridges or "tracing" on exposed vertical and bedding plane surfaces. Beales (1956, p. 867) states "the occurrence of relatively altered layers with much dolomitic mottling completely separated without apparent connection by relatively unaltered layers of the same microfacies, and both showing negligible porosity, strongly suggests that the magnesium ions were derived by diffusion from the overlying sea water rather than from escaping connate waters. Dolomitization may have stopped short of completion when compaction and recrystallization cut down the effective diffusion porosity of the sediment."

Dolomites.—Four main types of dolomite have been differentiated in the Wabamun group.[4]

(1) Cryptocrystalline and anhydritic dolomites, considered to be of primary or evaporitic origin. These dolomites appear translucent in thin fragments and commonly exhibit conchoidal fracture.

(2) Microsucrosic to chalky dolomites made up of euhedral crystals less than 1/16 mm. in size.

(3) Lithographic dolomites with anhedral crystals generally less than 1/30 mm. in size. The rocks appear dense, non-porous, locally argillaceous and exhibit conchoidal fracture.

(4) Fine to coarse dolomites with crystal size ranging from 1/16 to $\frac{1}{2}$ mm. or greater. These dolomites are either sucrosic (with intercrystalline porosity) or non-sucrosic, the latter composed of interlocked crystals.

The microsucrosic and at least some of the lithographic dolomites are believed to have formed by diagenetic dolomitization of calcareous muds or calcilutites. The coarser microsucrosic phases may represent dolomitized calcareous silts of either bioclastic or pelletoid origin. It is possible that some of the lithographic dolomites are of primary origin.

The coarser dolomites with crystal sizes greater than 1/16 mm. are considered to be of secondary origin where dolomitization occurred penecontemporaneously with deposition or at any time thereafter. These dolomites comprise the potential petroleum reservoirs. The sucrosic dolomites (as contrasted to the non-sucrosic type with interlocked crystals) exhibit varying amounts of intercrystalline porosity. However, both textural types may contain pinpoint and vuggy porosity. The effective porosity is subject to modification by infilling, particularly with calcite and anhydrite.

Fossils.—Crinoid fragments and brachiopods are the main fossils; gastropods, corals, bryozoans, cephalopods, ostracods, "algal" laminations, spicules, fish teeth, foraminifera, *etc.* are also reported from the Palliser formation. Fossils are concentrated in the uppermost Palliser and are less common, except in thin zones, in the middle and lower parts of the formation.

Method of Analysis

SUBDIVISION OF WABAMUN GROUP

The Wabamun group is bounded by two well defined and consistent horizons, the top of the silt zone (Graminia or Alexo) at the base and the Exshaw shale at the top. In this study, it is tentatively assumed that the Wabamun as here defined is a time-rock unit and that the bounding contacts approximate or closely approach time surfaces. Although a pre-Exshaw disconformity at the top of the Wabamum is reported locally, substantial erosion of Wabamun rocks before Exshaw deposition has not been demonstrated. The regional increase in thickness of the Wabamun appears to

[4] Dolomites similar to three of these types are described by Thomas and Glaister (1959, pp. 119–21) from the Elkton member of the Turner Valley formation (Mississippian) of southwestern Alberta. They state that "lime mud was easily dolomitized to produce a crypto- to microcrystalline, anhedral, interlocking type of dolomite with no effective porosity." The microsucrosic type is similar to their "micro-rhombic dolomites" that are considered "to be the dolomitized equivalents of comminuted, skeletal or non-skeletal grains in current agitated areas." Coarser crystalline dolomites in which original grain outlines are mainly or wholly destroyed are also described by Thomas and Glaister.

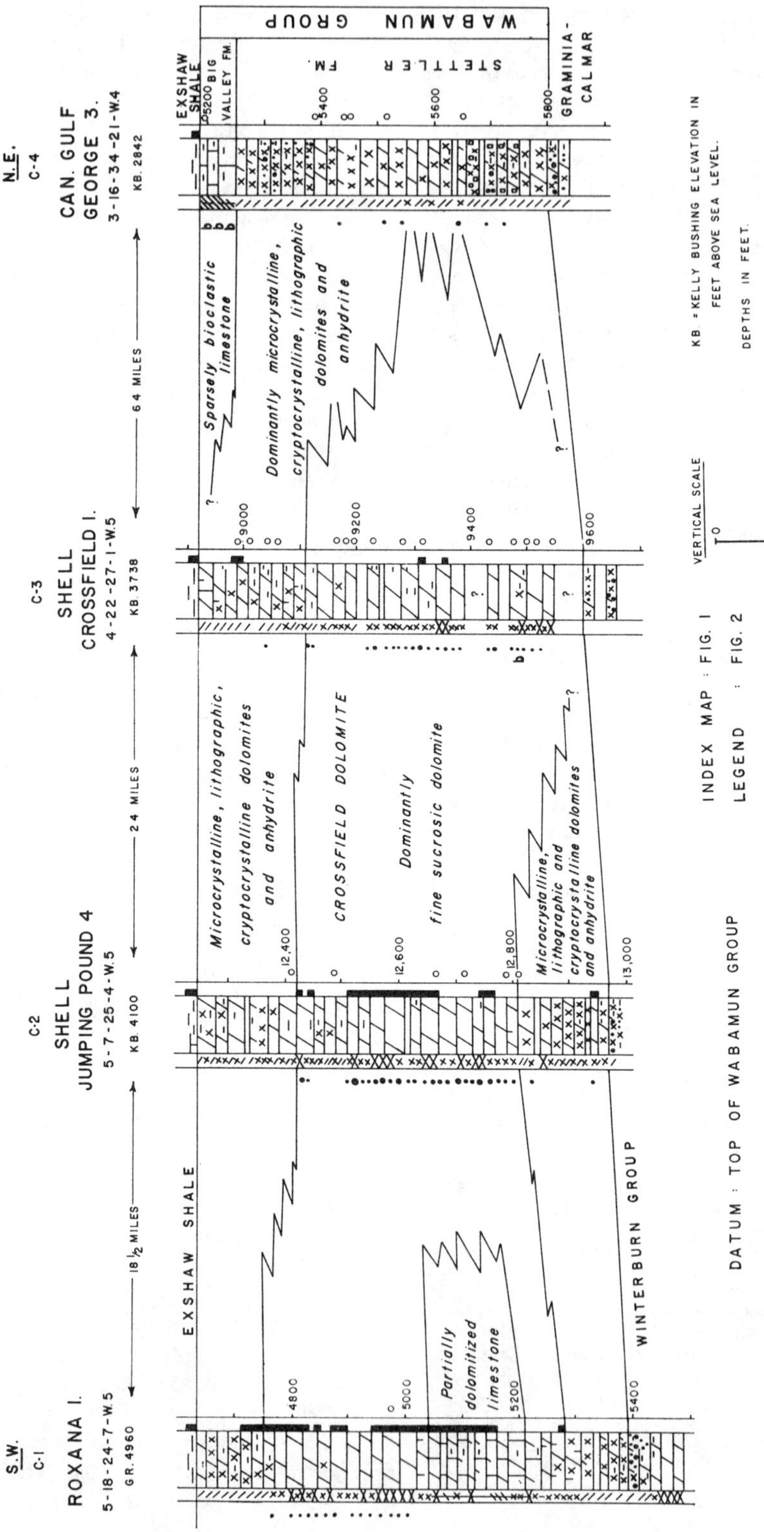

FIG. 5.—Southwest-northeast stratigraphic cross section of Wabamun group, from Moose Mountain area to Big Valley area.

be uniform, except in local areas where solution and collapse-brecciation with accompanying reduction in thickness has occurred.

For purposes of analysis, the Wabamun group is subdivided into four operational units (termed basal, lower middle, upper middle, and upper units) coinciding approximately with certain formal and informal stratigraphic subdivisions now in use (Fig. 2). The basal Wabamun includes the evaporite interval directly above the Graminia silt. The overlying carbonate (dolomite) section that extends to the top of the potentially productive Crossfield dolomite[5] comprises the lower middle unit. The upper middle Wabamun includes the post-Crossfield dolomite and pre-Big Valley beds representing the stratigraphically highest significant evaporites in the Devonian of the Stettler area. The informally named "Juno dolomite" commonly occurs at the top of this unit in the Stettler and Edmonton areas. The upper Wabamun corresponds with the type Big Valley formation and assumed lateral equivalents. This subdivision is based on the assumption that the regional westward increase in thickness is due primarily to differential subsidence during sedimentation. In areas where marker beds or distinct cyclic rock units are not present within the Palliser, such as in the uniform limestone sequence in the northwestern part of the map area, the operational units represent wedge-shape "slices" of which the relative thicknesses are directly proportional to those of the clearly defined units in the Calgary and Stettler areas. The boundaries of these rock "slices" are thought to approximate "phantom" time planes. Therefore, lithofacies maps of these four stratigraphic intervals should be meaningful for interpretation of the sedimentation history.

COMPUTATION OF STATISTICAL LITHOFACIES DATA

Lithologic indices or ratios were computed from thicknesses of the most significant rock groupings for each stratigraphic interval as follows.

$$\text{Evaporite Ratio (ER)} = \frac{\text{Primary evaporites (Halite, anhydrite, evaporitic dolomite)}}{\text{Non-evaporitic rocks}}$$

$$\text{Dolomite Ratio (DR)} = \frac{\text{Secondary dolomites}}{\text{Limestones}}$$

$$\text{Dolomite Crystallinity Ratio (XR)} = \frac{\text{Microcrystalline dolomites (less than } \tfrac{1}{16} \text{ mm.)}}{\text{Fine to coarse crystalline (greater than } \tfrac{1}{16} \text{ mm.) dolomites}}$$

$$\text{Calcarenite Ratio (CR)} = \frac{\text{Calcarenites} + \text{“sparse” calcarenites}}{\text{Calcilutites}}$$

$$\text{Bioclastic Ratio (BR)} = \frac{\text{Bioclastic calcarenites}}{\text{Pseudo-oölitic (pelletoid) calcarenites}}$$

Facies and Sedimentation

TOTAL WABAMUN GROUP

Isopachs.—The Wabamun group exhibits a regional increase in thickness west and southwest (Figs. 3, 9) from less than 500 feet near the Mississippian erosional limit in the Edmonton area of central Alberta, to approximately 1,800 feet in the Sunwapta Pass and Mount Assiniboine areas in the Rocky Mountains of southwestern Alberta (Severson, 1950; E. W. Best, personal communication; Usher, 1959, p. 34). Severson reports the Wabamun to be 1,900 feet thick in the Sunwapta Pass area, but the lower 180–250 feet of sandy limestone is here considered to belong to the underlying Alexo formation. East and northeast of the Mississippian zero edge, the isopach pattern is irregular as a result of differential pre-Cretaceous erosion. Because of these pre-Cretaceous erosional effects, the isopach patterns in Figure 3 and subsequent maps extend only a short distance east of the Mississippian limit.

In the plains area, the rate of westward thickening appears to be relatively uniform and gradual, except at the east near the Mississippian zero edge; here solution-brecciation and post-Mississippian truncation are responsible for the present irregular and generally closer spacing of isopachs

[5] Although the name "Crossfield dolomite" was originally used by the petroleum industry for the potential reservoir unit in the Wabamun group, the definition was first published by the Alberta Petroleum and Natural Gas Conservation Board in "Schedule of Wells Drilled for Oil and Gas in 1956" (p. 12) as "a biostromal member approximately 300 feet below the Exshaw; brown fine to medium crystalline dolomite with porosity." In this publication, the Crossfield was used only in the southern plains area.

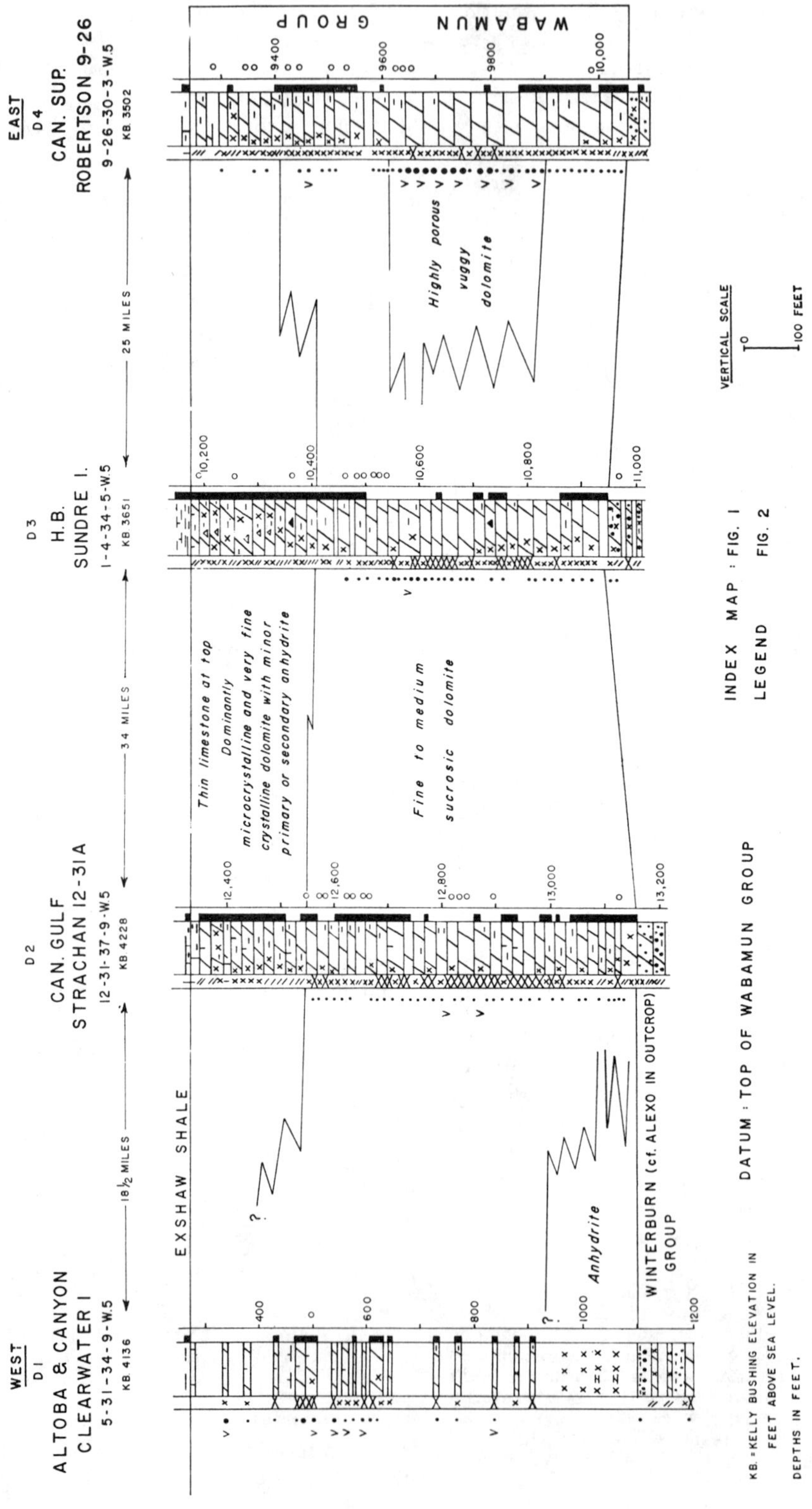

FIG. 6.—West-east stratigraphic cross section of Wabamun group, from foothills near Clearwater River to Harmattan-Elkton oil and gas field.

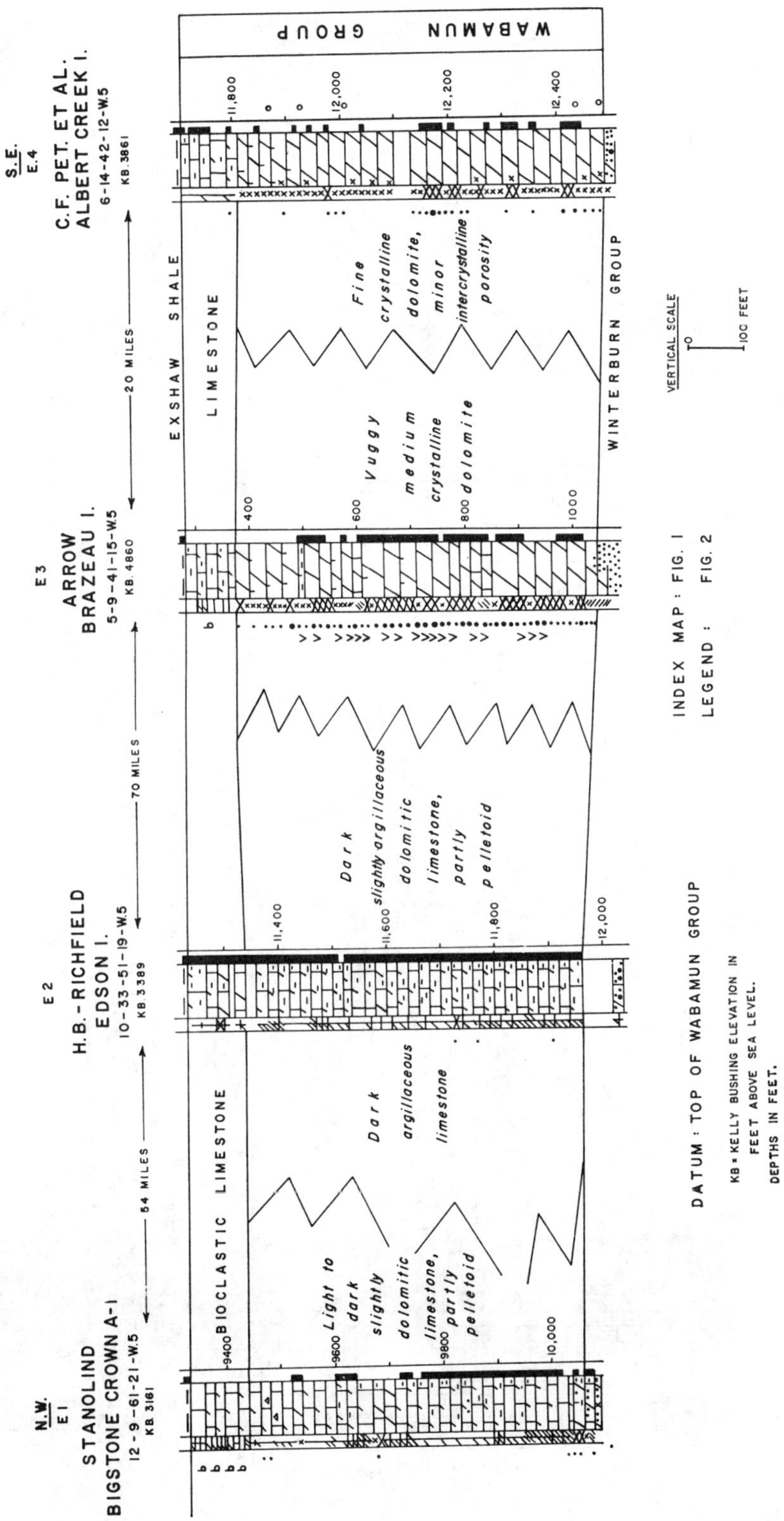

FIG. 7 —Northwest-southeast stratigraphic cross section of Wabamun group, from north of Beaver Creek and Pine Creek fields to Nordegg area.

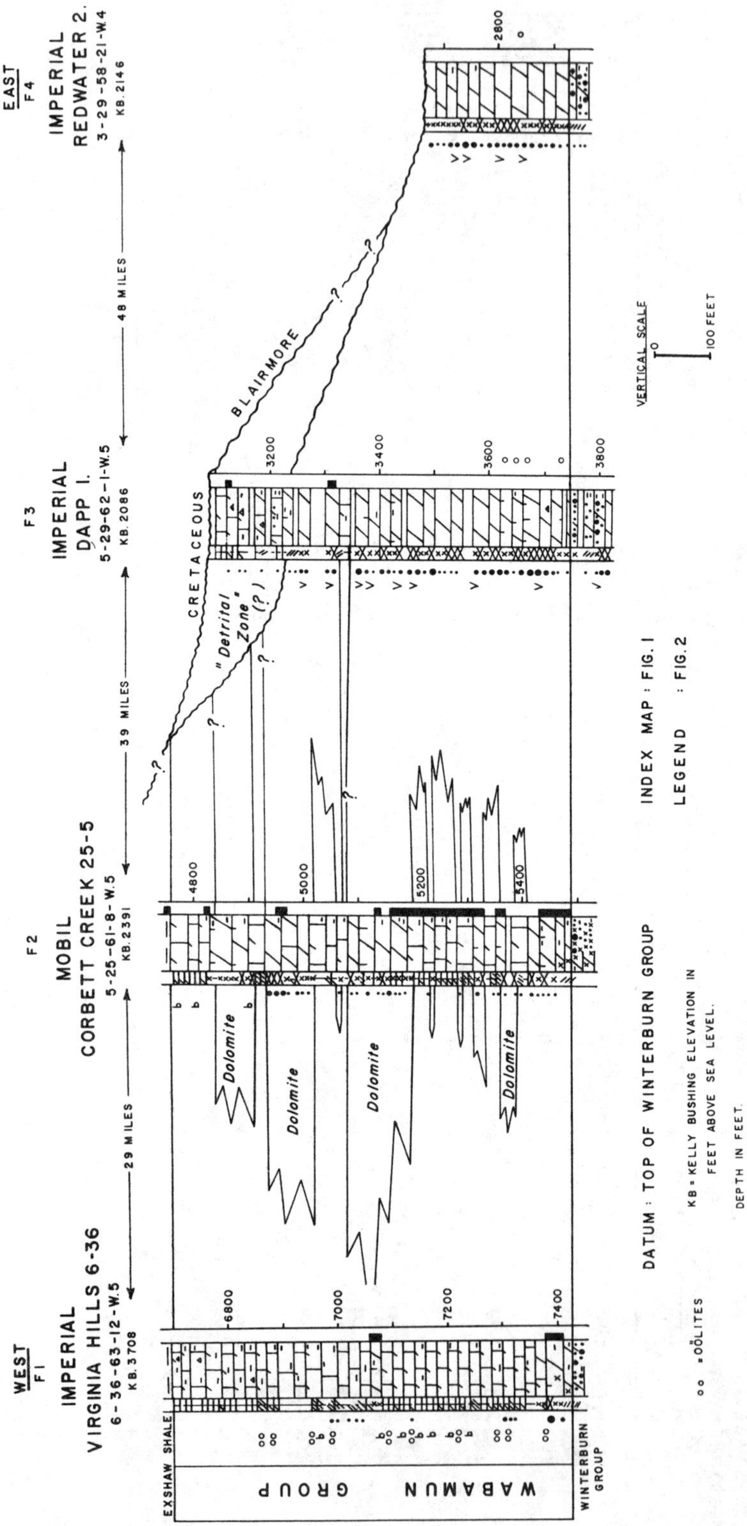

FIG. 8.—West-east stratigraphic cross section of Wabamun group, from Virginia Hills area to Redwater oil-field area.

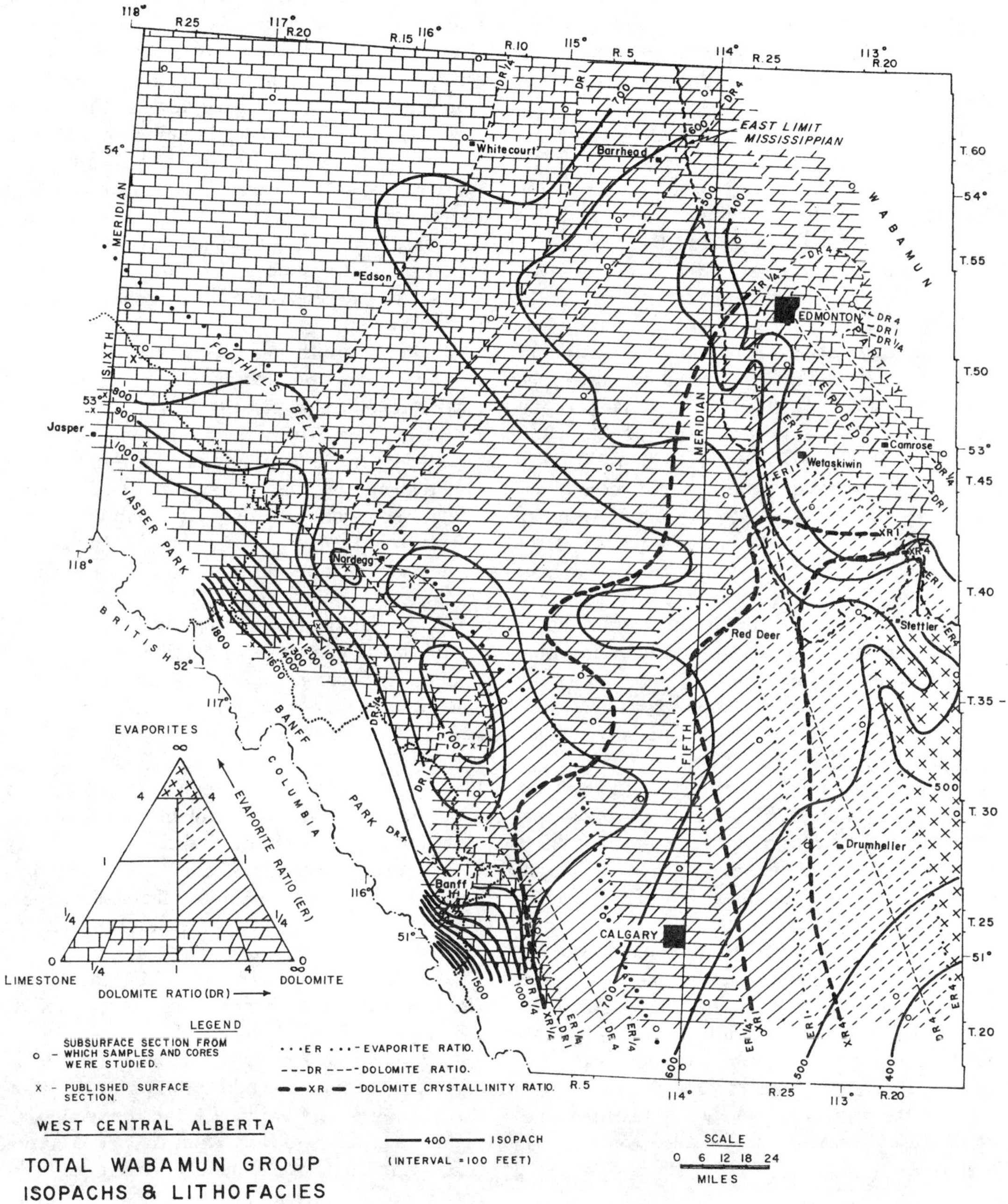

FIG. 9.—Isopachs and lithofacies pattern of total Wabamun group in west-central Alberta.

particularly evident in the Stettler area. In the Rocky Mountains, however, a marked increase in rate of thickening occurs west of the front ranges,[6] as noted especially in the Sunwapta Pass area west of Nordegg (based on published data by Severson, 1950) and south of Banff (Usher, 1959).

Anomalously thin Palliser surface sections are reported along Limestone Creek (Clearwater River area) in the Rocky Mountains between Nordegg and Banff (deWit and McLaren, 1950; McLaren, 1955). In the subsurface section in Altoba and Canyon Clearwater 1 well, a short distance east of Limestone Creek exposures, the Wabamun contains a prominent basal anhydrite that may have been removed by solution in the nearby outcrops. The interpreted equivalent interval is tentatively assumed to have been included in the uppermost part of the Alexo formation by deWit andMcLaren (1950, p. 47). They describe 12½ feet of gray limestone breccia underlying a 45-foot covered interval in the Alexo. In this study the top of the Alexo is placed at the laminated silty dolomites directly below the breccia. If this is correct, the relatively thin Palliser section may be certainly in part attributed to loss of section by post-depositional removal of evaporites by leaching.

The relatively thin Wabamun section west of Edmonton, for example, in Imperial-Canadian Superior Buck Creek 14–29 well, may also partly be caused by solution of evaporites. In this well, however, solution-breccias occur in the upper Wabamun at a stratigraphic level approximately equivalent to an anhydrite horizon in other wells on the south, for example, Imperial-Canadian Superior Pendryl 4–22.

Lithofacies.—The statistical lithologic triangle that forms the basis for the lithofacies patterns interrelates primary evaporites, secondary dolomites, and limestones. In addition to the evaporite ratio (evaporites to non-evaporitic rocks) and dolomite ratio (secondary dolomites to limestones), the ratio of extremely fine-textured dolomites to coarser dolomites is also contoured in Figure 9.

[6] In the mountains, these rates of change of both thickness and lithofacies are exaggerated by structural foreshortening in the belt of imbricate thrusts. Palinspastic base maps are required to show the original depositional patterns in such areas that have been subjected to later compressional folding and thrust faulting.

In the Wabamun group, evaporitic rocks are dominant in the Stettler-Drumheller area, east of Calgary and south of Wetaskiwin. In addition, a subsidiary evaporite facies belt, in which evaporites comprise up to 20 per cent of the Wabamun section, trends north-northwest along the foothills from southwest of Calgary to the Clearwater River. The main Stettler-Drumheller and subsidiary foothills evaporite provinces are separated by a peninsular area of secondary dolomites that trends north-northwest through High River, Okotoks, Calgary, Crossfield, Carstairs, Olds, Sundre, and Caroline and includes the bulk of the proved gas reserves in the Crossfield dolomite of the Wabamun group.

Dolomites are the main carbonate rocks southeast of a line from Nordegg to Barrhead, and locally dolomite is present to the exclusion of limestone. A relatively pure limestone facies belt occurs in the front ranges east of Banff and is bounded on the west by an almost completely dolomitized section on Sulphur Mountain. Usher (1959, p. 34) reports that in the Mount Assiniboine area, 24 miles south of Banff, "the Palliser formation reaches the unusual thickness of nearly 1,850 feet, and is there distinguished by including thin bedded cherty crinoidal limestones in its upper part (Costigan member?) and having massive nodular limestones low in the formation, as well as two black shaly limestone bands about the middle of the formation." The limestone facies east and south of Banff may represent a southern extension of the main limestone province along the western front ranges of the Rocky Mountains. However, in the eastern front ranges between the relatively non-dolomitized limestones west of Nordegg and those east of Banff, the Wabamun consists dominantly of dolomite at James Pass (McLaren, 1955) and Panther Dome (in Shell's Panther River 1 well, Lsd. 5, Sec. 19, T. 30, R. 10, W. 5).

North and east of the limit of Mississippian rocks the Wabamun exhibits abrupt lateral lithologic changes that are partly the result of post-depositional effects. North and east of Wetaskiwin, anhydrites are generally absent and the partly eroded section contains varying proportions of limestone. Although certain sections such as those in Leduc oil field contain interbeds of limestone in a dolomite sequence, other sections east and southeast of Edmonton locally consist wholly of limestone. In cores these limestones appear to be highly brecciated and exhibit some recrystallization, suggesting that brecciation by leaching of possible evaporites and limestones took place in addition to mechanical disintegration at the old erosion surface (unconformity).

Contours of the ratio of extremely fine-textured (microcrystalline, lithographic) dolomites to coarser (fine to coarse crystalline) dolomite are shown by heavy dashed lines in Figure 9. The pattern indicates that this crystallinity ratio decreases outward from the evaporitic facies, suggesting that the extremely fine-textured dolomites are genetically related to evaporites.

BASAL WABAMUN

The isopach and lithofacies pattern of the basal unit in the Wabamun (Fig. 10) resembles that of the total Wabamun group (Fig. 9). In the plains area the basal unit gently increases in thickness from 75 to 110 feet west and southwest. More pronounced increase in thickness occurs in the mountain area; the basal Wabamun is 250 feet thick in the Sunwapta Pass and Mount Assiniboine areas of southwestern Alberta.

Evaporites and Dolomites.—Evaporites are concentrated in two areas. The major evaporitic area appears elongate, extending from the southeastern corner of the map area northwest through the Stettler area and terminating southwest of Edmonton. However, evaporites in less amounts are distributed farther north.

Extremely fine-textured dolomite, of the type that is commonly closely associated with evaporites, extends northwest of the limit of significant anhydrite occurrence to the west of Edmonton, and to a minor extent, to the north edge of the map area north of Whitecourt. This suggests that incipient evaporitic conditions probably prevailed for a considerable distance northwest of Edmonton, as indicated by the thick dashed lines in Figure 10.

The apparent facies change from evaporites to dolomites north of Stettler and east of Wetaskiwin probably has been accentuated by pre-Mesozoic erosion and solution. However, in the Edmonton-Camrose area, the approximate conformance of the extremely fine-textured dolomites to the present evaporite concentration suggests that the original evaporites probably did not extend much farther north and east beyond that shown in Figure 10.

A subsidiary evaporite area occurs in the foothills west of Calgary and may extend northward to connect with the subsurface evaporites in the foothills of the Clearwater River area. No data are available between these two areas and the connection is thus questioned in Figure 10. At the north, in Altoba and Canyon Clearwater 1 well, the basal Wabamun contains more than 125 feet of anhydrite (cross section, Fig. 6), and the interpreted surface equivalent presently about 8 miles west along Limestone Creek contains possible solution breccias. This evaporitic facies may also extend east of the Altoba Clearwater 1 well to the Caroline area where Belyea (1958, Fig. 3) indicates anhydrites occurring in the basal Wabamun. These western evaporites appear to be essentially separated by a "peninsula" of secondary dolomite from the main Stettler-Drumheller evaporite province on the east.

Limestones.—In the basal Wabamun, relatively non-dolomitized limestones are confined to the extreme northwestern part of the plains and foothills areas west of Edson and to the Alberta Rocky Mountains between Jasper and Clearwater River and locally south and southeast of Banff.

Near the erosional limit of the Wabamun group in the Edmonton-Camrose area the basal Wabamun locally consists of limestone. As shown in Figure 10, northeast of Wetaskiwin there is an apparent abrupt gradation from evaporites and dolomites to dolomites and thence limestones. However, the character of the limestone section in the Edmonton-Camrose area suggests that these lateral lithologic changes are exaggerated by the effects of solution, brecciation, recrystallization, and reworking at and below the unconformity surface.

The calcarenite to calcilutite ratio contours in the northwestern part of the map area indicate relatively agitated water (shoal) facies in the Whitecourt-Kaybob area, whereas to the south and west in the Edson-Jasper area more quiet water limestones accumulated. In the dolomitized area at the southeast, most of the original limestone textures are obscured and the depositional pattern can not be interpreted. In general, the secondary dolomites exhibit maximum intercrystalline and local vuggy porosity near the west-

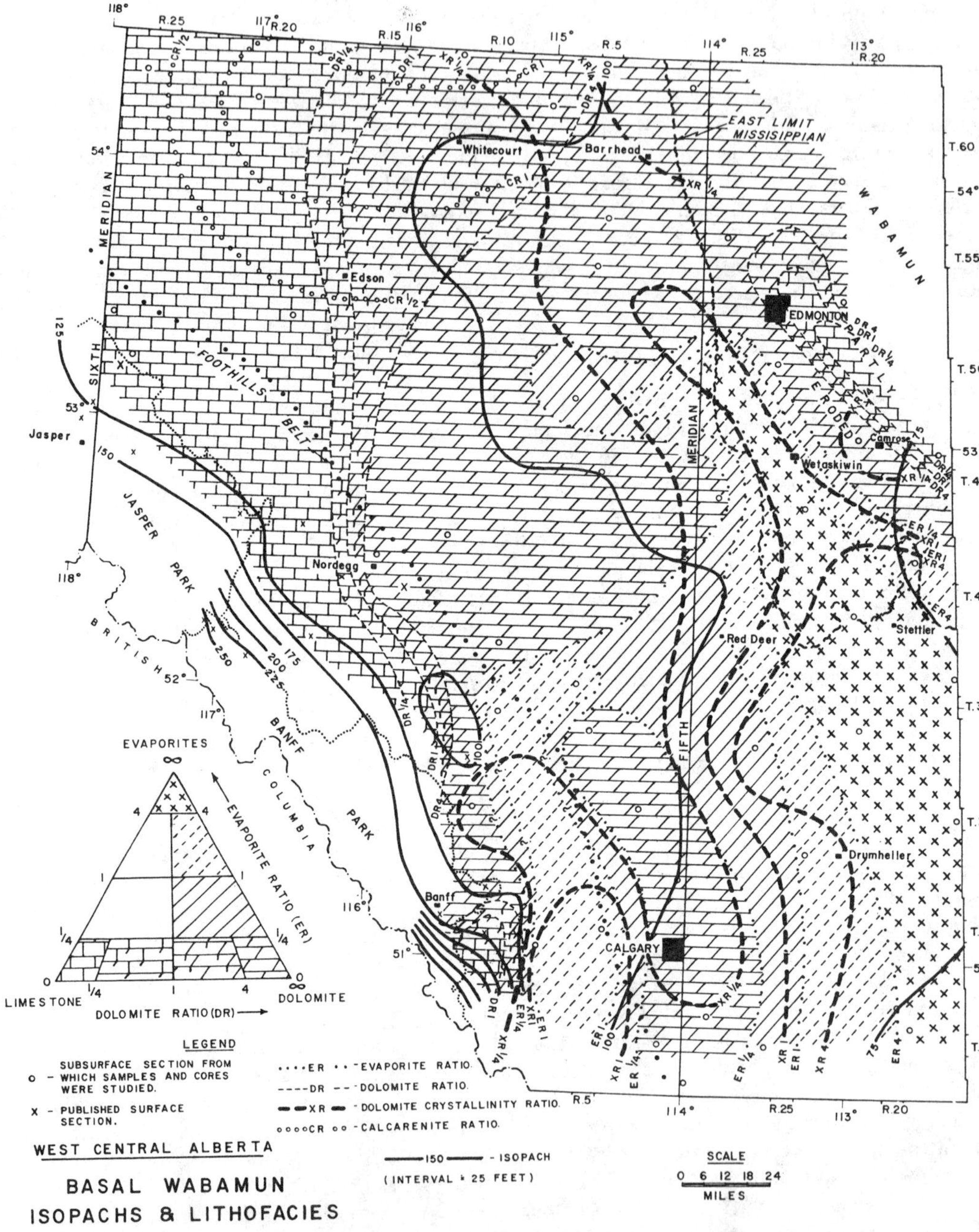

FIG. 10.—Basal Wabamun isopachs and lithofacies in west-central Alberta.

ward transition to limestones, for example, in the Whitecourt, Cynthia (Fig. 2), Dapp (Fig. 8), and Nordegg (Fig. 7) areas.

LOWER MIDDLE WABAMUN

Isopachs.—The lower middle Wabamun includes the productive Crossfield dolomite and equivalent beds and is the thickest operational unit. The isopachs shown in Figure 11 indicate that the thickness increases from about 300 feet in the plains area at the northeast, to more than 450 feet in the foothills. In the Rocky Mountains, the thickness increases to 500–590 feet in the Front Ranges along a line between Banff and north of Jasper, and thence abruptly to about 1,000 feet in Sunwapta Pass and Mount Assiniboine areas west of this line. This regional isopach pattern is modified by abrupt thickness changes in the foothills and mountains near the Clearwater River southeast of Nordegg. Relatively thin deposits occur due west of Edmonton and along a northwest trend from Olds (through the Wimborne, Innisfail, and Caroline areas). The latter thin deposits overlie Leduc reefs and are partly or wholly the result of continued differential compaction of the underlying Woodbend group in which the Leduc reef carbonates compacted less than off-reef shaly sediments. Other irregularities in thickness are caused by post-depositional differential solution of evaporites (Sutterlin, 1958), especially in the eastern part of the area.

Evaporites and dolomites.—Evaporites are concentrated in a belt trending south-southeast of Stettler (Fig. 11) and are negligible west of the fifth meridian. The termination of evaporites north of the Stettler area is caused primarily by leaching or pre-Mesozoic erosion. The evaporite ratio is notably decreased in the northernmost well (Gulf's Porter 16) that contains a relatively complete section of Wabamun in turn overlain by Mississippian beds. However, the slightly thinner Wabamun section in this well suggests that some of the evaporites have been removed by subsurface leaching processes. Although both the ratios of evaporites to non-evaporitic rocks and of extremely fine-textured dolomites to coarser dolomites in Gulf's Porter 16 well indicate that the evaporites were deposited in less proportion north of Stettler, the latter ratio (dolomite crystallinity) suggests that evaporites originally may have accumulated north of their present edge, in the Camrose area southeast of Edmonton. However, unlike that of the basal Wabamun unit, incipient evaporitic conditions did not extend northwest of Edmonton during lower middle Wabamun sedimentation.

West of the fifth meridian, the unit contains a few thin basal anhydrite beds along the foothills belt west of Calgary, north to the Clearwater River. These evaporites represent a cyclic transition from the basal Wabamun anhydrite to overlying carbonate deposition in this area.

The dolomitization pattern in the foothills and mountain areas resembles that of the basal Wabamun, except that in the northwest part of the map area high dolomite content does not extend as far west. Thus, in the basal Wabamun dolomites predominate over limestones east of a line trending due north of Nordegg (to the west of Whitecourt), whereas this line trends north-northeast of Nordegg to west of Barrhead in the overlying lower middle Wabamun unit.

Marked variation in dolomite types occurs in the lower middle Wabamun east of the non-dolomitized limestone province. Within and peripheral to the highly evaporitic facies in the southeastern part of the map area, the unit consists of light-colored (cream and light tan) non-permeable cryptocrystalline, lithographic and microsucrosic dolomites and anhydrite, with interbeds of very fine to fine sucrosic dolomites exhibiting pin-point, intercrystalline and fine vuggy porosity (sections A-5 and A-6, Fig. 2). Toward the periphery of the evaporitic facies, the fine sucrosic secondary dolomites increase in relative abundance at the expense of anhydrites and extremely fine-textured dolomites; this change appears first in the lower part of the unit (section A-5, Fig. 2; section C-4, Fig. 5).

In the dolomitized province west of the evaporitic rocks, two main facies are developed. The more common facies consists of a series of fine to medium sucrosic dolomites generally exhibiting intercrystalline porosity in varying amounts and minor vuggy zones (e.g., in Shell's Crossfield 1 and Shell's Jumping Pound Unit 4, Fig. 5). However, along the western margin of the completely dolomitized facies, certain sections exhibit thick zones of excellent vuggy porosity, probably of original organic (reefal?) origin, for example, in Altoba and Canyon Clearwater 1 (Fig. 6), Arrow Brazeau 1 (Fig. 7), and Imperial Dapp 1 (Fig. 8). Presumably, this dolomite margin may mark an

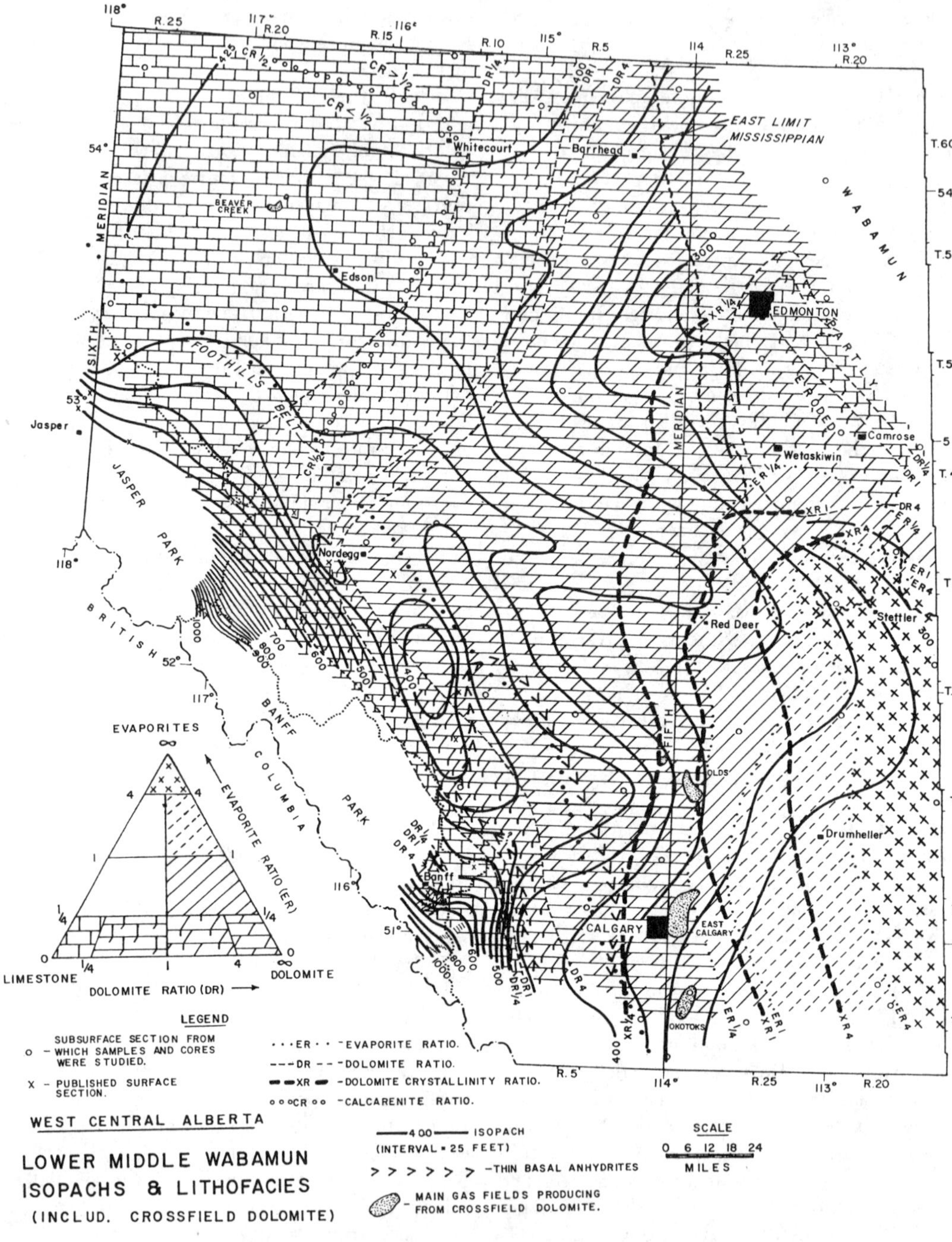

FIG. 11.—Lower middle Wabamun isopachs and lithofacies in west-central Alberta.

approximate Wabamun reef trend. Vuggy dolomites also occur in local sections that are not situated on the dolomitized margin, for example, in Canadian Superior's Robertson 9–26 (Fig. 6). According to H. R. Belyea (personal communication) some of these vuggy sections appear to be associated with underlying Leduc reefs.

West of the margin of complete dolomitization, the facies change to limestone is transitional and the section consists of limestone and dolomite interbeds. This transitional belt is about 30 miles wide in the plains at the northwest (Fig. 8) but in the mountains at the southwest (Nordegg, Banff areas), the facies change appears to be relatively abrupt, being exaggerated by structural foreshortening.

The carbonate rocks are generally darker-colored (brown and gray) in the western part of the map area and lighter-colored (cream, tan, light brown) in the eastern part. In the Calgary area the transition from dark- to light-colored rocks is observed between Okotoks and Jumping Pound at the west and Twin Dome and Crossfield at the east (Figs. 4, 5). In the northwest part of the map area, the transition occurs between the Simonette and Whitecourt areas (Fig. 2).

Limestones.—It is tentatively interpreted that the limestone province at the northwest is continuous southward along the Rocky Mountains from Jasper to a presently narrow structurally foreshortened belt (20 miles wide) east and south of Banff. The lower middle Wabamun is completely dolomitized on Sulphur Mountain near Banff, and thus locally, the relatively non-dolomitized limestone facies is restricted at the west; east toward the Roxana 1 well northwest of Moose Mountain the facies change to pure dolomites is not as abrupt, as shown by the transitional belt of interbedded limestones and dolomites in Figure 5.

The apparently anomalous dolomite and limestone facies in the Edmonton-Camrose area where the Wabamun has been breached by pre-Cretaceous erosion, is partly the result of post-depositional processes such as leaching, brecciation, reworking, and recrystallization.

The calcarenite ratio contour of one-half in the limestone province at the northwest indicates that sharply defined depositional shoals were not developed. The relative calcarenite content increases slightly toward the north edge of the map, and in partly dolomitized limestone facies at the east (DR $\frac{1}{4}$–1). Although not shown in Figure 11, the relative proportion of bioclastic to pelletoid calcarenites increases toward a north-south line passing through Whitecourt, approximately along the western limit of significant dolomitization (DR $\frac{1}{4}$). This line may mark the approximate margin of a depositional shoal.

UPPER MIDDLE WABAMUN

Isopachs.—This unit represents the post-Crossfield and pre-Big Valley interval. Westward, the unit gradually increases in thickness from 100 to 150 feet in the plains, and thence more abruptly to about 200 feet in the Jasper and Banff areas and 360 feet in Sunwapta Pass and Mount Assiniboine areas (Fig. 12). In the Clearwater River area north of Banff, the unit is anomalously thin, as were the underlying units. West of Edmonton, the regionally thinner deposits may be related to subsurface removal of evaporites and solution-brecciation in the upper part of this unit (Imperial-Canadian Superior Buck Creek 14–29, Fig. 2). Isopach irregularities in the southeastern part of the map area are also partly or wholly the result of similar solution-brecciation processes (Sutterlin, 1958).

Evaporites and dolomites.—The lithofacies pattern of the upper middle Wabamun closely resembles that of the directly underlying unit (Figs. 11 and 12). The highest concentration of evaporites extends south of Stettler and significant evaporite content is confined essentially east of the fifth meridian in the southeastern part of the map area. The termination of evaporitic facies north of Stettler has been caused primarily by pre-Mesozoic erosion and solution of evaporites. However, because extremely fine-textured dolomites extend north beyond the limit of anhydrites near Stettler it is inferred that the anhydrites also originally extended north beyond their present limit of distribution. Furthermore, extremely fine-textured dolomites that are conspicuous west and northwest of Edmonton may represent western equivalents of anhydrites of the original extension of the evaporite province due north of Stettler.

Minor evaporites also occur in the foothills west of Calgary where the evaporite ratio has a value of approximately one-eighth. Furthermore, as in lower units, this minor evaporite development is separated from the Stettler province by an essentially non-evaporitic belt in the Calgary

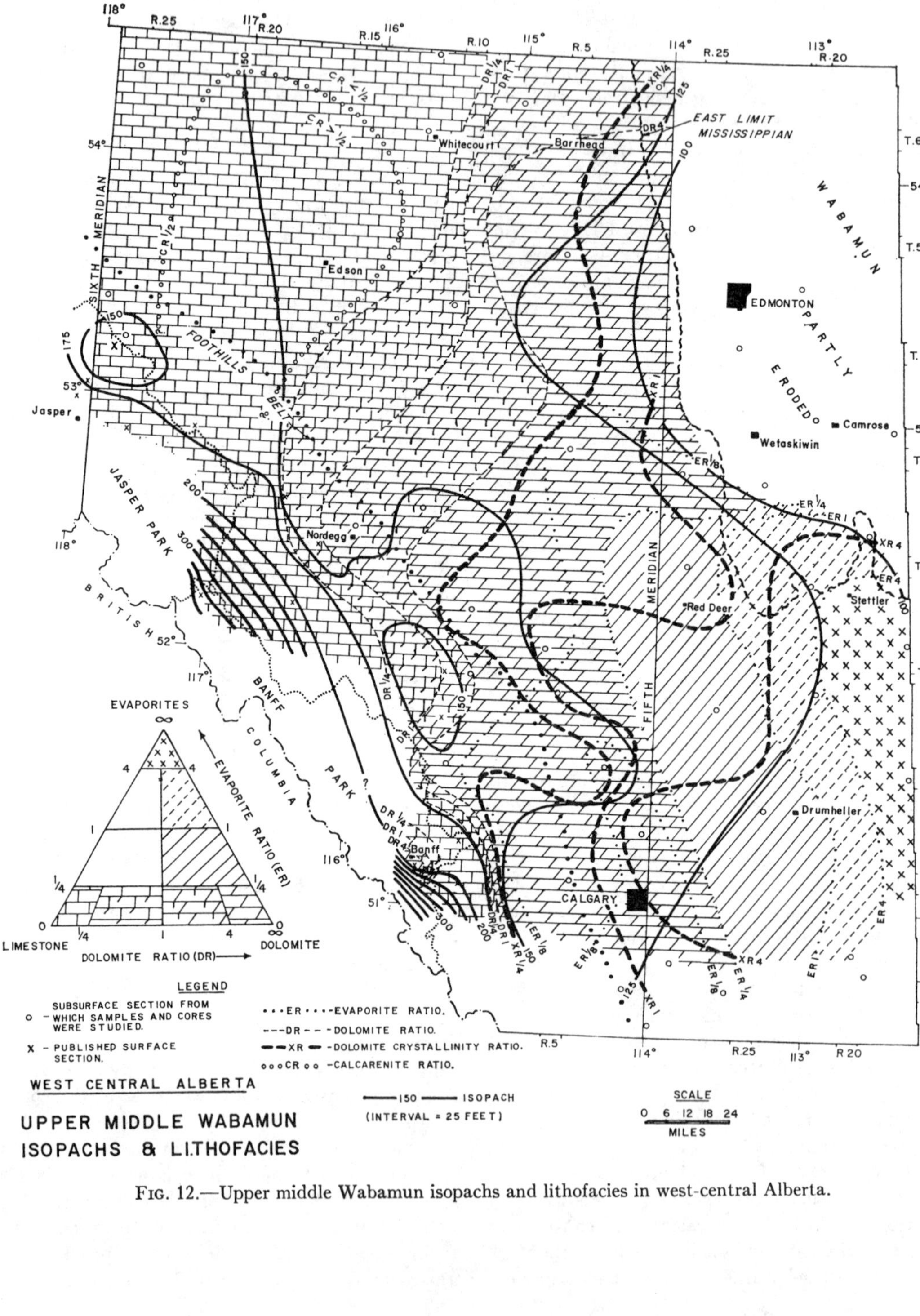

FIG. 12.—Upper middle Wabamun isopachs and lithofacies in west-central Alberta.

Crossfield-Carstairs area. The extremely fine-textured dolomites exhibit more widespread distribution west of the Stettler evaporite province in the upper middle Wabamun than in lower units. In addition, the irregularity of the dolomite crystallinity contours in the southern area suggests that incipient evaporite conditions developed in localized "pans" characteristic of shelf areas with minor topographic relief. Undoubtedly, the differentiation of depositional environments and topography in the underlying unit, including possible reef masses and shoal and intershoal areas on the Wabamun bank, was expressed during subsequent Wabamun deposition. Evaporitic and associated lithographic or microsucrosic dolomites would be expected to accumulate in topographic depressions between shoals.

In the highly evaporitic area near Stettler, the upper middle Wabamun commonly contains less anhydrite and more interbedded fine-textured dolomites than the beds directly underneath (see Gulf's Porter 16 well, Fig. 2). Although the evaporites are more widespread than in the lower middle Wabamun, they do not represent as intensely restricted conditions in the Stettler area, suggesting that the topographic influence of the shoals was reduced by deposition in the intershoal areas. The evaporites of the upper middle Wabamun, extending west to near the present western limit of complete dolomitization, are underlain and overlain by secondary dolomites (e.g., in California Standard's Lacombe Province 65–10, Fig. 2) that are also of widespread distribution and suggest the influence of a cycle of regional restriction, preceded and followed by relatively free circulation. This unit locally contains nodular chert (in Imperial-Canadian Superior Buck Creek, 14–29, Fig. 2), silt or arenaceous material, and breccias of solution origin.

Limestones.—The non-dolomitized limestones are distributed similarly to those in the underlying unit, that is, in the northwestern part of the map area and along the Rocky Mountains, except in the vicinity of Banff (Sulphur Mountain). The transition from dolomite to limestone in the upper middle Wabamun as compared with underlying units is less abrupt in the Nordegg area.

A local concentration of calcarenites occurs in the general Edson area, although in general, the relatively low proportion of calcarenites in the limestone province at the northwest suggests that these were deposits of the bank interior at some distance from the bank margin. The apparent eastern edge of the belt with higher calcarenite content coincides with the dolomite ratio contour of $\frac{1}{4}$. This suggests that selective dolomitization of calcarenites may be responsible for the apparent eastward decrease in relative amount of calcarenites as compared with calcilutites.

UPPER WABAMUN

This uppermost unit includes the Big Valley formation of the plains and the upper part or the whole of the Costigan member of the Palliser formation in the mountains. Westward, the unit increases in thickness from 50 to 75 feet in the plains area and attains a thickness of more than 150 feet in Sunwapta Pass and Mount Assiniboine areas (Fig. 13).

The general lithofacies pattern exhibits considerable departure from that of underlying Wabamun units. The former Stettler evaporite province of lower units is not evident and the same area is characterized by limestones and shales. However, a smaller evaporitic province reappears in the upper Wabamun west of Calgary.[7]

Dolomites.—A belt of dominant dolomitization surrounds the evaporitic area and extends northward along a north to north-northwest trend west of the fifth meridian, terminating north of Pembina oil field. In the Nordegg to Banff area, the western limit of dominantly dolomitized limestones occurs approximately along the Front Ranges. At the west, this dolomitized belt is bounded by a relatively narrow band of significant but not dominant dolomitization (dolomite ratio of $\frac{1}{4}$ to 1); west of the latter and north of the Clearwater River a huge limestone area with only negligible dolomitization is present. South of the Clearwater River, significant dolomite content occurs throughout the mountains in the Banff area, but is most conspicuous on Sulphur Mountain near Banff.

[7] Penner (1958 , pp. 268–69) believes that the absence of a normal Big Valley limestone and shale section at the top of the Wabamun group in the Calgary area suggests tectonic movements which caused either erosion or non-deposition of the Big Valley prior to Mississippian deposition. Accordingly, these uppermost Wabamun evaporites in the Calgary area would be correlative with the upper middle unit evaporites of the Stettler-Drumheller area and thus would not be a facies of the Big Valley formation as tentatively assumed in the present study. Further data are required to establish the correct interpretation.

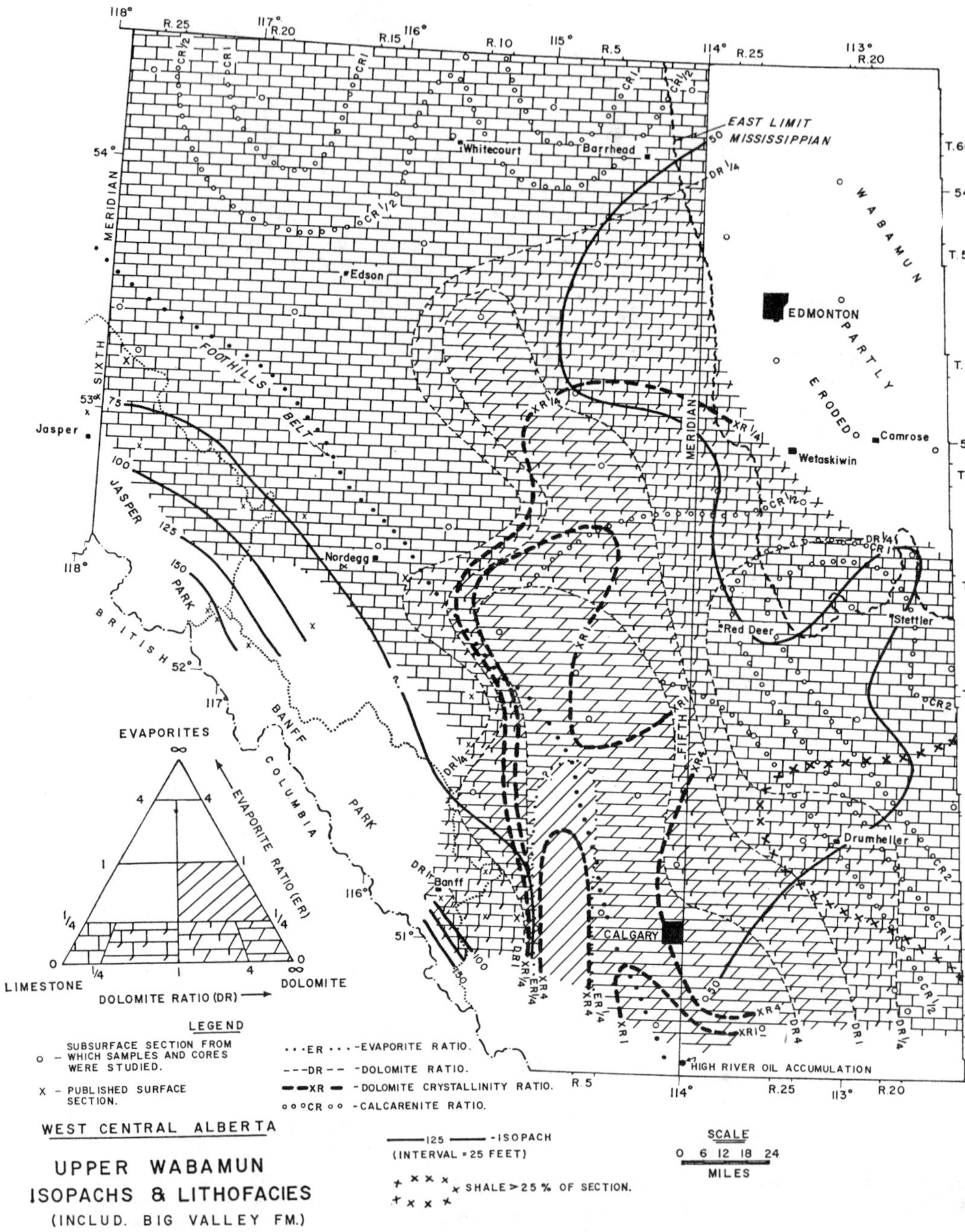

FIG. 13.—Upper Wabamun isopachs and lithofacies in west-central Alberta.

The dolomite crystallinity ratio (proportion of extremely fine-textured to coarser dolomites) contours suggest that the highly dolomitized facies represents the maximum possible limits of the final evaporitic deposition in the Devonian. The slightly higher salinity of marine waters in this area would make the calcium carbonate deposits most readily susceptible to penecontemporaneous dolomitization.

Limestones.—In the non-dolomitized areas, the ratio of calcarenites to calcilutites is highest in the Stettler area where fossiliferous or bioclastic limestones constitute more than 67 per cent of the limestone section. This calcarenite facies appears to adjoin the eastern side of the dolomitized belt. However, the relative decrease in calcarenite content west toward the dolomitized area may be somewhat misleading, for it is possible that originally calcarenites extended westward into this belt. Because these calcarenites were more permeable as compared with the calcilutites, it is suspected that many of the secondary dolomites actually represent calcarenite beds. A conspicuous increase in proportion of calcarenites also occurs in the extreme northern part of the area, north of Edson and west of Barrhead. Calcarenites are concentrated in two lobate areas, west and east of Whitecourt, that extend north of the map area. In the central parts of the lobes, the calcarenite ratio exceeds a value of one, that is, calcarenites, dominantly bioclastic, form more than 50 per cent of the limestones.

Argillaceous rocks occur in increasing proportion southeast of Stettler, indicating that quiet-water clay deposition alternated with deposition of more agitated-water sparsely bioclastic and fossiliferous limestones. Gray and green shales and gray argillaceous fossiliferous limestones containing abundant brachiopods and crinoid fragments are characteristic sediments. However, farther eastward in Saskatchewan, the upper Wabamun, as well as underlying beds, change facies to redbeds, anhydrites, and red and green argillaceous, silty, and sandy dolomite.

In summary, final Wabamun sedimentation was characterized by a return to more widespread conditions of relatively free circulation where prolific invertebrate life was fostered in all but the southern part of the area where locally slight evaporitic conditions may have persisted essentially to the end of Devonian deposition.

Historical Interpretation

At the beginning of Wabamun deposition, the tectonic framework consisted of a broad stable shelf in central and south-central Alberta, bounded on the southwest by a mild tectonic hingeline along the present Rocky Mountain Front Ranges. West of this line, the rate of subsidence and sedimentation increased up to two-fold. This broad pattern represents mainly a continuation of that of preceding deposition of the Alexo-Graminia-Calmar argillaceous, silty, arenaceous and carbonate rocks and anhydrites (Sutterlin, 1958).

The presence of pelletoid or pseudo-oölitic limestone in the lower and middle Wabamun and Palliser, and ripple marks and mud cracks in the Costigan member of the Palliser indicate shallow-water deposition throughout. During basal Wabamun sedimentation, pelletoid calcarenites and calcilutites accumulated on a widespread bank in the western part of the area, whereas evaporites (anhydrite, primary(?) dolomite, and locally halite) were precipitated on the east. Intense evaporitic precipitation took place in the Stettler-Drumheller area at the east, and in the present foothills area from west of Calgary northward to the Clearwater River. Surrounding the evaporite areas, the calcarenites and calcilutites were subject to probable penecontemporaneous dolomitization because of increased salinity above the depositional interface. The calcareous muds were diagenetically dolomitized to very fine-textured microsucrosic dolomites, whereas the calcarenites changed to very fine to fine crystalline or sucrosic rock that has preserved some of the calcarenite texture. In the central part of the bank at the northwest, calcilutites were deposited in greatest abundance and dolomitization was relatively minor. Toward the bank margins, the proportion of calcarenites and dolomitization increased, and locally, vuggy biostromal reefs, formed near the eastern margin of the bank. As interpreted by Sutterlin (1958) the broad limestone bank, whose eastern margin was dolomitized, acted as a sill area which was responsible for restricting the marine waters to the southeast. Presumably the restriction was sufficient to promote evaporite precipitation and penecontemporaneous dolomitization. Sutterlin also inferred that there was rapid accumulation of carbonate rocks on the bank as compared with relatively slow evaporite sedimentation beyond the bank at the southeast.

It is interpreted that after deposition of the basal Wabamun there was a slight increase in subsidence, and an approach to normal marine conditions was attained over much of the map area. However, in the Stettler-Drumheller area evaporites continued to precipitate. West from the Stettler-Drumheller area toward Nordegg and Barrhead, a decrease in relative restriction is interpreted, although the salinity was sufficiently high to promote essentially complete dolomitization of calcareous sediments. On the bank proper, calcilutites and pelletoid limestones were characteristic deposits, whereas bioclastic limestones were of more local distribution, being probably concentrated along bank margins and edges of prominent shoals within the bank (e.g., in the Whitecourt area). Extremely fine-textured non-porous dolomites change facies westward from the evaporite province to more permeable fine sucrosic dolomites (Crossfield dolomite) and locally vuggy reef(?) facies.

During upper middle Wabamun sedimentation, the carbonate sediments transgressed farther east toward the evaporite province and accordingly the completely dolomitized belt was confined to a narrow belt near the present western limit of occurrence of the very fine-textured dolomites and associated anhydrites. By latest Wabamun time, normal circulation was re-established over most of west-central Alberta, except west of Calgary and Olds at the southwest where evaporitic conditions and accompanying dolomitization persisted essentially to the end of Wabamun sedimentation. Fossiliferous and bioclastic limestones accumulated over widespread areas, including the former Stettler-Drumheller evaporite province. Also, argillaceous material was deposited in increasing quantities southeastward in the Stettler-Drumheller area, suggesting a shelf environment of alternating quiet and slightly agitated normal marine waters. Pelletoid limestones and non-fossiliferous calcilutites that were subsequently dolomitized probably accumulated only in local restricted areas associated with the evaporites in the area west of Calgary and Olds.

Oil and Gas Occurrences in Wabamun

Significant Discoveries

In the Edmonton area of central Alberta, numerous indications of oil have been encountered in Wabamun rocks at the pre-Cretaceous unconformity. These oil shows were found since 1947 incidental to exploratory drilling for deeper Upper Devonian Leduc reefs and proved to be marginal or non-commercial. In the southern part of the map area where the Wabamun is overlain by Mississippian beds, oil was encountered in a thin porous unit in the uppermost Wabamun at High River and in Royalite-Anglo Eyremore 1 well (Lsd. 4, Sec. 21, T. 18, R. 18, W. 4 M.), south of Bassano (Fig. 1). The latter well yielded 2,700 feet of oil (41.6° A.P.I.) on drill-stem test, but efforts toward completion in the Wabamun were not successful.

In May, 1951, a commercial gas accumulation in the Crossfield dolomite of the Wabamun group was discovered at Okotoks, south of Calgary, in Shell's Mackid 1 well (Lsd. 1, Sec. 19, T. 21, R. 28, W. 4 M.) which yielded 10 MMcf. of sour gas per day (Fig. 1). Subsequently, large gas reserves were discovered and developed along this trend to the north of Okotoks, resulting in the Olds and East Calgary fields. The discovery well for Olds field, Shell's Olds 1 (Lsd. 16, Sec. 11, T. 32, R. 1, W. 5 M.) was completed May, 1952, with a potential of 2.5 MMcf. per day of wet gas from the Crossfield dolomite. In April, 1954, the East Calgary field discovery well, Socony, C.P.R. Kathryn 3–2 (Lsd. 2, Sec. 3, T. 26, R. 28, W. 4 M.), was completed for 1.1 MMcf. of sour gas per day from the same zone. The size and potential of the East Calgary gas field[8] was indicated by a followup well in 1955, Socony's Calgary 36–10 (Lsd. 10, Sec. 36, T. 24, R. 29, W. 4 M.), located 8 miles southwest of the discovery well, which yielded 23 MMcf. of sour gas per day from the Crossfield dolomite. In 1956, the Beaver Creek gas field was discovered with the drilling of Hudson's Bay, Stanolind, Fina Beaver Creek 1 (Lsd. 1, Sec. 32, T. 56, R. 19, W. 5 M.) which on drill-stem test yielded gas at the rate of 12.8 MMcf. per day from the Wabamun group. This well is 24 miles northwest of Edson.

In 1958, gas was found in the Wabamun group at three exploratory tests in the foothills and western part of the plains. In February, the British American, Triad, Royalite, Sun Lambert

[8] Preliminary data published by Mason and Riddell (1959) show the Wabamun (Crossfield dolomite) gas accumulation at East Calgary field to have a productive area of 35,400 acres and disposable gas reserves of 334 billion cubic feet.

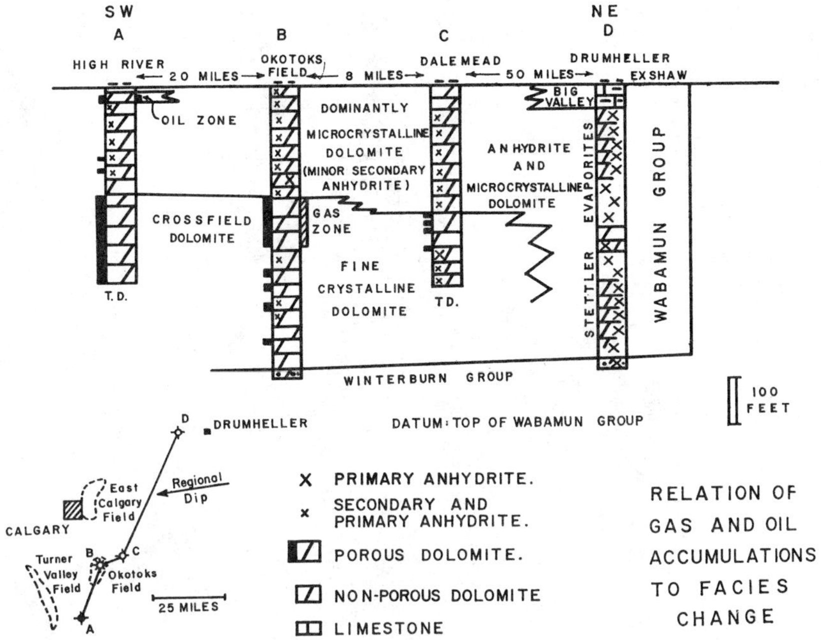

FIG. 14.—Generalized southwest-northeast stratigraphic cross section of Wabamun group, from High River to Drumheller. Relation of gas and oil accumulations to facies change is indicated. Datum, top of Wabamun group.

Creek 6–4 well (Lsd. 6, Sec. 4, T. 51, R. 21, W. 5 M.), 18 miles southwest of Edson, yielded gas on drill-stem test of the Wabamun at the rate of up to 10.6 MMcf. per day, but the well was abandoned later. In September, Pan-American's A-1 Berland River 10–10 (Lsd. 10, Sec. 10, T. 58, R. 25, W. 5 M.), 28 miles west-northwest of Beaver Creek field, tested 1.9 MMcf. of gas per day in the Wabamun (at 270–360 feet below the top of the group). Gas containing hydrogen sulphide is reported to occur in the Crossfield dolomite in Shell's Panther River 1 well (Lsd. 6, Sec. 19, T. 30, R. 10, W. 5 M.) located about 30 miles north-northeast of Banff (Hunt, 1959).

Subsequently, in 1959 and 1960, Wabamun gas discoveries were reported in Shell's Simonette 12–26 (Lsd. 12, Sec. 26, T. 62, W. 5 M.), California Standard's Westerose South 12-2 (Lsd. 12, Sec. 2, T. 45, R. 1, W. 5 M.), and Cities Service *et al.* Spruce Grove 1-12 (Lsd. 1, Sec. 12, T. 52, R. 27, W. 4 M.), the last well yielding only 362 Mcf. of gas per day and 310 feet of gassified oil on drill-stem test of the Wabamun. The Westerose South 12-2 well flowed 5 MMcf of gas per day from a thin net pay zone in the Wabamun, although the underlying Leduc reef reportedly contained 672 feet of gas pay section.

TYPES OF TRAPS

Although certain Wabamun petroleum occurrences, such as those in the foothills belt, are partly or entirely structurally controlled,[9] most oil and gas accumulations in the area of study discovered to date appear to be essentially stratigraphically controlled. Two distinct types of stratigraphic accumulations are recognized, namely, facies and unconformity traps.

Facies traps.—In the Okotoks-East Calgary-Olds area, commercial accumulations of gas occur in fine to medium sucrosic dolomites that exhibit varying amounts of intercrystalline or pinpoint porosity and are overlain by extremely fine-textured non-permeable dolomites and anhydrites (Figs. 4, 5, 14). Northeast and east of these fields, up regional dip, the permeable secondary dolomites in the middle Wabamun (termed Crossfield

[9] Overlying Leduc reefs, such as Big Lake and Golden Spike, draping of the porous Wabamun zone above the reef probably contributes to the formation of the trap for hydrocarbons.

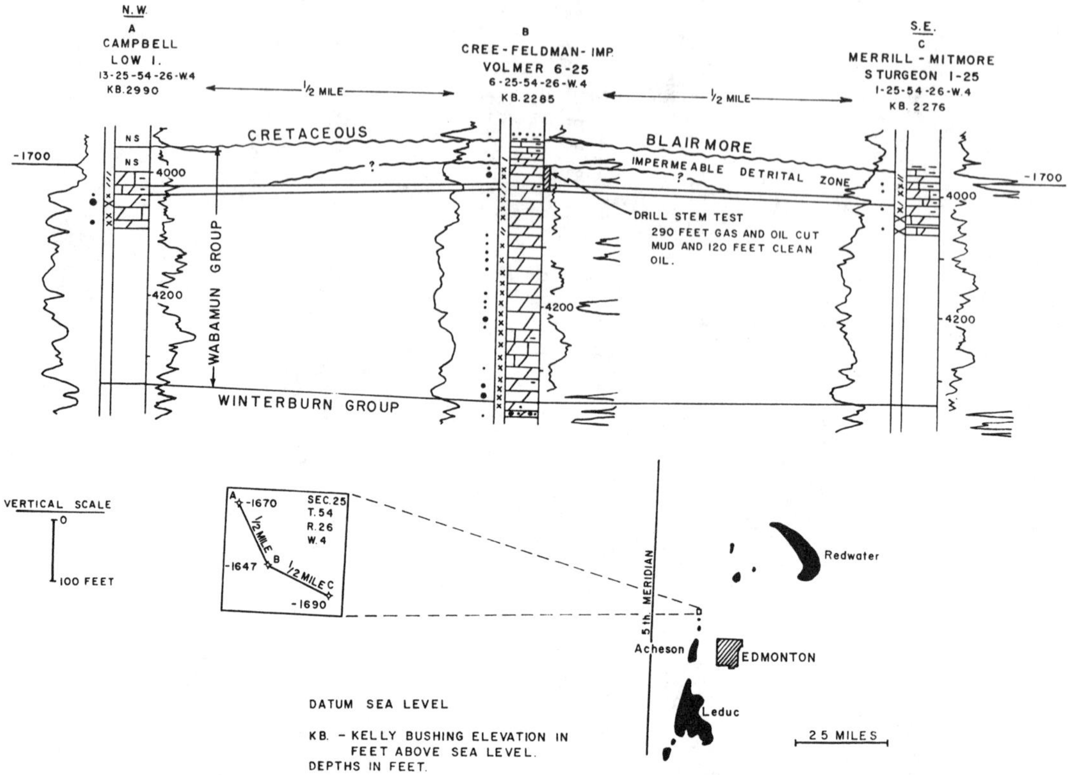

FIG. 15.—Northwest-southeast cross section of Wabamun group near Edmonton showing small oil accumulation in dolomite below the pre-Cretaceous unconformity. Datum, sea-level.

dolomite) are represented by relatively non-permeable extremely fine-textured (lithographic, microsucrosic) dolomites, the latter containing increasing amounts of interbedded anhydrite northeast and east toward the Stettler-Drumheller evaporite province. In a few places within the potentially productive dolomite facies, anhydrite or calcite infilling of pore space occurs and, as a result, non-commercial gas wells or barren sections are locally encountered. The secondary infilling of potentially favorable dolomite facies and the sinuous nature of the line of facies change to non-permeable extremely fine-textured dolomites are responsible for localizing gas accumulations along facies and regional structural strike that trends north-northwest in the Okotoks-Calgary-Olds area.

In the High River area south of Okotoks, a minor accumulation of oil occurs in the uppermost part of the Wabamun group. The oil is trapped similarly within a thin (10-foot) porous secondary dolomite (Figs. 4, 14) bed overlain and underlain by relatively impermeable microsucrosic and lithographic dolomites and anhydrite; the porous dolomite changes facies to relatively non-permeable dolomite toward the northeast (approximately up regional dip).

Unconformity traps.—Several marginal or minor accumulations of oil have been discovered in the general Edmonton area of central Alberta, where permeable Wabamun dolomites have been truncated and covered by Cretaceous beds. The accumulations appear to be localized in topographic hills or ridges (Fig. 15) that were formed by irregular erosion of Wabamun beds where the Mississippian cover had been removed. These topographic features were covered by Lower Cretaceous sands and shales. Where these hills or ridges are made up of permeable dolomites and directly overlain by Cretaceous shales, a potential oil trap occurs. Brecciation and secondary infilling of porous beds at the unconformity commonly cause considerable irregularity in the shape of the reservoir. In addition, the oil accumulation is generally underlain by water without an intervening effective non-permeable zone or floor and difficulty has been encountered in producing oil because of water problems. Specialized completion techniques where acid is used with caution are required to develop these as commercial fields.

REFERENCES

ANDRICHUK, J. M., 1951, "Regional Stratigraphic Analysis of Devonian System in Wyoming, Montana, Southern Saskatchewan, and Alberta," *Bull. Amer. Assoc. Petrol. Geol.*, Vol. 35, No. 11, pp. 2368–2408.

———, AND WONFOR, J. S., 1954, "Late Devonian Geologic History in Stettler Area, Alberta, Canada," *ibid.*, Vol. 38, pp. 2500–36.

BAILLIE, A. D., 1955, "Devonian System of Williston Basin," *ibid.*, Vol. 39, No. 5, pp. 575–629.

BEACH, H. H., 1943, "Moose Mountain and Morley Map-Areas, Alberta," *Geol. Survey Canada Mem. 236.*

BEALES, F. W., 1953, "Dolomitic Mottling in Palliser (Devonian) Limestone, Banff and Jasper National Parks, Alberta," *Bull. Amer. Assoc. Petrol. Geol.*, Vol. 37, pp. 2281–93.

———, 1956, "Conditions of Deposition of Palliser (Devonian) Limestone of Southwestern Alberta," *ibid.*, Vol. 40, No. 5, pp. 848–70.

BELYEA, H. R., 1955, "Cross Sections Through the Devonian System of the Alberta Plains," *Geol. Survey Canada Paper 55-3.*

———, 1957, "Correlation of Devonian Subsurface Formations, Southern Alberta," *ibid.*, *Paper 55-38.*

———, 1958, "Devonian Formations between Nordegg Area and Rimbey-Meadowbrook Reef Chain, Alberta," *Guidebook*, 8th Ann. Field Conf., Alberta Soc. Petrol. Geol., pp. 74–106.

———, AND MCLAREN, D. J., 1956, "Devonian Sediments of Bow Valley and Adjacent Areas," *ibid.*, 6th Ann. Field Conf., pp. 66–91.

CHOQUETTE, A. L., 1955, "The Blue Ridge Member of the Graminia Formation," *Jour. Alberta Soc. Petrol. Geol.*, Vol. 3, No. 5, pp. 70–73.

CRICKMAY, C. H., 1952, "Discrimination of Late Upper Devonian," *Jour. Paleon.*, Vol. 26, No. 4, pp. 585–609.

DEWIT, R., AND MCLAREN, D. J., 1950, "Devonian Sections in the Rocky Mountains between Crowsnest Pass and Jasper, Alberta," *Geol. Survey Canada Paper 50-23.*

EDIE, R. W., 1956, "Origin and Characteristics of Evaporitic Dolomite," *Jour. Alberta Soc. Petrol. Geol.*, Vol. 4, No. 1, pp. 16–23.

FOLK, R. L., 1959, "Practical Petrographic Classification of Limestones," *Bull. Amer. Assoc. Petrol. Geol.*, Vol. 43, No. 1 (January), pp. 1–38.

FOX, F. G., 1951, "Devonian Stratigraphy of Rocky Mountains and Foothills between Crowsnest Pass and Athabasca River, Alberta, Canada," *ibid.*, Vol. 35, pp. 822–43.

GEOLOGICAL STAFF, IMPERIAL OIL LIMITED, WESTERN DIVISION, 1950, "Devonian Nomenclature in Edmonton Area, Alberta, Canada," *ibid.*, Vol. 34, No. 9, pp. 1807–25.

HARKER, P., AND MCLAREN, D. J., 1958, "The Devonian-Mississippian Boundary in the Alberta Rocky Mountains," *Jurassic and Carboniferous of Western Canada*, Amer. Assoc. Petrol. Geol., pp. 244–59.

HERBALY, E. L., 1959, "Geometry of Oil and Gas Accumulations Associated with the Edmonton Reef Chain in Central Alberta," *Jour. Alberta Soc. Petrol. Geol.*, Vol. 7, No. 7, pp. 141–48.

HUNT, C. W., 1959, "Developments at Panther River, 1956 to 1959," *ibid.*, pp. 157–60.

ILLING, L. V., 1954, "Bahaman Calcareous Sands," *Bull. Amer. Assoc. Petrol. Geol.*, Vol. 38, pp. 1–95.

LAW, J., 1955, "Geology of Northwestern Alberta and Adjacent Areas," *ibid.*, Vol. 39, No. 10, pp. 1927–78.

MASON, A. D. M., AND RIDDELL, C., 1959, "East Calgary Gas Field," *Guidebook*, 9th Ann. Field Conf., Alberta Soc. Petrol. Geol., pp. 152–57.

MCLAREN, D. J., 1954, "Upper Devonian Rhynchonellid Zones in the Canadian Rocky Mountains," *Western Canada Sedimentary Basin*, Amer. Assoc. Petrol. Geol., pp. 159–81.

———, 1955, "Devonian Formations in the Alberta Rocky Mountains between Bow and Athabasca Rivers," *Geol. Survey Canada Bull. 35.*

———, 1959, "The Role of Fossils in Defining Rock Units with Examples from the Devonian of Western and Arctic Canada," *Amer. Jour. Sci.*, Vol. 257 (December), pp. 734–51.

PATTERSON, A. M., 1955, "The Devonian of Jasper Park," *Guidebook*, 5th Ann. Field Conf., Alberta Soc. Petrol. Geol., pp. 117–27.

PATTON, W. J. H., 1958, "Mississippian Succession in South Nahanni River Area, Northwest Territories," *Jurassic and Carboniferous of Western Canada*," Amer. Assoc. Petrol. Geol., pp. 309–26.

PENNER, D. G., 1958, "Mississippian Stratigraphy of Southern Alberta Plains," *ibid.*, pp. 260–88.

RAASCH, G. O., 1956, "Late Devonian and/or Mississippian Faunal Succession in Stettler Area, Alberta," *Jour. Alberta Soc. Petrol. Geol.*, Vol. 4, No. 5, pp. 112–18.

SEVERSON, J. L., 1950, "Devonian Stratigraphy, Sunwapta Pass Area, Alberta, Canada," *Bull. Amer. Assoc. Petrol. Geol.*, Vol. 34, No. 9, pp. 1826–49.

SHIMER, H. W., 1926, "Upper Paleozoic Faunas of the Lake Minnewanka Section, near Banff, Alberta," *Geol. Survey Canada Bull. 42*, pp. 1–89.

SUTTERLIN, P. G., 1958, "Uppermost Devonian (Post-Woodbend) Studies in Southern Alberta," *Ph.D. Dissertation*, Northwestern University. 108 pp.

TAYLOR, P. W., 1957, "Revision of Devonian Nomenclature in the Rocky Mountains," *Jour. Alberta Soc. Petrol. Geol.*, Vol. 5, No. 8, pp. 183–95.

THOMAS, G. E., AND GLAISTER, R. P., 1959, "Facies and Porosity Relationships in the Mississippian Elkton Carbonate Cycle of Southwestern Alberta," *Guidebook*, 9th Ann. Field Conf., Alberta Soc. Petrol. Geol., pp. 113–21.

USHER, J. L., 1959, "The Geology of the Western Front Ranges South of Bow River, Alberta," *ibid.*, pp. 23–35.

WARREN, P. S., 1927, "Banff Area, Alberta," *Geol. Survey Canada Mem. 153.* 94 pp.

WONFOR, J. S., AND ANDRICHUK, J. M., 1956, "The Wabamun Group in the Stettler Area, Alberta," *Jour. Alberta Soc. Petrol. Geol.*, Vol. 4, No. 5, pp. 99–111.

WORKMAN, L. E., 1953, "Winterburn Silt Concentration," *Newsletter, ibid.*, Vol. 1.

———, 1955, "Northern Limit of Potlatch Anhydrite," *Jour.*, *ibid.*, Vol. 3, No. 1, p. 10.

Reprinted from:
BULLETIN OF THE AMERICAN ASSOCIATION OF PETROLEUM GEOLOGISTS
VOL. 51, NO. 10 (OCTOBER, 1967), PP. 2033–2043, 22 FIGS.

USE OF SEDIMENTARY STRUCTURES IN DETERMINATION OF DEPOSITIONAL ENVIRONMENTS, MESAVERDE FORMATION, WILLIAMS FORK MOUNTAINS, COLORADO[1]

CHARLES D. MASTERS[2]
Denver, Colorado

ABSTRACT

Rock-stratigraphic units composing the Mesaverde Formation of northwestern Colorado were deposited in offbeach, beach, lagoon, swamp, and floodplain environments of deposition. The vertical distribution of these rock-stratigraphic units reflects the lateral distribution of environments at the time of deposition. In a similar manner, the vertical distribution of sedimentary structures within the rock-stratigraphic units reflects the lateral distribution of processes at the time of deposition. Therefore, an understanding of modern processes and the distribution of their resultant sedimentary structures will permit prediction, through the principle of uniformitarianism, of the vertical distribution of stratification within the depositional products of ancient environments.

It is possible to subdivide the products of the various nearshore marine and nonmarine deposits into a series of subenvironments based on types of stratification. In a regressive suite of rocks, beginning with the offbeach environment, the rocks are characterized by parallel laminae and low-angle cross-laminae reflecting deposition from suspension and from low-energy traction currents. Landward, through the beach environment, varying processes and energy levels result in a sequence of sedimentary structures characteristic of the offshore-beach, submarine-bar, and foreshore-beach subenvironments. The beach zone may be overlain by rocks of the lagoonal, swamp, or floodplain environments. The lagoonal environment comprises sedimentary structures indicative of processes in tidal-delta, tidal-channel, lagoon-pond, and tidal-flat subenvironments. The swamp environment is characterized predominantly by lithology—coal and carbonaceous sediments—rather than by stratification. The floodplain environment comprises stratification characteristic of the channel and interchannel subenvironments.

INTRODUCTION

A fundamental problem to be solved in oil exploration is that of the lateral distribution, beyond the borehole, of the potential reservoir rocks. One way to predict that distribution is to determine the depositional environments of the rock-stratigraphic units and their vertical sequence in the borehole. Figure 1 shows a diagrammatic west-east cross section of the Mesaverde rocks in the Sand Wash basin of northwestern Colorado. This clastic wedge of sediments is bounded on top and bottom by marine shales, and may be subdivided into three asymmetric cycles of deposition composed of rocks deposited in a variety of nearshore-marine and nonmarine environments. The first cycle consists of dominantly mainland-beach rocks overlain by rocks of floodplain origin, whereas the upper two cycles comprise barrier-island deposits overlain by rocks of lagoonal origin. From any particular outcrop section or borehole through this wedge of sediments, certain elements of the lateral distribution of the rock-stratigraphic units are predictable, if the sequence of depositional environments represented in the rock column, the source direction of the sediments, and their depositional strike are known. For example, the wedges of marine shale and coarser clastics will maintain an approximately equal thickness parallel with depositional strike, but will thicken or thin normal to depositional strike; sandstone units deposited in a fluvial environment should be expected to have only limited lateral continuity, whereas beach-sandstone units may be traced for great distances parallel with depositional strike and may be expected to become either higher or lower in the section if traced normal to depositional strike. The problem discussed in this paper is the determination of depositional environments.

[1] Manuscript received and accepted, May 29, 1967.

[2] Pan American Petroleum Corporation.

The data and concepts presented in this paper compose a part of the writer's Ph.D. dissertation, Yale University, 1965.

The writer gratefully acknowledges the help of John E. Sanders, Hudson Laboratories, Dobbs Ferry, New York, and Richard P. Sheldon, U. S. Geological Survey, Denver, Colorado, in all stages of the dissertation work. Pan American Petroleum Corporation contributed generous financial support during the course of the work.

DESCRIPTION OF PROBLEM AND METHOD OF APPROACH

In recent years, a great deal of attention has focused on the determination of the depositional

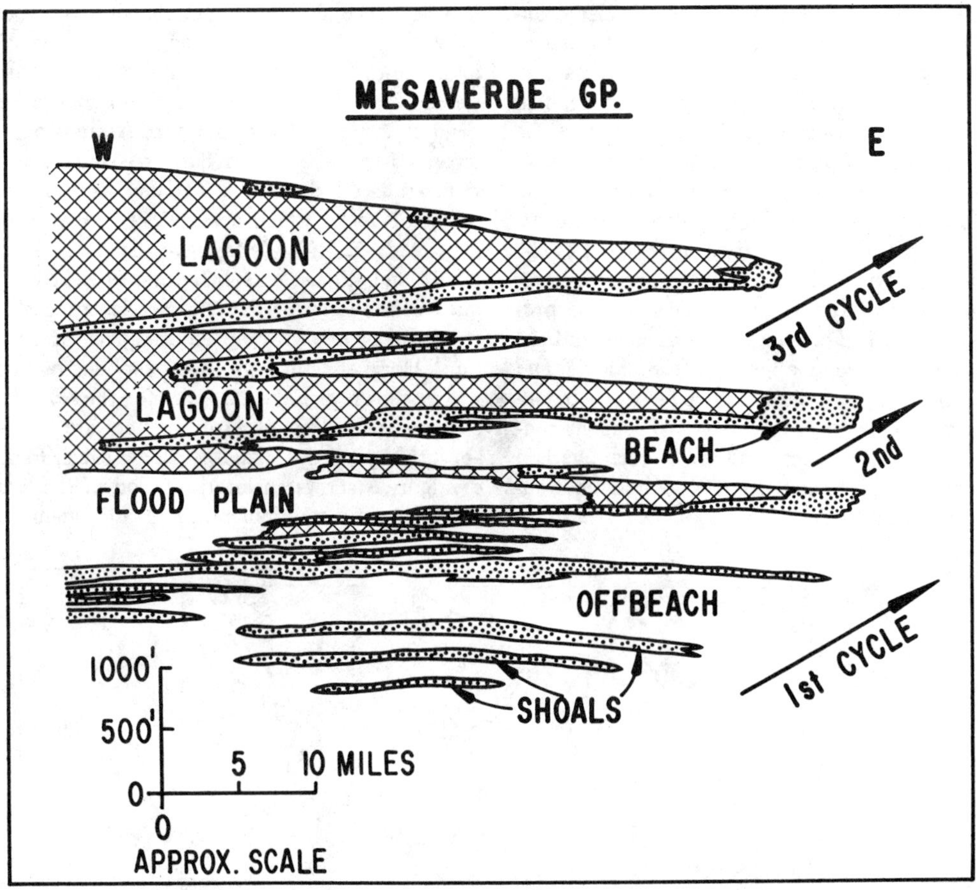

FIG. 1.—Diagrammatic cross section of Mesaverde Formation in Sand Wash basin, northwestern Colorado.

environments of ancient rocks. Substantial progress, in particular, has been made in paleoecology, but many important rock columns contain an inadequate fauna, and other methods of environmental determination are necessary. This writer's thesis is that depositional environments can be recognized by the analysis of vertical sequences of stratification types. Stratification develops in a rock as a function of physical and organic processes acting on grains and laminae within a particular environment. Following Walther's Law (Walther, 1893–94) the vertical sequence of stratification types is predictable by reference to the expectable lateral distribution of these physical and organic processes within the environment at the time of deposition. Thus, as the vertical sequence of environmental products mirrors the lateral distribution of environments at the time of deposition (an expression of Walther's Law), so will the vertical sequence of stratification types mirror the lateral distribution of processes at the time of deposition.

The concepts involved in Walther's Law and the writer's corollary of it are shown in Figure 2. Panel 1 of Figure 2 is a diagrammatic sketch of a modern nearshore suite of depositional environments comprising the offbeach, beach, lagoon, swamp, and floodplain. Each environment may be subdivided further on the basis of distinctive processes occurring within it, which are reflected in the geologic record by recognizable stratification types. For example, the lagoonal environment may include stratification peculiar to the tidal-delta, tidal-channel, tidal-flat, and lagoon-pond subenvironments; the beach environment may include stratification peculiar to the offshore-beach, submarine-bar, and foreshore-beach subenvironments. As a result of progradation, the rock-stratigraphic records of these subenvironments and their contained stratification become arranged in vertical superposition (Fig. 2, Panel 2). (The time line of deposition, which relates the rock-stratigraphic units to the lateral sequence of depositional environments (Fig. 2, Panel 1), crosses the diagram from upper left to lower right.) The transition between offbeach shale and offshore-beach sandstone appears at the base of the prograded sequence. The submarine-bar zone and the foreshore-beach zone appear at the middle and top, respectively, of the beach sequence. Lagoonal rocks, comprising a variety of subenvironments, overlie the beach, and are in turn overlain by rocks of the swamp and floodplain environments of deposition.

A sequence similar to that shown diagrammatically (Fig. 2), crops out along U.S. Highway 40 in northwestern Colorado on the west flank of the Tow Creek anticline. Panel 1 of Figure 3 is a panoramic view of that outcrop. A sandstone, interpreted to have been a barrier island, is on the right. It is overlain by a lagoonal sequence composed of rocks which are believed to have been deposited in tidal-delta, tidal-channel, lagoon-pond, and tidal-flat subenvironments. This sequence is overlain by a thin section of carbonaceous rocks of a swamp environment of deposition. Here the analogy with the diagrammatic sketch shown in Figure 2 ends, for the carbonaceous rocks in this outcrop section are overlain abruptly by marine shale of a new cycle. Panel 2 of Figure 3 shows an interpretation of the lateral sequence of environments and the manner in

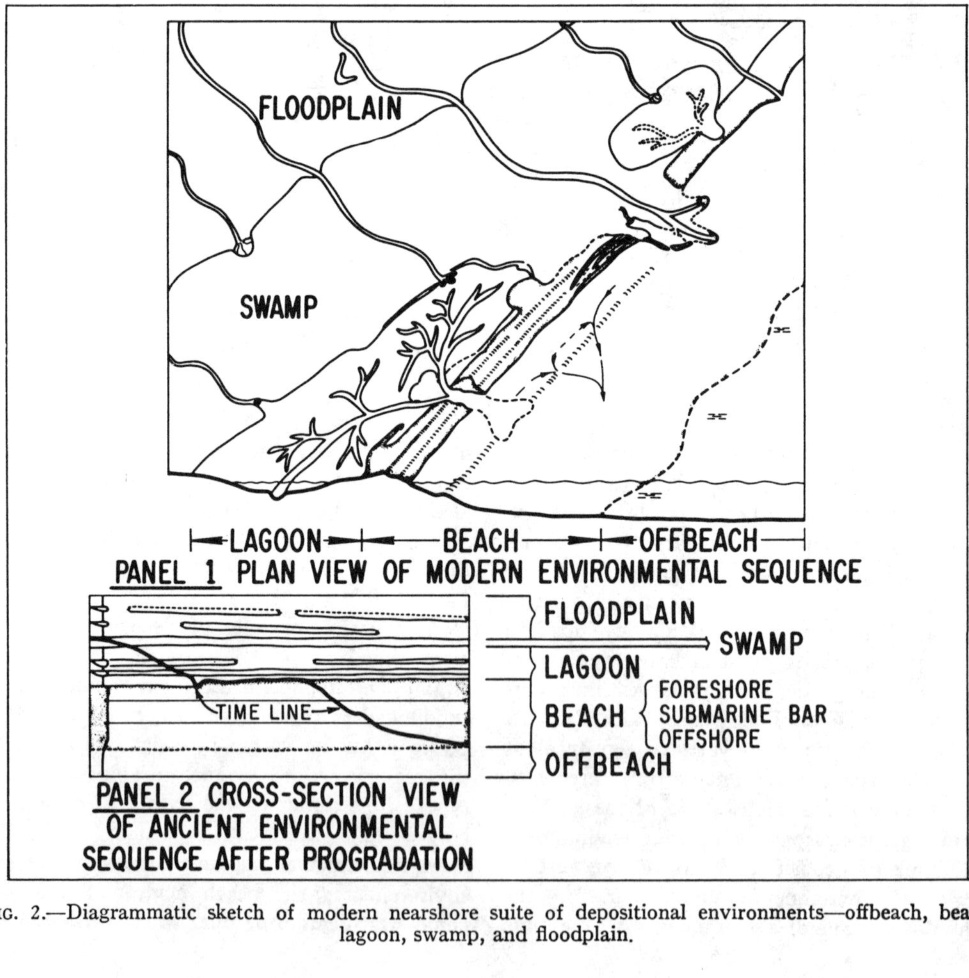

FIG. 2.—Diagrammatic sketch of modern nearshore suite of depositional environments—offbeach, beach, lagoon, swamp, and floodplain.

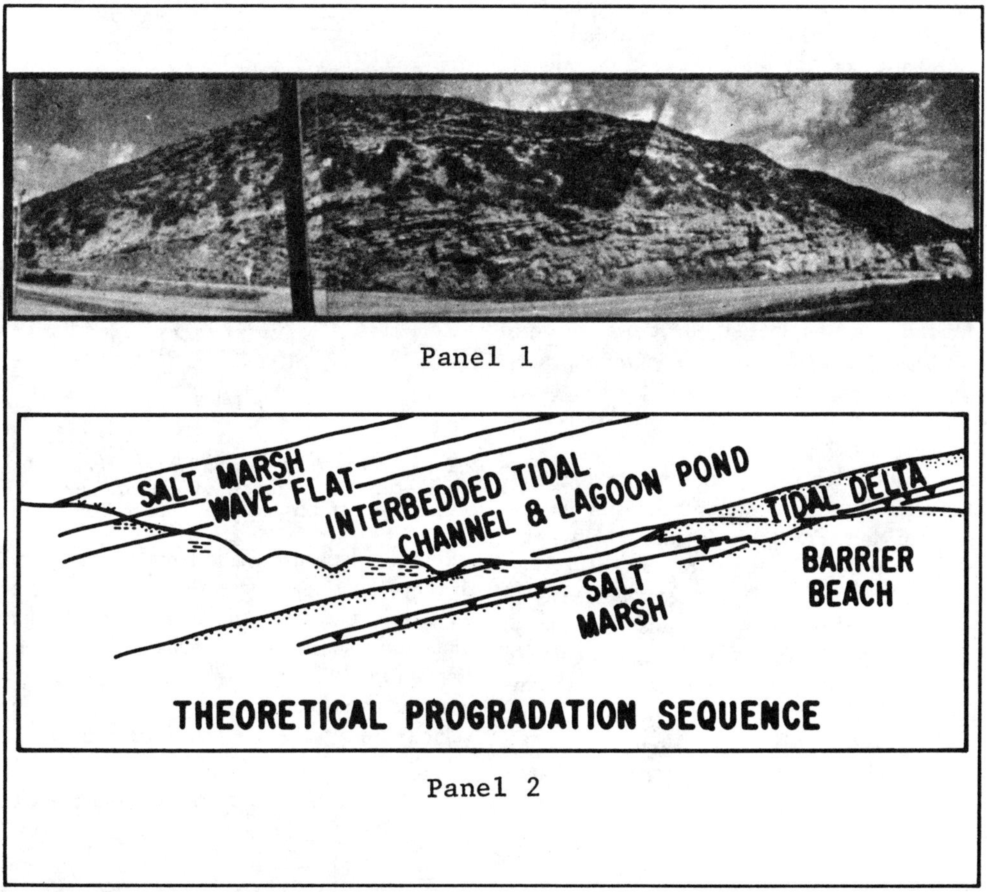

FIG. 3.—(Panel 1) Panorama of outcrop along U. S. Highway 40, northwestern Colorado; (panel 2) Interpretation of sequence of depositional environments, and hypothetical profile representing topography at time of deposition.

which the various environmental products became superimposed. The remainder of this paper is concerned with the use of sedimentary structures for interpreting these depositional environments. Many of the documentary photographs were taken along the section of outcrop shown in the panorama.

DESCRIPTION OF CERTAIN BASIC PROCESSES OF STRATIFICATION

Stratification types which permit the recognition of these various environments and subenvironments develop from both primary and secondary processes. The primary processes include unidirectional and oscillating traction currents, or sediment fallout from suspension. The secondary processes include slumping, differential loading of the sediments, and the action of burrowing organisms.

In recent years, the stratification resulting from unidirectional currents has become well documented (Allen, 1965; Harms *et al.*, 1963; Simons *et al.*, 1961). Commonly, asymmetric ripples develop under these conditions (Fig. 4, Panel 1), with the size of the ripple and the angle of the foreset slope being controlled, among other things, by interrelations between current velocity and water depth (Menard, 1951). The ripple migrates in the direction of the current by erosion on the back slope and deposition on the lee. However, if sufficient sediment fallout takes place concomitant with migration, the upstream ripple migrates up the back slope of the downstream ripple producing the phenomenon of ripple-drift lamination (Sorby, 1859), as shown in longitudinal section (Fig. 4, Panels 2, 3). If sediment fallout is high, the back slope of the downstream ripple is protected from erosion by sedimentation and the entire ripple form is preserved, resulting in false bedding of alternating fine and coarse sediments (Type III ripple-drift lamination of Walker, 1963; see Fig. 4, Panel 4). The cross-section view of ripple-drift lamination is determined by the shape of the migrating ripple form (Fig. 5). Migrating cuspate (curved) ripple forms will generate festoon cross-lamination in cross section, whereas the cross section view of laminae generated by migrating straight-crested ripple forms shows subparallel laminae. The longitudinal view of both types of ripples is the same.

Another important phenomenon of unidirectional current flow is the lateral migration of the current with time. Panel 1 of Figure 6 shows the shape of the bed of the River Klaralven in Sweden (Sundborg, 1956). The greatest current velocity is in the thalweg and diminishes laterally toward each bank. Panel 2 shows diagrammatically the stratification and erosion effects of the varying current velocities. In the deep water on the side of the gentle slope, the current has sufficient velocity to transport relatively coarse grains and deposit them in large-scale ripples. Farther up the slope, decreasing velocity and water depth give rise to sediment transport of relatively finer grain sizes and to deposits of smaller-scale ripple forms. Continuing up the slope to a point near the bank margin, further decrease in current velocity results in the transport and deposition,

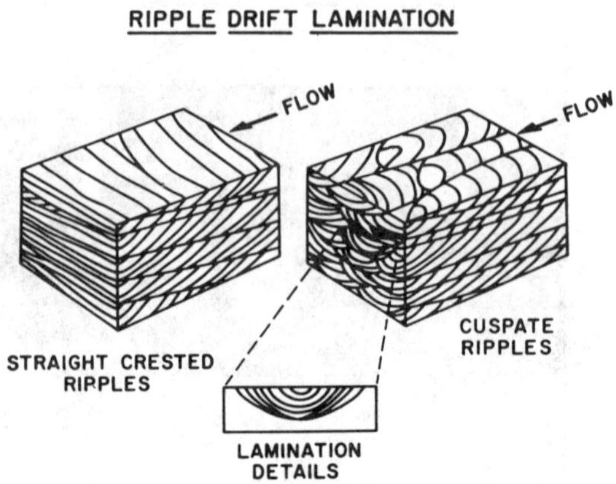

FIG. 5.—Diagrams showing orientation of laminae resulting from migration of cuspate and straight-crested ripple forms (modified from Allen, 1963).

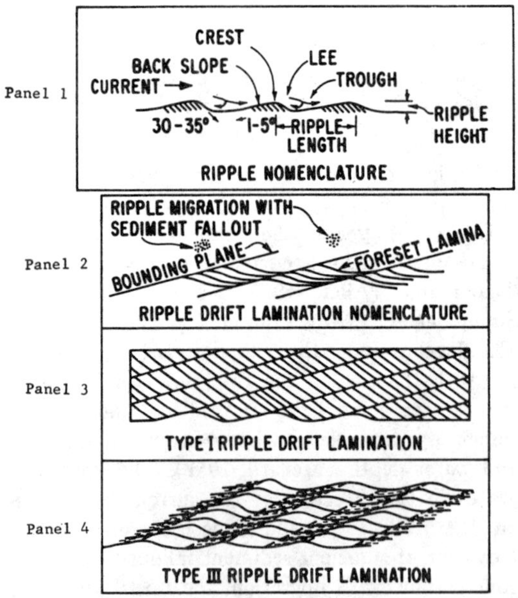

FIG. 4.—Diagrams showing nomenclature of ripples and ripple-drift lamination.

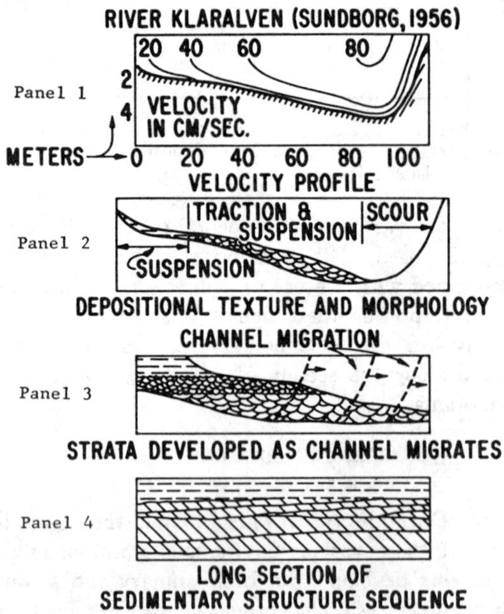

FIG. 6.—Diagrams showing development of fining-up sequence of stratification. See text for explanation of each panel.

FIG. 7.—Thinly-laminated siltstone and shale of offbeach environment. Rippled bed in center of photograph reflects gentle traction-current process.

from suspension fallout, of mud and silt. On the opposite side of the stream, in the thalweg, however, where the current is maximum, the bank is undercut by erosion resulting in the channel and the entire point-bar sequence migrating from left to right. This migration causes the horizontally laminated silt and mud to overlap the small-scale ripples, and the small-scale ripples to overlap the large-scale ripples, producing a "fining-up" sequence of stratification (Allen, 1963).

These processes of sedimentation represent only some of the processes that produce stratification in sedimentary rocks. However, these traction-current processes are fundamental to sedimentation in all nearshore environments and must be understood in order to interpret the sequence of stratification types found in the geologic record.

PRESENTATION AND INTERPRETATION OF DATA

Stratification found in progradational rocks of the Mesaverde Formation represents processes active in the offbeach, beach, lagoon, swamp, and floodplain environments. The vertical sequence of stratification in any one of these environments reflects the lateral sequence of processes, from seaward to landward, active at the time of deposition. Though certain patterns of stratification may characterize a particular environment, nature is too variable for a set sequence of processes to be responsible for clastic deposition in any particular environment. The photographs, discussion, and interpretations that follow present typical stratification types found in various subenvironments. The sequence of photographs reflects, successively, processes active from the far offshore environments to the landward environments.

OFFBEACH

Sedimentary rocks of the offbeach environment (Fig. 7) include thinly laminated siltstone, and shale deposited mainly from suspension. In places, gentle currents have winnowed the bottom sediments and produced a ripple form, thereby producing lenses of cross-laminated sandstone or siltstone.

BEACH

The massive sandstone of the beach-environment deposition is the focal point of any shallow-water, nearshore-marine stratigraphic study. Commonly, the beach environment may be subdivided into several subenvironments representing discrete processes of deposition: the offshore-beach zone, the submarine-bar zone, and the foreshore-beach zone. In places, stratification characteristic of tidal inlets may be found.

Offshore-beach zone.—Rocks deposited in the offshore-beach subenvironment include, at their base, a transition zone from the offbeach environment, and reflect vertically upward the progressively higher energy of the active-wave zone. The basal-transition zone represents the migration of the sand-mud line during progradation (Fig. 8). Here and there, however, the contact at the base of the beach is sharp or even channeled, reflecting storm or other active-current processes during progradation (Fig. 9). Farther landward, in the offshore-beach subenvironment, current en-

FIG. 8.—Gradational contact zone between beach sandstone and offbeach shale. The contact reflects migration, during progradation, of sand-mud line of deposition (see Fig. 2).

ergy from oscillating wave motion may generate a symmetrical ripple form (Fig. 10). The crests of the ripples, because of wave refraction, are most commonly oriented nearly parallel with the shoreline.

Submarine-bar zone.—Under certain wave and bottom conditions (King, 1959; Bascom, 1964), the central part of the offshore-beach zone may change to a breaker zone. Submarine bars (King, 1959) develop at the breaker zone and unidirectional rip currents form between the submarine bar and the foreshore of the beach. Stratification characteristic of the submarine-bar zone includes landward-dipping foreset laminae of submarine-bar deposition, as well as large-scale festoon cross-laminae of rip-current origin, which shows current direction to have been approximately parallel with, or normal to, the depositional strike (Fig. 11).

The origin of stratification that is characteristic of large-scale unidirectional currents in the central part of the beach-sand zone need not always have been the same. Unidirectional currents of tidal-inlet origin, because of their capacity to cut deeply into the beach-sand section, are also responsible for some of the stratification there. For example, Figure 12 shows a "fining-up" sequence of stratification in the central part of a beach sandstone, interpreted to have been deposited by a migrating, unidirectional current. The dip of the foreset laminae suggests a landward-oriented current, and this direction of

FIG. 10.—Oscillation ripples formed in offshore-beach zone.

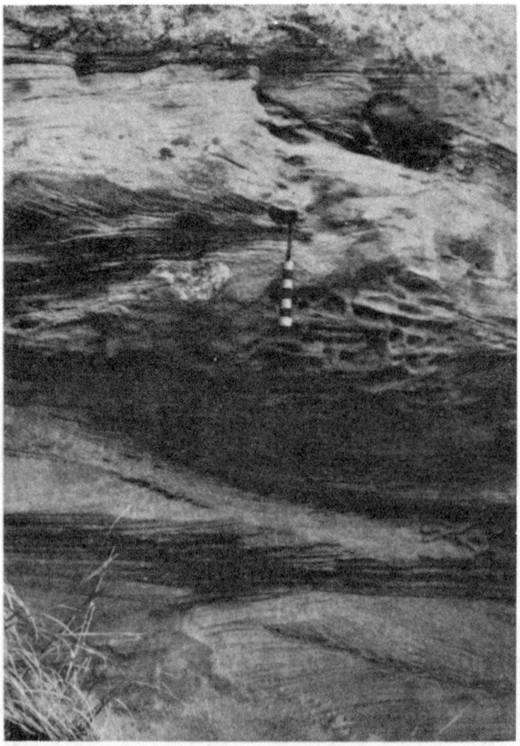

FIG. 11.—Large-scale festoon cross-laminae in central part of beach sandstone. Interpreted to have been caused by unidirectional rip-current flow.

FIG. 9.—Contact zone between beach sandstone and offbeach shale is abrupt and channeled. Channel results from high-energy scouring current action at sand-mud line.

FIG. 12.—Fining-up sequence of stratification in central part of beach sandstone showing current direction normal to shoreline and toward land. Laminae interpreted to represent lateral migration or tidal-inlet current (see Fig. 6).

FIG. 13.—Low-angle, foreshore cross-laminae dipping from left to right in offshore direction.

movement could not be attributable to a rip current. This writer believes, therefore, that a tidal-inlet current origin is a more rational explanation for the observed stratification. This interpretation is supported by data from Hoyt and Henry (1965) who reported large-scale ripples in tidal inlets along the Georgia coast that probably would produce stratification similar to that shown in the lower part of the "fining-up" sequence (Fig. 12).

Foreshore-beach zone.—At the top of a beach sandstone, one should expect to find low-angle, landward- or seaward-dipping cross-laminae, representing backshore or foreshore stratification. In fact, the writer found dominantly foreshore stratification (Fig. 13) at the top of the beach sandstones. This, however, may have been the result of too few observations because the massive sandstone occurs in steep cliffs making it physically difficult to examine the top strata.

LAGOON

The lagoonal environment comprises tidal-delta, tidal-channel, lagoon-pond, and tidal-flat subenvironments.

Tidal delta.—A tidal delta is deposited just inside of the inlet of a lagoon, and in the geologic record should lie directly above the barrier-island sandstone. Description of the stratification in modern tidal deltas is not known to the writer. However, based on stratigraphic position and inferences regarding the processes involved in the stratification, the rock shown in Figure 14 is interpreted as having been deposited in a tidal delta. The stratification in the rock consists of multiple sets of foreset laminae (dipping predominantly toward the left) bounded by bedding planes dipping in the same direction, indicating repeated currents of comparable dimensions depositing the successive sets of foreset laminae. At places in

FIG. 14.—Tidal-delta sandstone. Successive sets of foreset laminae dip predominantly right to left.

FIG. 15.—Tidal-channel scour filled with finely laminated, suspension-laid sediments of lagoon pond.

the outcrop, bipolar orientation of cross-laminae implies a relation to the in-and-out movements of tidal currents.

Tidal channel.—Tidal-channel deposits may be recognized as lenticular sandstone bodies, commonly displaying a "fining-up" sequence of stratification and located within a lagoonal environment. Because the high-velocity tidal currents function within the quiet-water environment of the lagoon pond, it is common to find that the channel is filled with quiet-water, suspension-laid sediments typical of the lagoon pond (Fig. 15).

Lagoon pond.—Sediments of lagoon pond environment may show fine laminations (Fig. 16) or these same fine laminations may be disturbed by

FIG. 16.—Closeup of finely laminated, suspension-laid sediments of lagoon pond.

FIG. 17.—Rocks of lagoon pond mottled by action of burrowing organisms.

FIG. 18.—Contorted laminae in rocks of lagoon pond caused by differential loading from overlying sand units.

FIG. 19.—Low amplitude, straight-crested ripples of tidal-flat environment. Brunton compass is held parallel to ripple-crests; varying crestal orientation of ripples in this environment is characteristic.

burrowing organisms (Fig. 17) or contorted by load flowage caused by differential compaction of the overlying sediments (Fig. 18).

Tidal flat.—Tidal flats are located at the margin of the lagoon pond in shallow water. There, the gentle waves of the lagoon winnow the sediment to clean silt or sand and generate long-crested ripple forms. Commonly, diverse orientations of the ripple crest may be observed, reflecting varying wind patterns or the diverse current di-

Fig. 20.—Cross section of tidal-flat ripples. Note rounded crests reflecting fluctuations in water depth.

rections associated with ebb and flood tides (Fig. 19). A cross section (Fig. 20) shows the ripples to be symmetrical and to possess both sharp and round peaks. Gilbert (1899) suggested that the rounding of ripple crests is due to fluctuations in water depth.

SWAMP

Rocks characteristic of salt marsh and swamp environments develop between the lagoonal and floodplain environments and are characterized by carbonaceous sediments and coal beds. Commonly it is convenient to include these organic deposits with either the associated lagoonal or floodplain rocks.

FLOODPLAIN

The floodplain environment comprises interchannel and channel deposits. The interchannel mudstone commonly is structureless, thus providing essentially no diagnostic structural criteria. The channel sandstone commonly is lenticular and displays internal cutting and filling. Many

Fig. 21.—Cut-and-fill structure typical of floodplain-channel sandstones.

Fig. 22.—Fining-up sequence of cross-laminae typical of floodplain channel processes. See text and Figure 6 for explanation of origin.

channels are lined with fragments of mudstone (Fig. 21). A "fining-up" sequence of stratification is common in the channel environment and can be found on all different scales of development; the large-scale examples are difficult to photograph, but Figure 22 shows a small-scale example with relatively large foreset laminae at the base, grading up into a smaller scale ripple-drift lamination. The strata grade to siltstone and mudstone at the top.

Conclusions

The foregoing series of diagrams and photographs show how a vertical sequence of sedimentary structures can be used to interpret depositional environments. Just as the products of the different depositional environments are arranged vertically in the geologic record, reflecting the lateral distribution of environments at the time of deposition, so also do the sedimentary structures arranged vertically within these rock-stratigraphic units reflect the lateral distribution of processes at the time of deposition. An understanding of the lateral distribution of processes and their resultant sedimentary structures in modern depositional environments, therefore, will aid in interpreting depositional environments in the geologic record. In turn, the recognition of these environments and their vertical distribution will aid in predicting the lateral distribution of reservoir rocks beyond the borehole.

References Cited

Allen, J. R. L., 1963, Asymmetrical ripple marks and the origin of water laid cosets of cross strata: Liverpool and Manchester Geol. Jour., v. 3, pt. 2, p. 187–236.

——— 1965, A review of the origin and characteristics of Recent alluvial sediments: Sedimentology, v. 5, no. 2, p. 89–191.

Bascom, W. H., 1964, Waves and beaches: Garden City, New York, Doubleday and Co., Inc., 267 p.

Gilbert, G. K., 1899, Ripple marks and cross bedding: Geol. Soc. America Bull., v. 10, p. 135–140.

Harms, J. C., D. B. MacKenzie, and D. G. McCubbin, 1963, Stratification in modern sands of the Red River, Louisiana: Jour. Geology, v. 71, no. 5, p. 566–580.

Hoyt, J. H., and V. J. Henry, Jr., 1965, Significance of inlet sedimentation in the recognition of ancient barrier islands, *in* 19th Field Conf., Sedimentation of Late Cretaceous and Tertiary outcrops, Rock Springs uplift: Wyo. Geol. Assoc., p. 191–194.

King, C. A. M., 1959, Beaches and coasts: London, Edward Arnold, 403 p.

Menard, H. W., Jr., 1951, Sediment movement in relation to current velocity: Jour. Sed. Petrology, v. 20, no. 3, p. 148–160.

Simons, D. B., E. V. Richardson, and M. L. Albertson, 1961, Flume studies using medium sand (0.45 mm): U. S. Geol. Survey Water Supply Paper 1498-A, 76 p.

Sorby, H. C., 1859, On the structures produced by the currents present during deposition of stratified rocks: Geologist, v. 2, p. 137–147.

Sundborg, A., 1956, The River Klaralven—a study of fluvial process: Geografiska Ann., v. 38, no. 2, p. 127–316.

Walker, R. G., 1963, Distinctive types of ripple drift cross lamination: Sedimentology, v. 2, no. 3, p. 173–188.

Walther, Johannes, 1893–94, Einleitung in die Geologie als historische Wissenschaft; Beobachtung uber die Bildung der Gesteine und ihrer organischen Einschlusse: Jena, G. Fischer, Bd. 1.

Reprinted from:
BULLETIN OF THE AMERICAN ASSOCIATION OF PETROLEUM GEOLOGISTS
VOL. 51, NO. 10 (OCTOBER, 1967), PP. 2044–2055, 14 FIGS.

RELATIONSHIP OF LOWER CRETACEOUS DEPOSITIONAL ENVIRONMENT TO OIL ACCUMULATION, NORTHEASTERN POWDER RIVER BASIN, WYOMING[1]

ROBERT W. STAPP[2]
Denver, Colorado

ABSTRACT

The Lower Cretaceous rocks produce oil from stratigraphic traps in the northeast Powder River basin. Density of subsurface data allows reconstruction of paleodepositional environments. Through detailed subsurface studies it is possible to relate time-equivalent sandstone bodies to depositional patterns. Structure has little or no bearing on petroleum accumulation in this area.

Identification of paleodepositional environments by means of subsurface studies depends on precise electric-log correlations and a geomorphic interpretation.

The depositional trends of the Lower Cretaceous Fall River sandstone bodies and the so-called "Muddy" sandstone beds are determined readily by selective isopachous mapping. Care must be exercised to recognize unconformities, disconformities, or other breaks in an apparently continuous cycle of deposition. Isopachous maps are constructed by using a lithologic time marker as the upper boundary and either a lithologic time marker or an unconformity as the lower boundary. These continuous cycles of deposition indicate a genetically related depositional environment.

The Fall River contains three sandstone bodies that are mappable as separate units in a complex marginal-marine shoreline environment. These sandstone beds thicken locally and, in places, merge into a single unit. The Fall River does not have channel characteristics in this area.

The "Muddy" contains two sandstone bodies previously named Newcastle and Dynneson. These sandstone units have distinct depositional patterns and the Newcastle shows a pronounced dendritic drainage pattern.

Oil production is from the eastern updip edges of maximum sandstone development in both the Fall River and "Muddy" sandstone units. By detailed subsurface mapping it is possible to relate time-equivalent sandstone bodies to the depositional environments which control oil accumulation.

INTRODUCTION

The fact is recognized that the future of the exploration geologist depends to a large extent on his skill in finding petroleum in stratigraphic traps. The purely structural approach to prospecting is rapidly becoming a thing of the past. Prospecting for stratigraphic oil presents to the exploration geologist the maximum opportunity for a "total" understanding of the geologic section being studied. Viewed from an economic standpoint, the rewards for discovering stratigraphic entrapments like East Texas, Patrick Draw, and Coyote Creek (although differing greatly in size) are certainly worthwhile.

STUDY AREA

The Powder River basin is in northeastern Wyoming and southeastern Montana (Fig. 1). It is bounded on the east by the Black Hills, on the west by the Big Horn Mountains, and on the south by the Laramie and Hartville uplifts.

The area discussed is in the northeast part of the basin where the Lower Cretaceous rocks have regional dip of about 1° toward the west-southwest. The current Permian Minnelusa play makes the northeast part of the Powder River basin the most active exploration area in the Rocky Mountain region today.

The study area covers 20 townships (720 sq mi)—T. 48 N. to T. 52 N., and R. 67 W. to R. 70 W. It covers parts of Campbell, Crook, and Weston Counties, Wyoming.

Within the study area, the Lower Cretaceous rocks produce oil at a profit from a variety of stratigraphic traps. Fields such as Coyote Creek will produce more than 30 million bbl of oil from a wholly stratigraphic trap. Depths to the base of the Cretaceous range from 500 to 8,500 ft.

There have been approximately 1,250 tests drilled through the Cretaceous rocks within the study area (Fig. 2). Location of cross-section lines used to establish correlations are shown. The cross sections are not included in the illustrations because the critical detail is destroyed in the reduction process.

[1] Manuscript received and accepted, May 29, 1967. Modified from paper presented at the 16th annual meeting of the Rocky Mountain Section of the Association, Denver, Colorado, October 25, 1966. Received "Outstanding Paper Commendation" from the Rocky Mountain Section.

[2] Senior exploration geologist, Phillips Petroleum Company. The writer thanks Phillips Petroleum Company for permission to publish.

STUDY TECHNIQUES

Prospecting for stratigraphic traps requires an understanding of the paleodepositional environments of the rocks under consideration.

The Lower Cretaceous rocks discussed are shown on the type log for the area (Fig. 3). They are discussed from the bottom upward because that was the order of deposition, and reconstruction of the paleodepositional environments is more easily demonstrated.

The requirements for reconstruction of these depositional environments from subsurface data are:

1. Adequate density of subsurface control.
2. Precise electric-log correlations.
3. Selective isopachous mapping of genetically related time units of one cycle of sedimentation. The upper boundary of these cycles *must* be a lithologic time marker and the lower boundary may be either a lithologic time marker or an unconformity.
4. Sample analysis.

The interpretation of the paleodepositional environment as shown by isopach and lithologic data is based on a thorough understanding of geomorphology and sedimentation. Care must be exercised in recognizing unconformities, disconformities, or any break in a continuous cycle of deposition.

DEPOSITIONAL ENVIRONMENTS

The basal part of the Lower Cretaceous is represented by the Fuson-Lakota sequence which is highly lenticular and lithologically varied. It is primarily of continental origin and there is only scattered, mostly uneconomic, production from these rocks. Because this paper deals with the relationship of depositional environment to oil accumulation, these rocks are not discussed.

The close of Fuson-Lakota deposition was marked by a widespread transgression that essentially represents a time line. This transgression is the initial incursion of the interior Cretaceous

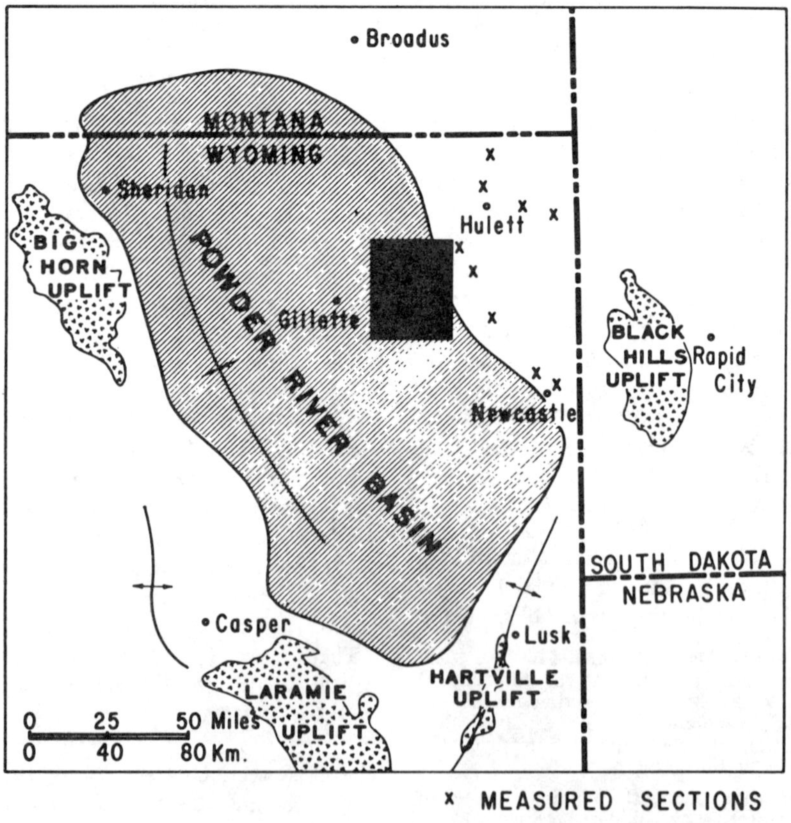

FIG. 1.—Location map, Powder River basin, northeastern Wyoming.

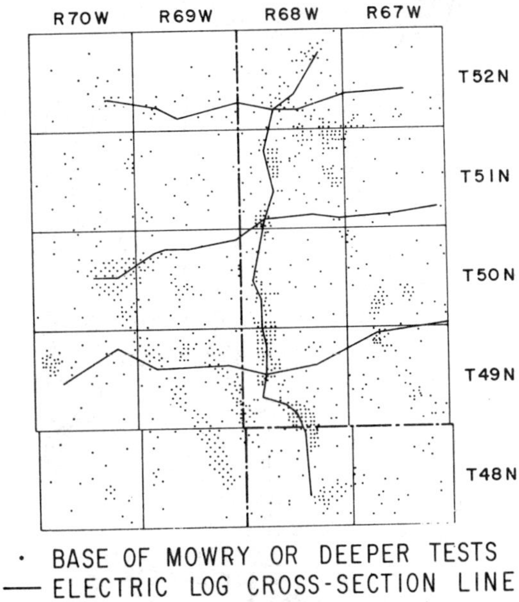

FIG. 2.—Within study area approximately 1,250 tests have been drilled through Cretaceous. Cross-section lines shown are used to establish correlations.

sea. The resulting deposit of very dark gray, very micro-micaceous shale and light-gray siltstone laminations is easily recognized on electric logs. This regional transgression separates the continental deposits of the Fuson-Lakota from the overlying marginal-marine deposits of the Fall River.

The Fall River contains three sandstone beds that are mappable as separate units in a complex marginal-marine shoreline environment.

Overlying the Fall River is the marine Skull Creek Shale which represents a further transgression of the Early Cretaceous sea. Within the lower part of the Skull Creek Shale is a siltstone zone time line called the "Dakota silt" by petroleum geologists.

At the end of Skull Creek deposition the study area emerged, the shoreline regressed, and the Newcastle drainage pattern was etched on the Skull Creek mudflats. The Newcastle dendritic drainage pattern was then inundated by another transgression during which the blanket-type sandstone named "Dynneson" by Wulf (1962, p.

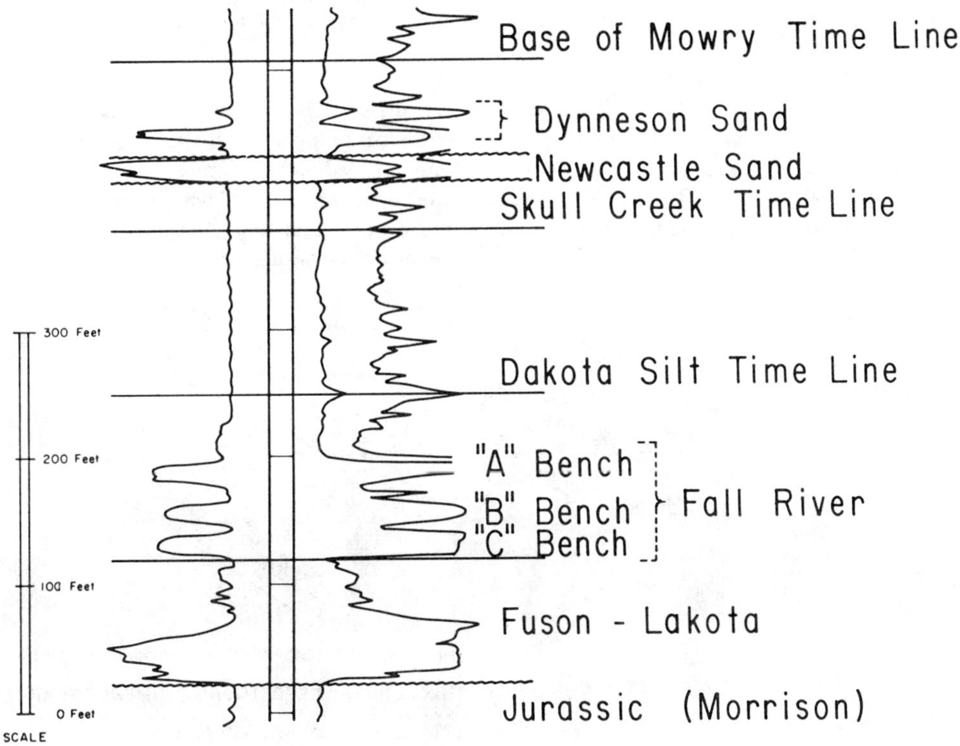

FIG. 3.—Composite electric log of Lower Cretaceous, northeast Powder River basin, showing nomenclature and correlations used.

1397) was deposited. Transgression and marine sedimentation continued throughout deposition of the overlying Mowry Shale.

Structure

A bentonite in the basal part of the Mowry provides an excellent lithologic time marker which is used as datum for a contour map of the present-day structure (Fig. 4).

The Lower Cretaceous rocks dip west-southwest at approximately 100 ft per mile in most of the study area. Eastward the dip rate increases to approximately 1,000 ft per mile on the Little Missouri monocline. Minor structural anomalies are present at Donkey Creek, Timber Creek, Raven Creek, and Robinson Ranch, but generally there is very little interruption of the regional dip.

All Lower Cretaceous oil production is shown on the structural contour map. It is apparent that structural features play an insignificant part in the accumulation of oil.

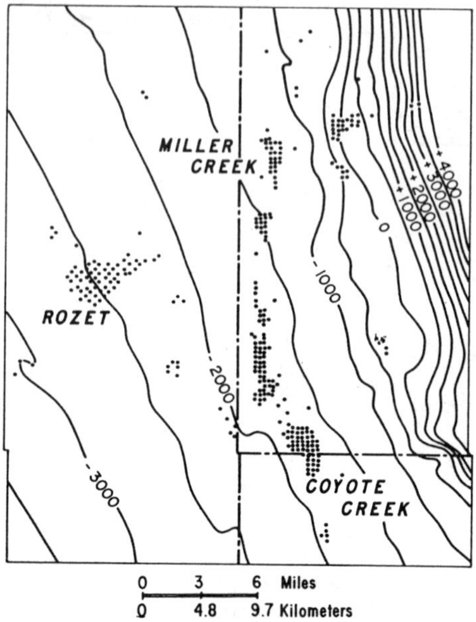

FIG. 4.—Structural contour map; datum is base of Mowry Shale. Monoclinal dip plays insignificant part in oil accumulation. For orientation, three oil fields that produce from Lower Cretaceous rocks are named.

Mapping Methods

Fall River

By applying the requirements for reconstructing paleodepositional environments from subsurface data, a selective isopachous map was made of the section containing the Fall River sandstones. The upper boundary is the "Dakota silt" and the lower boundary is the regional transgression at the base of the Fall River. This isopach interval contains the three Fall River sandstones that are genetically related time units. No unconformities or disconformities break the continuous cycle of sedimentation. This genetically related unit of sedimentary rocks ranges in thickness from 100 to 150 ft, with variations being the result of sandstone development. A channel deposit may be identified by its downcutting into

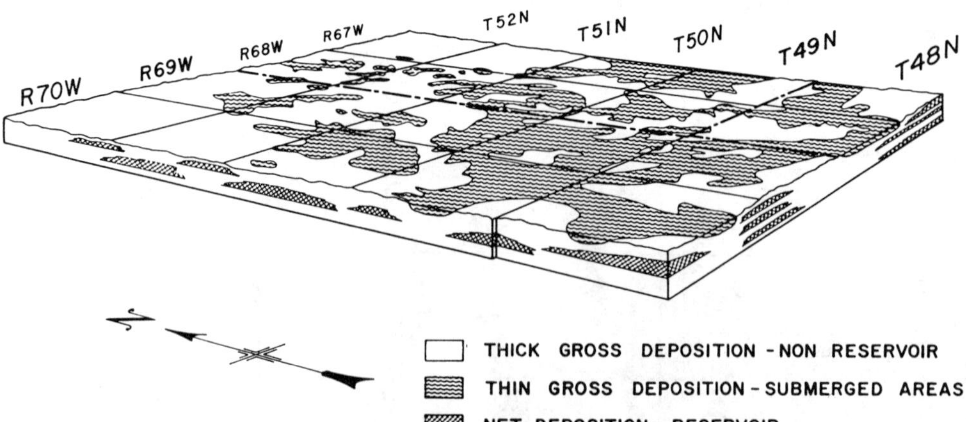

FIG. 5.—Block diagram showing complex marginal-marine shoreline environment at close of Fall River deposition.

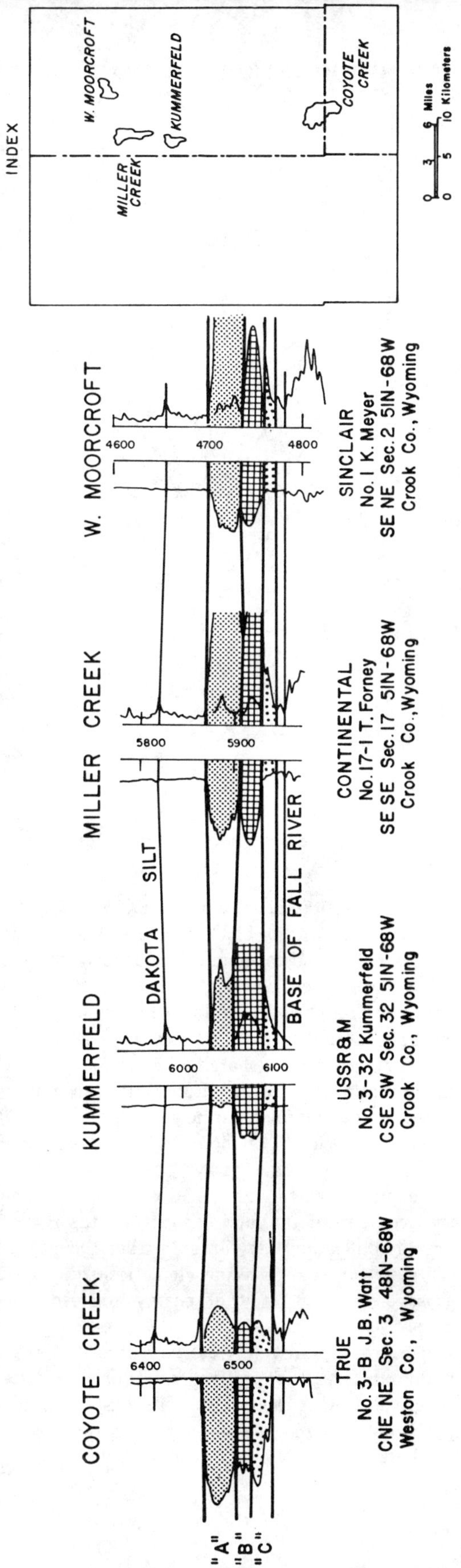

the underlying strata. Detailed electric-log correlations throughout the study area show no thickening of the Fall River Sandstone due to channeling or downcutting. Each of the three Fall River sandstone units demonstrates variations in thickness in an upward direction with the base remaining relatively flat. The three oil-bearing sandstones have been arbitrarily called "A," "B," and "C" sands in descending order for the purpose of this study.

The "A" and "B" sands of the Fall River are developed throughout the area as gross units (Figs. 9 and 10). The lowermost "C" sand is developed as a gross unit primarily in the southern part of the study area (Fig. 8).

It is within the thick gross sandstone developments that the net reservoir sandstone is present. The term "net reservoir sand" describes sandstone that is capable of commercial production based on porosity and permeability values shown on electric logs, cores, drill-stem tests, and sample descriptions. These net reservoir sandstone units produce oil at their eastern updip limits.

The block diagram (Fig. 5) represents the complex marginal-marine shoreline environment that existed at the close of Fall River deposition, and at the beginning of the marine inundation and deposition of the Skull Creek Shale. It is an "aerial" view looking from the southwest to the northeast across the study area. The white areas represent the thick developments of the gross Fall River sandstone bodies. The wave-pattern areas represent the earliest Skull Creek marine inundation over areas of thin Fall River. These thin developments of the Fall River are generally shaly, tight, nonreservoir lithologic types and were topographically low. It is within the thick intervals shown as low-relief, slightly emergent, coastal features that the net reservoir sandstones develop. The net reservoir sandstone units are shown in the cross-hatched pattern on the block diagram and their distribution is discussed individually.

In places the three Fall River sandstone beds merge to form what appears to be a single, thick, porous net sandstone body that locally cannot be separated into distinct units. This occurs in parts of the Coyote Creek field. In other places only the upper "A" and middle "B" sands merge to

FIG. 6.—Fall River correlations in productive areas.

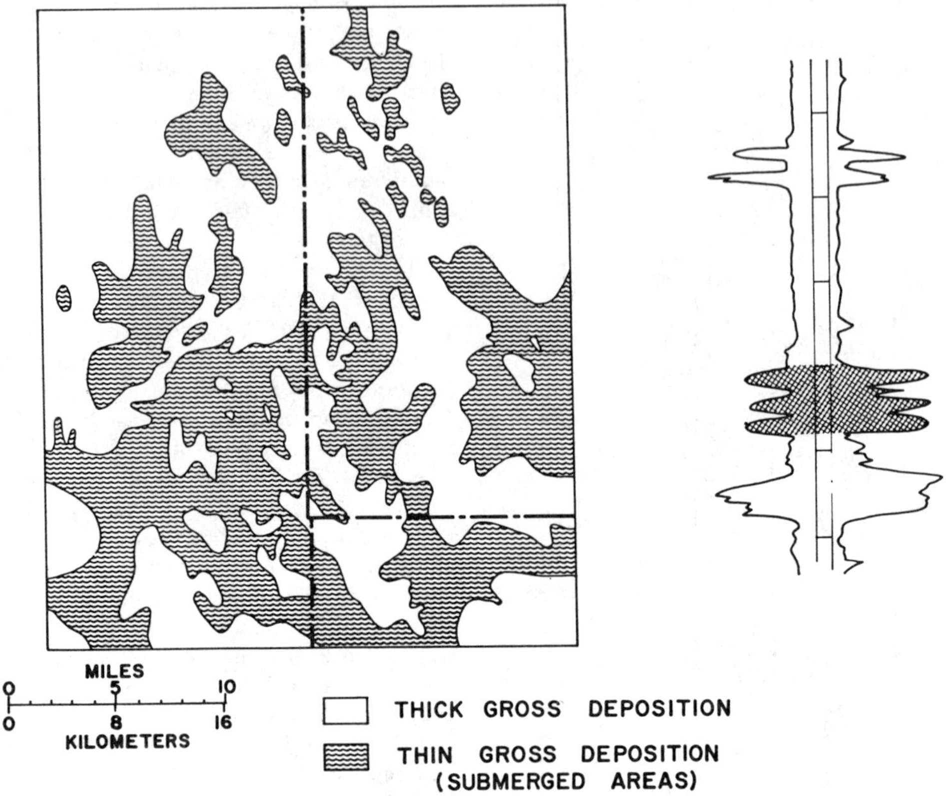

FIG. 7.—Planar map view of complex marginal-marine shoreline environment at close of Fall River deposition.

form a thick, porous, net sandstone. Examples of this are present at Donkey Creek, Kummerfeld, Miller Creek, and West Moorcroft. In other places the three sandstones thicken individually and develop net reservoir porosity with no apparent vertical "stacking." Examples of Fall River correlations in some of the productive areas are shown in Figure 6.

The lithologic types are essentially identical in the different areas where any or all of the three Fall River sandstone units develop porosity and permeability values of reservoir quality. The sandstones consist of light gray to white, very fine- to medium-size quartz grains which are well sorted, sub-angular, and cemented with secondary silica. There are some scattered black shale and carbonaceous inclusions. Complete loss of porosity and permeability can occur abruptly in a lateral direction; the thick, clean, reservoir sandstone passing into a thin, shaly siltstone. The interval change from thick to thin is at the top of the sandstone, the base remaining essentially flat.

Figure 7 represents the more conventional planar map view of the block diagram. The porous net reservoir sandstones are developed within the white areas which represent thick Fall River deposition. The paleodepositional environment during Fall River deposition in this area was a complex marginal-marine shoreline. Each of the three sandstones has its own environmental range associated with a marginal-marine shoreline.

Factors such as offshore currents, longshore currents, tides, storms, wave base, proximity to shore, and sediment supply had varying effects on the geometry of the sand deposits. At the risk of over-simplification one can state that the porous net reservoir sandstones develop within the thick gross units as a result of energy and sediment supply.

Areas of thickest net reservoir porosity values within the three Fall River sandstones coincide remarkably with the areas of thickest gross isopachs between the "Dakota silt" and the base of the Fall River.

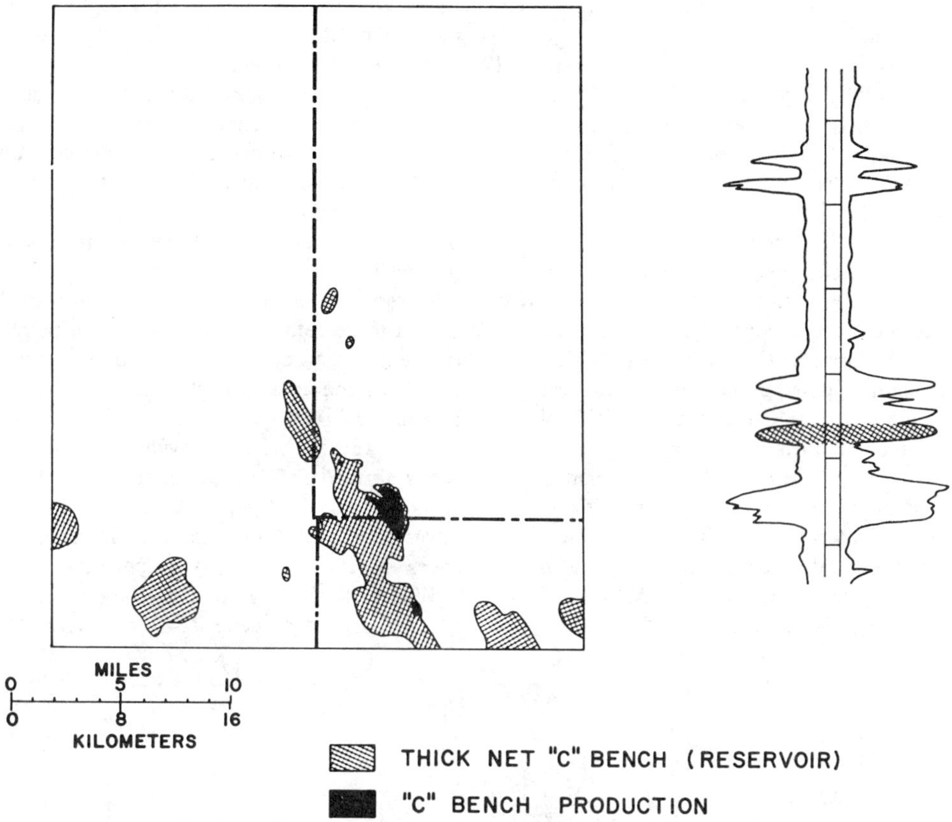

FIG. 8.—Distribution of thick net (reservoir) Fall River "C" bench sandstone and productive areas.

Figures 8, 9, and 10 show the thick net-reservoir porosity values and the production from them in the individual Fall River sandstone units. The thick net-reservoir porosity values shown on the Fall River illustrations are outlined generally by a minimum isopach of 20 ft. There is thick gross development of the Fall River where Campbell, Crook, and Weston Counties meet (Fig. 7). It is within this thick interval that the lowermost "C" bench sand develops the net-reservoir porosity value from which there is production at Coyote Creek.

Fall River "C" bench.—The "C" bench sand of the Fall River is developed in the southern part of the study area and the thick net-reservoir porosity values are shown in the cross-hatched pattern (Fig. 8). Production at Coyote Creek and South Coyote Creek is at the eastern, updip edge of a large north-northwest-trending porous sandstone deposit. One well produces from the "C" sand in a stratigraphic trap on the south plunge of the Donkey Creek structural anomaly.

In the north half of the study area the time-equivalent of the "C" bench is essentially a marine, silty shale. As seen in Figure 8, the "C" bench sand development in the southern part of the study area represents the nearshore-sand deposits of a mildly regressive shoreline environment in the primarily transgressive Lower Cretaceous sequence.

The first appearance of the Early Cretaceous sea is marked by the regional transgression at the base of the Fall River. Deposition of the Fall River "C" bench just above this shoreline transgression, with no break in sedimentation, means that the "C" bench is of marginal-marine origin. It is the first development of the complex, marginal-marine shoreline deposition of the Fall River.

The sediment supply for the "C" bench sand is from the southeast. Other authors have postulated a deltaic environment for the Fall River extending approximately 30 mi toward the southeast. This could have provided an adequate source of sediment for the development of the near-shore "C" bench.

Much has been published regarding the Fall River rocks by Waagé (1959, p. 1397), Reeside (1957, p. 510), Robinson *et al.* (1964, p. 23), and others; there is agreement that a marginal-marine environment existed in this area during Fall River deposition.

Fall River "B" bench.—The "B" bench developed throughout the study area is the result of continuing regression of the shoreline with abundant sediment supply (Fig. 9). The overall regression is toward the northwest. Thick net porosity values allow production at the eastern updip limits at Miller Creek, Kummerfeld, North Donkey Creek, and Coyote Creek.

Fall River "A" bench.—The "A" bench develops thick net porosity values throughout the study area similar to the "B" bench (Fig. 10). However, the areal extent of net porosity in the "A" bench is not as widespread. With the foreknowledge that the marine Skull Creek Shale is just above the "A" bench, with no break in sedimentation, it is apparent that another marine transgression followed.

Fall River summary.—Identifying specific environments for particular sandstone beds in a complex, marginal-marine shoreline environment is difficult at best. Such identification is primarily an interpretation of the geometry of a sandstone body as it is related to its own cycle of sedimentation. The accuracy of the specific environmental interpretation of the geometry of a sandstone body can be refined only to a certain point. This is the point where subsurface data cease to be definitive and a calculated guess must be made as to the origin of specific sandstone bodies based on the rather broad knowledge that the Fall River formed in a marginal-marine environment. It is obvious that oil accumulation in the Fall River sandstone bodies is at their eastern updip edges of thick net-reservoir porosity values.

Within the study area a complete range of types of sandstone deposits—those typically asso-

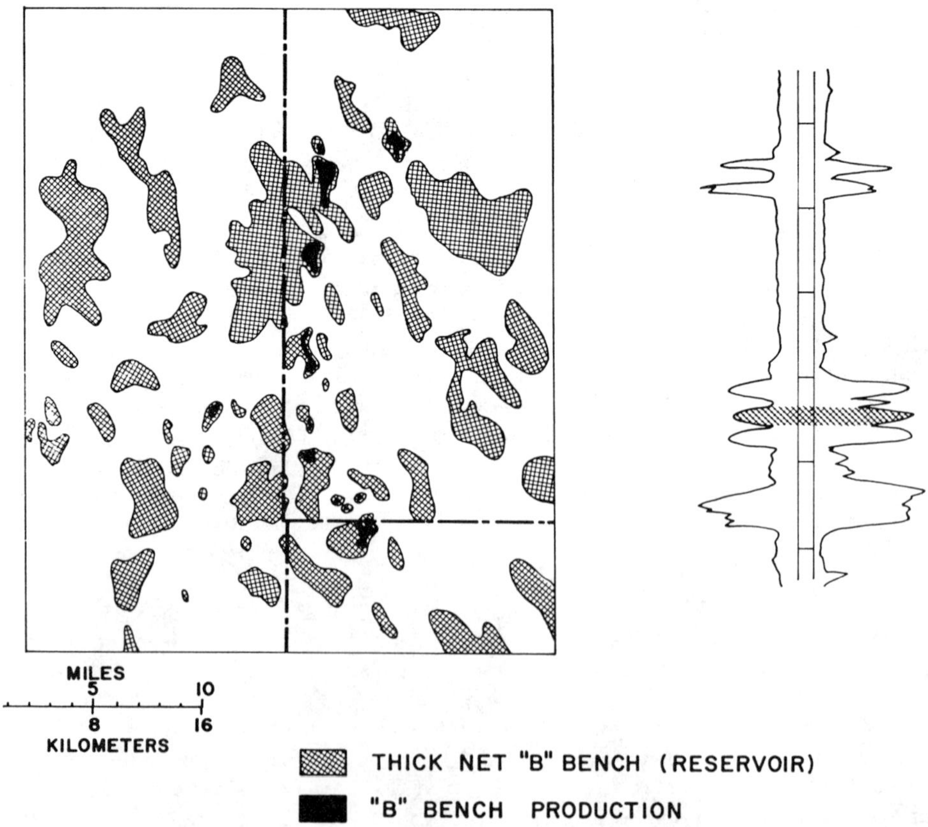

FIG. 9.—Distribution of thick net (reservoir) Fall River "B" bench sandstone and productive areas.

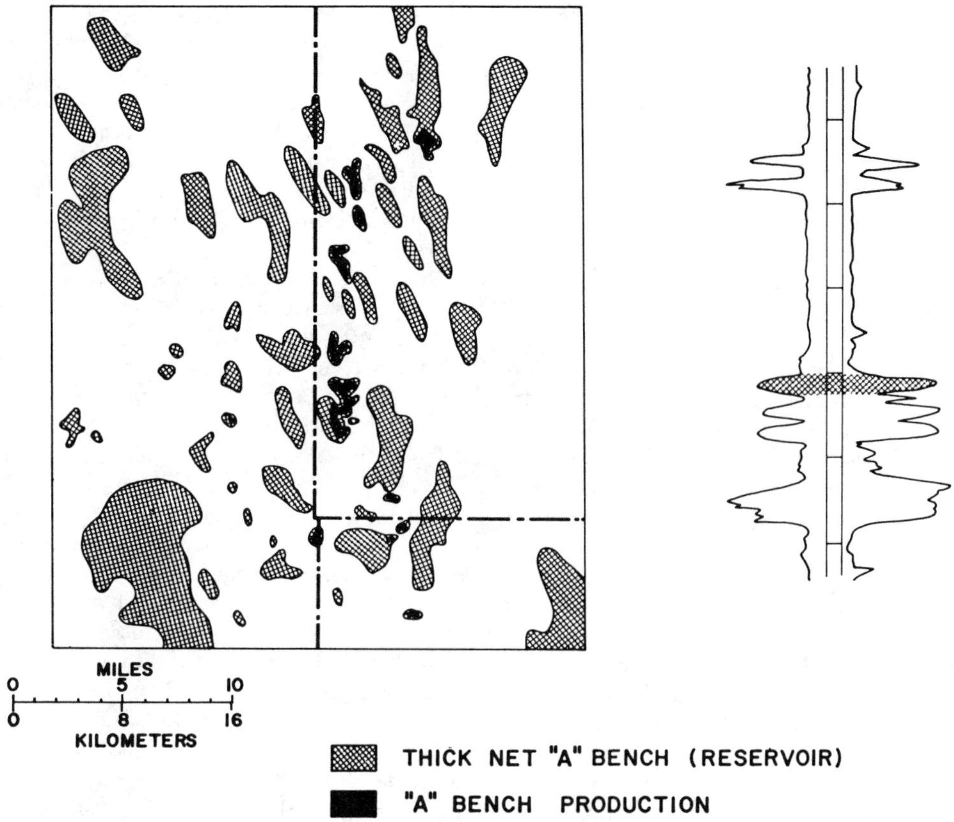

FIG. 10.—Distribution of thick net (reservoir) Fall River "A" bench sandstone and productive areas.

ciated with a very shallow-water marine and transitional shoreline environment—probably is to be found. Types which may be expected include features such as cheniers (which are relic beaches formed during shoreline regression), thick beach sand (formed during a shoreline stillstand), tidal-pass sand deposits, tidal deltas, and the various stages of offshore-bar developments, together with associated lagoons and intertidal flats.

A stillstand or unchanging balance of energy and sediment supply during the deposition of any of the shoreline features could result in a vertical merging of the three Fall River benches into a single thick bench. This condition explains the thick sandstones from which Fall River oil is produced in many places.

The marine Skull Creek sea transgressed the study area from the northwest and approximately 200 ft of marine shale was deposited before the area once more became emergent. This emergence or withdrawal of the sea northwestward created a widespread mudflat on which the agents of erosion began to work.

NEWCASTLE

The dendritic drainage system of the Newcastle formed on the Skull Creek mudflat. Channeling is obvious at the base of the Newcastle, and the thickest Newcastle sandstone developments are at the expense of the Skull Creek Shale. The top of the Newcastle is a lithologic time line. An isopachous map of the gross Newcastle between the time line at the top and the channel unconformity at the base, defines the dendritic drainage pattern in Figure 11. The general direction of stream flow is northward. The relatively uncompactible, thick Fall River Sandstone provided a topography which continued to be reflected to a lesser degree throughout the deposition of the overlying Skull Creek Shale. This gentle topography at the close of the time of Skull Creek deposition controlled, in places, the Newcastle drainage pattern. The thick Fall River Sandstone at Miller Creek and Kummerfeld caused an eastward deflection in the subsequent Newcastle stream channel. The Newcastle channel was superimposed on thin areas of the Fall River Sandstone, that provide the stratigraphic traps which produce oil at Miller Creek and Kummerfeld. Within this drainage, net reservoir-quality sandstones were developed. In the southern part of the area, the Skull Creek Shale was being drained, and there was no source of coarse clastic sediment for the development of net reservoir sandstone. In the northeastern part, however, a drainage channel is present that extends toward the east. This channel crops out in the northeastern edge of the study area; projection of the channel eastward across the Black Hills into South Dakota provides an ample supply of clastic sediment from the massive Dakota Sandstone.

The conventional planar map view of the block diagram is shown on Figure 12. On this map the net deposition of reservoir quality sandstone is shown in the cross-hatched pattern and the scattered production from these sandstone beds is shown in black. Storms, or extreme high tides from the sea, not far north, could have redistributed local sands in the southern part of the area. The Newcastle produces oil at the eastern updip edges of porous sandstone units in fields such as Donkey Creek, East Rozet, O'Connor, and others.

DYNNESON

The close of Newcastle deposition was marked by the final transgression of the Early Cretaceous sea. The Dynneson Sandstone was deposited at the beginning of this transgression.

The distribution of the gross Dynneson unit is shown on the block diagram in Figure 13. The southeastern part of the study area was high and did not receive Dynneson sediments. The water depth increased northwestward, and it was within this deeper water, below effective wave base, that the deposition of reservoir-quality net sands occurred.

Figure 13 indicates that the southeastern limit of the reservoir sandstone was primarily a function of water depth and, therefore, effective wave base. Sediment supply was from the underlying Newcastle sandstone and transport was primarily by offshore currents.

The conventional planar map view of the Dynneson block diagram indicates areas of production (*e.g.*, Rozet and West Moorcroft, Figs. 4, 6, 14). Production is at the eastern updip edge of porous net sandstone development.

At Rozet the southeastern limit of porous sandstone is caused by the decrease in water depth and its interrelation with wave base and offshore currents. The northern limit of porous sandstone at Rozet is also the result of a decrease in water depth. It is believed that the difference in water depth was caused by differential compaction in the Skull Creek Shale above a thick Fall River Sandstone. The funnel-like constriction on the east in the Dynneson porous sandstone provides the trap from which oil is produced at Rozet.

Marine sedimentation of the overlying Mowry Shale was continuous, completing the Early Cretaceous deposition.

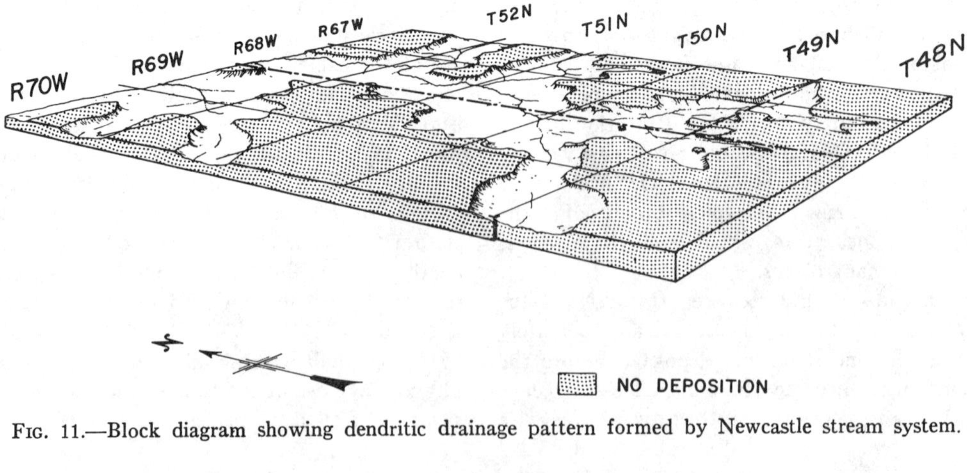

FIG. 11.—Block diagram showing dendritic drainage pattern formed by Newcastle stream system.

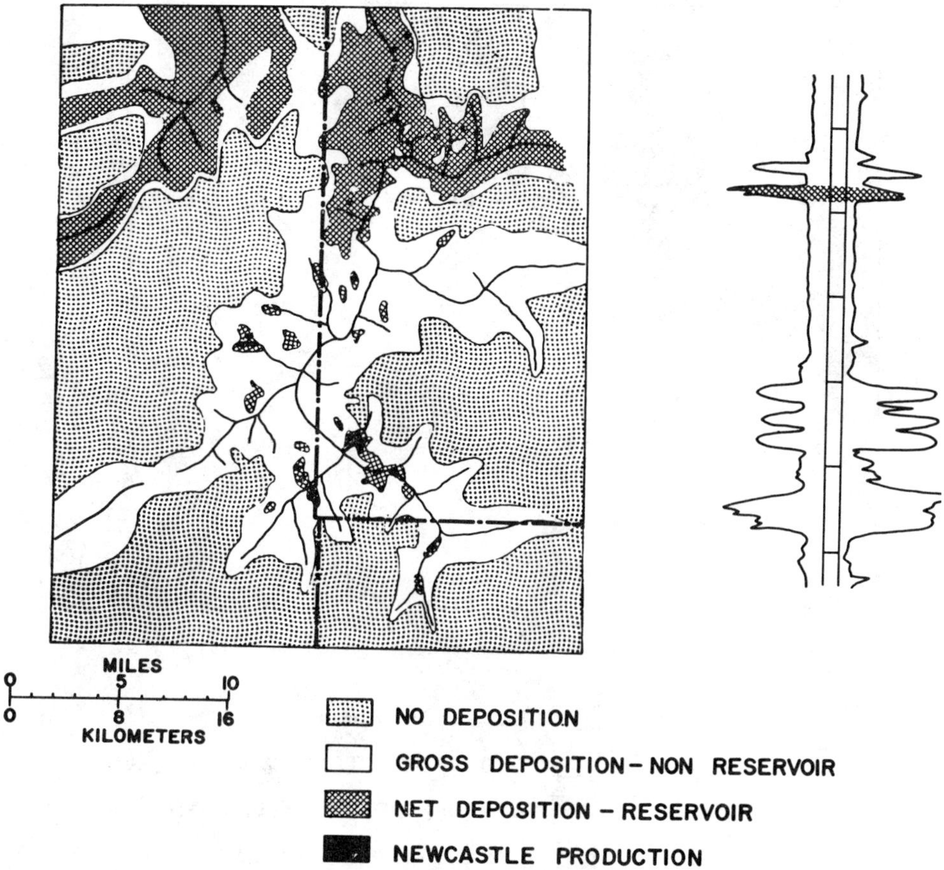

FIG. 12.—Planar map view of Newcastle drainage system and distribution of net (reservoir) Newcastle sandstone showing productive areas.

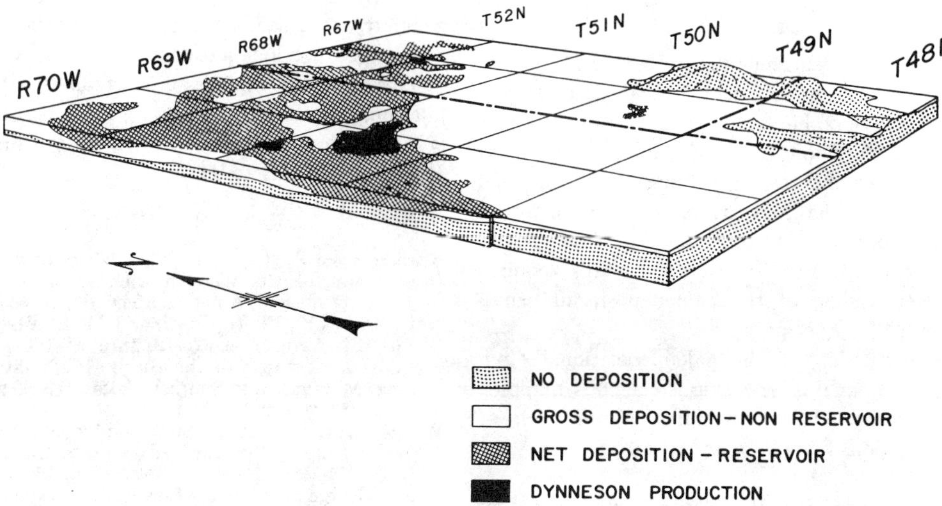

FIG. 13.—Block diagram showing distribution of gross Dynneson sandstone with development of net (reservoir) sandstone and productive areas.

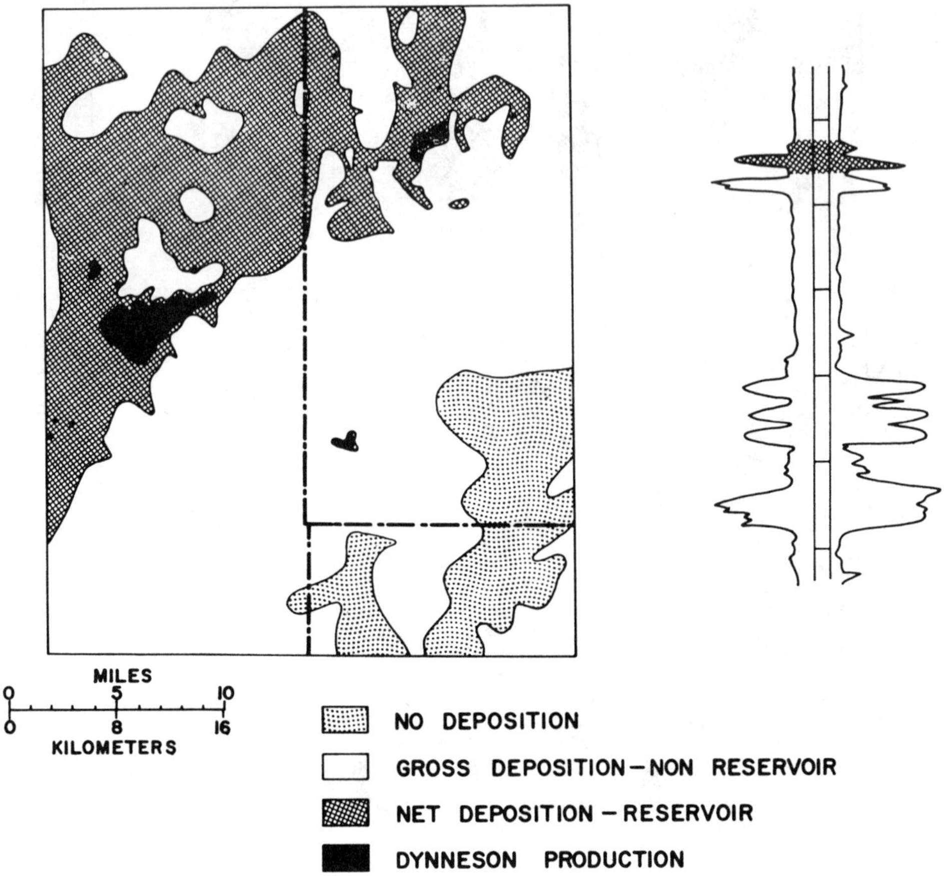

FIG. 14.—Planar map view of Dynneson sandstone (both gross and net) and productive areas.

Conclusions

The Lower Cretaceous production in the northeastern Powder River basin is from a wide variety of stratigraphic traps. Oil accumulation is present in the updip edges of the offshore, blanket-type sandstone of the Dynneson, the channel sandstone of the Newcastle, and the complex marginal-marine sandstone of the Fall River. Prospecting for such stratigraphic traps requires an understanding of the paleodepositional environments of the rocks.

Reconstruction of the paleodepositional environment is done by selective isopachous mapping of genetically related time units of one cycle of sedimentation. Sample analysis is a necessary requirement.

The identification of specific environments of deposition is primarily an interpretation of the geometry of a sandstone body as it relates to its own cycle of sedimentation.

This approach requires much tedious data compilation, but the end result is well worth the time spent. Many fields, such as Coyote Creek and Rozet, are waiting to be found.

References Cited

Reeside, John B., Jr., 1957, Paleoecology of the Cretaceous seas of the Western Interior of the United States: Geol. Soc. America Mem. 67, p. 505–541.

Robinson, C. S., W. J. Mapel, and M. H. Bergendahl, 1964, Stratigraphy and structure of the northern and western flanks of the Black Hills uplift, Wyoming, Montana, and South Dakota: U.S. Geol. Survey Prof. Paper 404, 134 p.

Waagé, Karl M., 1959, Stratigraphy of the Inyan Kara Group in the Black Hills: U.S. Geol. Survey Bull. 1081-B, p. 11–90.

Wulf, George R., 1962, Lower Cretaceous Albian rocks in northern Great Plains: Am. Assoc. Petroleum Geologists Bull., v. 46, p. 1371–1415.

Reprinted from:
BULLETIN OF THE AMERICAN ASSOCIATION OF PETROLEUM GEOLOGISTS
VOL. 52, NO. 9 (SEPTEMBER, 1968), PP. 1689–1699, 13 FIGS., 2 TABLES

APPLICATION OF SEDIMENTARY DIRECTIONAL FEATURES AND SCALAR PROPERTIES TO HYDROCARBON EXPLORATION[1]

GARY A. McDANIEL[2]
Oklahoma City, Oklahoma 73102

ABSTRACT

Ripple marks, cross-bedding, and groove casts are examples of directional features of sediments. Systematic downcurrent variations in grain size or mineral composition are examples of scalar properties. Although directional features normally are considered to be the tools of the surface geologist, these features also can be used by the subsurface geologist by means of the oriented core and the dipmeter. Recent literature abounds with reports of new types of scalar properties.

Directional features and scalar properties of sediments are used to determine paleocurrent direction. The integration of paleocurrent data with environmental information yields a paleogeographic map, from which prediction of favorable areas for wildcat exploration and the exploitation of wildcat discoveries can be made.

A *directional features → paleocurrent direction + depositional environment = paleogeographic map* study of the early Desmoinesian Hartshorne Formation in the Arkoma basin, Oklahoma, indicates that the productive Hartshorne Sandstone in the discovery gas well of the South Pine Hollow field of Pittsburg County, Oklahoma, is probably a deltaic distributary channel sandstone with a NE-SW alignment. This type of study could have been used to exploit the discovery well; the information needed to reach the same conclusion by subsurface geology was not available until some time after the initial discovery.

INTRODUCTION

Historically, exploration programs in a new province progress through three phases. The initial phase is surface-structural. When information generated by phase 1 is coupled with geophysical data, the programs gradually shift into phase 2, subsurface-structural. As subsurface information becomes even more plentiful, exploration programs become oriented toward phase 3, the search for that most elusive of all hydrocarbon deposits, the stratigraphic trap.

Most future domestic reserves will be found in subtle stratigraphic traps. Only by the utilization and integration of every available exploration tool can the success ratio necessary to sustain the petroleum industry be attained in stratigraphic exploration.

One of the more useful tools of the stratigraphic explorationist is the paleogeographic map. With this map he can delineate favorable areas for wildcat exploration and can exploit wildcat successes. To construct a paleogeographic map the explorationist first must have knowledge of the depositional environment of the stratigraphic unit to be studied. Second, he must orient this environment with paleocurrent data. Paleocurrent data can be determined directly by the use of sedimentary directional features such as ripple marks and cross-bedding, or indirectly by study of sedimentary scalar properties. A scalar property, of which grain-size variation is an example, shows a systematic downcurrent change.

The discussion of paleocurrent directional indicators and their application to hydrocarbon exploration is presented by the writer in two parts: (1) a brief review of directional features and scalar properties and (2) a case study wherein directional features are used to construct paleogeographic maps which are used to predict field development from a wildcat discovery. The case-study area, 154 sq mi (398.9 sq km) in the Arkoma basin of eastern Oklahoma (Fig. 1), has been designated as the South Pine Hollow area.

During the last 10 years the Arkoma basin has become a major gas-producing province. One of the significant wildcat discoveries was made by Carter Oil Company in 1959. In a well in the South Pine Hollow area, Sec. 24, T5N, R12E (Fig. 2), drilled to a depth of 3,445 ft (1,052 m), the lower Hartshorne Sandstone (Pennsylvanian, lower Desmoinesian Stage) was found at a depth of 3,186 ft (971.1 m) and was 215 ft (65.5 m) thick. The microlog indicated a net 122 ft (37.2 m) of pay. The well subsequently was completed for a calculated open flow of 5.65 MMcf (159.9 thousand cu m) of gas a day with a shut-in pressure of 500 psi (35.2 kg/cm^2).

[1] Manuscript received, February 4, 1968; accepted, March 29, 1968. Given at the 2d Biennial Meeting of the Mid-Continent Section of the Association, Wichita, Kansas, September 1967.

[2] Geologist, Midwest Oil Corporation.

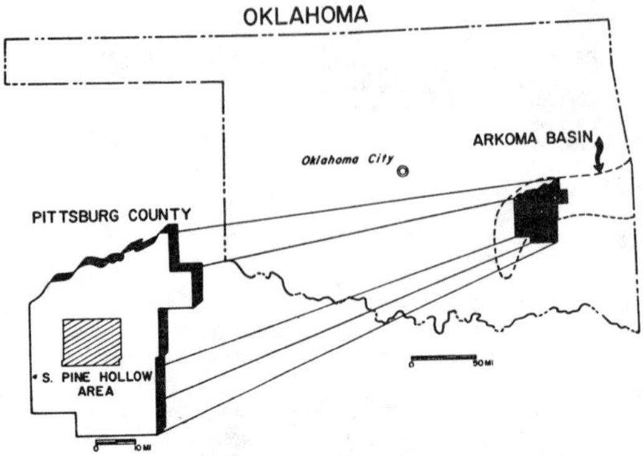

FIG. 1.—Location map, South Pine Hollow area, Oklahoma.

TABLE I. SEDIMENTARY DIRECTIONAL FEATURES

	Unidirectional	*Bidirectional*
Surface		
A. Asymmetrical ripple marks	X	—
B. Symmetrical ripple marks	—	X
Erosional Surface		
C. Rib and furrow	X	—
D. Fossil lineation	X	X
Internal		
E. Festoon cross-bedding	X	—
F. Planar cross-bedding	X	—
G. Parting lineation	—	X
H. Fossil lineation	X	X
Bottom (Sole Markings)		
I. Flute casts	X	—
J. Striations and groove casts	X	X

At the time of the discovery there were only two Hartshorne Sandstone penetrations in the South Pine Hollow area. All of the wells, including Carter's, were drilled near the major surface structural feature, the axis of the McAlester anticline–Penitentiary fault. Approximately 10 mi (16.1 km) northeast of the discovery well, the Hartshorne Sandstone is exposed (Fig. 2).

After this wildcat discovery explorationists were called on to answer questions and make evaluations and recommendations. They were asked whether or not the accumulation was controlled structurally, whether the thick lower Hartshorne Sandstone was a bar, channel, or blanket deposit, and which of the eight possible offsets should be drilled. A paleogeographic map, based on paleocurrent information derived from directional features and/or scalar properties, could have provided the answers.

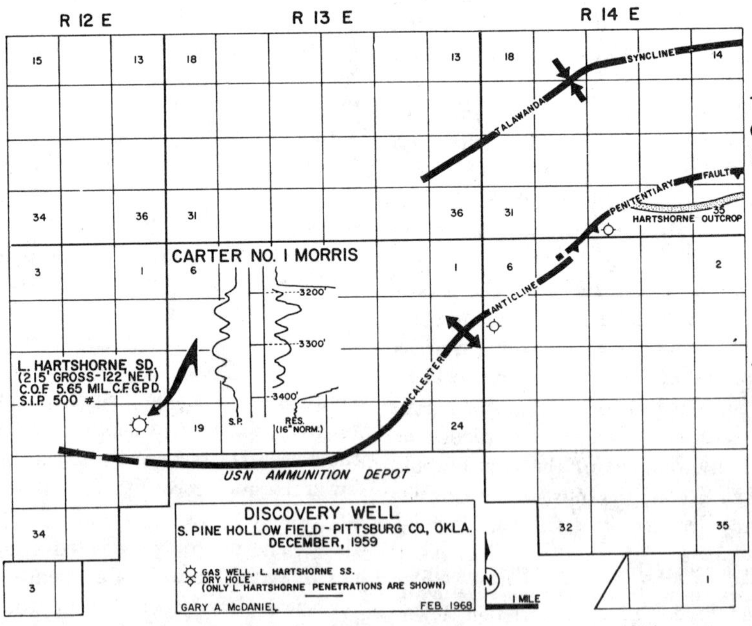

FIG. 2.—Discovery well, South Pine Hollow field, Oklahoma. Major structural axes within South Pine Hollow area are shown.

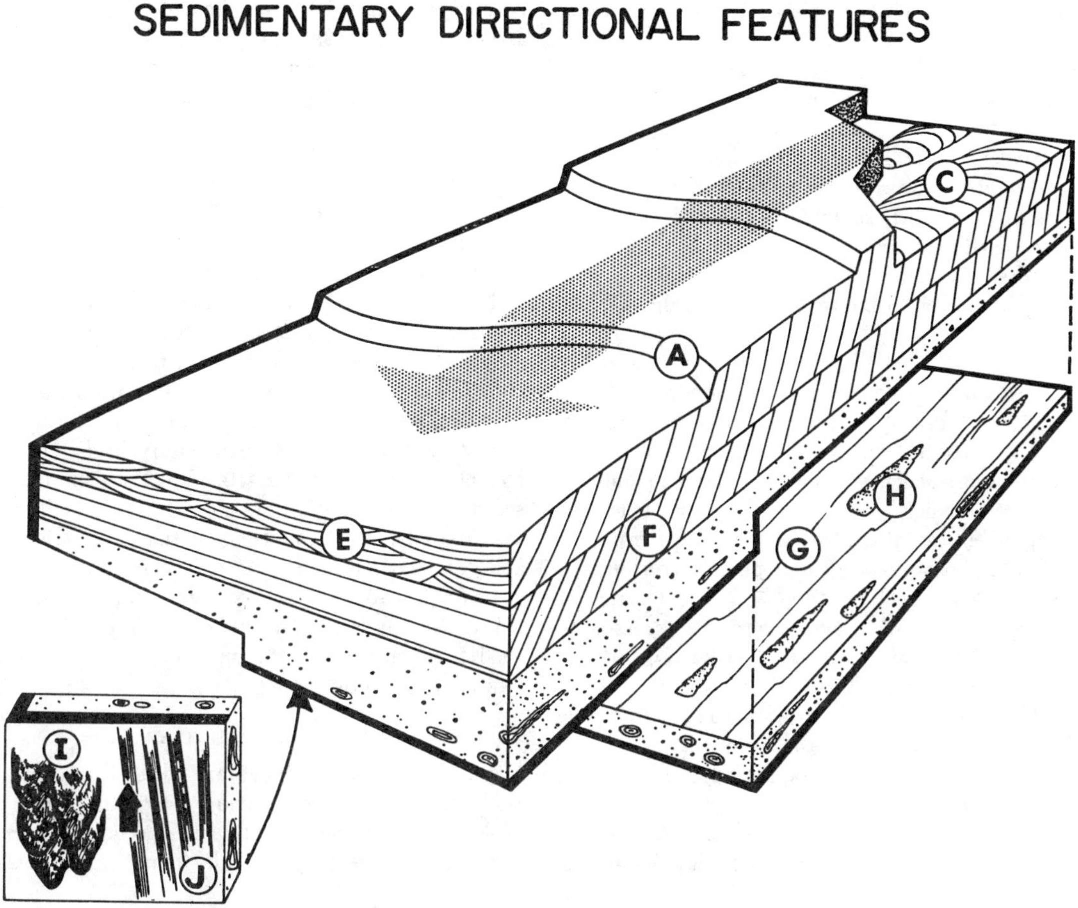

FIG. 3.—Sedimentary directional features. Letters refer to Table I. *A*, Asymmetrical ripple mark; *C*, rib and furrow; *E*, festoon cross-bedding; *F*, planar cross-bedding; *G*, parting lineation; *H*, fossil lineation; *I*, flute casts; *J*, striations and groove casts.

SEDIMENTARY DIRECTIONAL FEATURES

The subject of directional features is covered thoroughly by Pettijohn (1962, p. 1470–1478). The following discussion is based on Pettijohn's article supplemented by personal observations.

Sedimentary directional features have been classified with respect to their occurrence in a sedimentary bed as *surface, erosional surface, internal,* and *bottom* (Table I). Figure 3 is a block diagram which illustrates the directional features from Table I.

SURFACE FEATURES

Asymmetrical ripple marks.—Asymmetrical ripple marks (Fig. 3, *A*) are linear features which form transverse to the paleocurrent. They are unidirectional and can be found in fluvial, marine, and eolian environments. A unidirectional feature indicates the source direction of the paleocurrent; a bidirectional feature indicates the direction of the flow alignment but not the source.

Symmetrical ripple marks.—Symmetrical ripple marks are formed by oscillatory waves and are bidirectional.

EROSIONAL SURFACE FEATURES

Rib and furrow.—Ribs and furrows (Fig. 3, *C*) are the erosional surface expression of festoon cross-bedding. The furrows open down the paleoslope, and are thus unidirectional features. Ribs are the raised linear surfaces between adjoining sets of furrows.

Fossil lineation.—The alignment of the long axes of fossils by paleocurrents is called "fossil lineation." It can be either a unidirectional or bidirectional feature. Figure 3, *H*, indicates the source of the flow because the small end of the fossil is upcurrent. In this example fossil lineation is a unidirectional feature.

Fossil lineation can be either an erosional surface feature or an internal feature (Table I).

INTERNAL FEATURES

Festoon cross-bedding.—Cross-bedding is a product of gravity sliding and may be found in fluvial, marine, and eolian environments. The term "cross-bedding" should be applied only to sediments which have a deviation from true bedding in excess of 10°. There are two types of cross-bedding—festoon and planar (torrential). Festoon cross-bedding (Fig. 3, *E*) is a series of curved surfaces, concave upward. The cross-laminations are tangential to the base of the bed and are truncated at the top.

Planar cross-bedding.—In planar or torrential cross-beds (Fig. 3, *F*) the laminae are parallel with one another and meet the adjoining beds at a sharp angle.

Parting lineation.—This feature (Fig. 3, *G*) consists of irregular, elongate, and parallel or subparallel markings found on the bedding planes of a rock where it is split or parted. The method of formation is unknown, but the markings form parallel with the depositional current and are a bidirectional indicator.

BOTTOM FEATURES (SOLE MARKINGS)

Flute casts.—Flute casts (Fig. 3, *I*) are elongate, spatulate, bulbous elevations which have a beaklike upcurrent termination; the downcurrent end merges with the bedding plane. These unidirectional features are thought to be products of scouring by eddies on the mud surface.

Striations and groove casts.—Striations and groove casts (Fig. 3, *J*) are raised linear features found on the bottom of sedimentary beds. They commonly are present in sets, with the dominant set being crossed at a low angle by a secondary set. Their origin is sand accumulation in striations and grooves on a mud surface. The striations and grooves probably are caused by currents moving debris such as large sand grains, shale chips, and shell fragments across the mud surface. They are bidirectional features.

SUBSURFACE APPLICATION

Oriented core.—Sedimentary directional features are used most commonly by the surface geologist, but if the formation has no outcrop, or if the outcrop is geographically remote from the

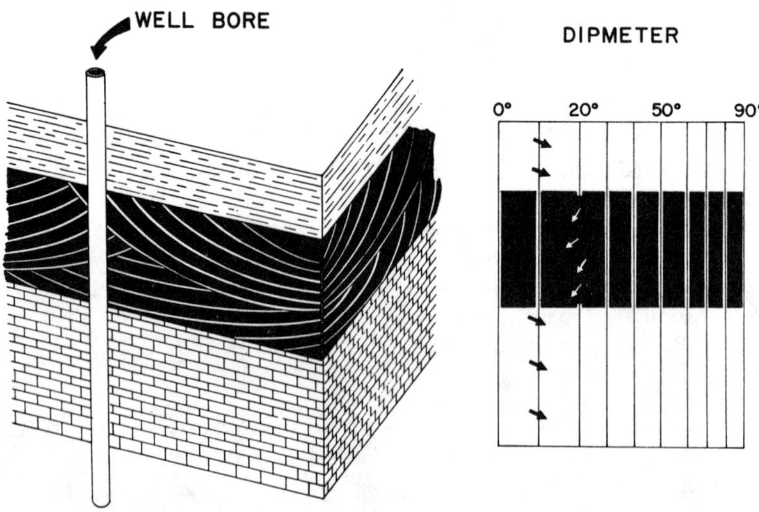

FIG. 4.—Interpretation of transport direction of cross-bedded sandstone by dipmeter.

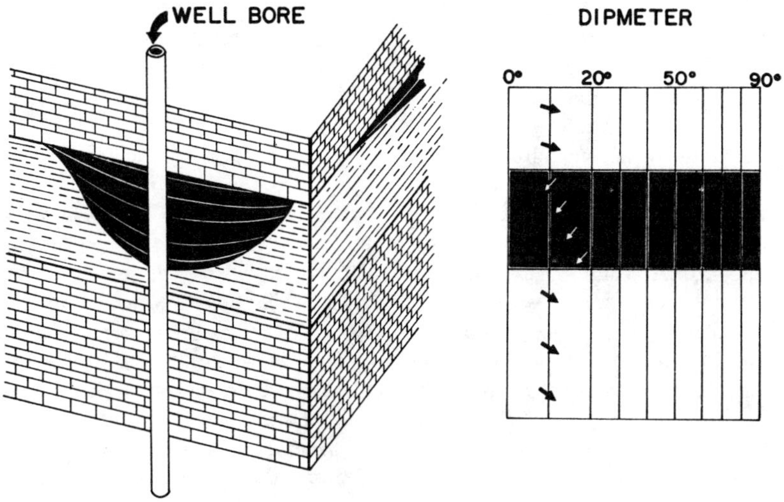

FIG. 5.—Interpretation of transport direction of channel sandstone by dipmeter.

study area, the subsurface geologist can utilize sedimentary directional features found in an oriented core. Cross-bedding, parting lineation, fossil lineation, striations and groove casts, and flute casts can be identified in cores.

Stratigraphic dipmeter.—In addition to the oriented core, the dipmeter may be used by the subsurface geologist to gather paleocurrent information. This is an interpretational technique using the standard dipmeter.

Figure 4 illustrates a festoon cross-bedded sandstone. Regional dip is indicated by the overlying shale and the underlying limestone to be approximately 10° southeast. Dip directions within the sandstone are 20° slightly south of west, an almost 90° deviation from the regional dip. The change in dip direction and the increase in the dip are attributed to the cross-bedding laminae. Removal of the regional structural influence from the cross-bedding dips would result in a determination of the paleocurrent flow direction.

A second dipmeter example (Fig. 5) illustrates a channel sandstone. Regional dip is 10° southeast as shown by the overlying and underlying limestone beds. The dipmeter readings in the channel sandstone show a southwest dip, and an increase in dip from the top to the bottom of the channel. The variation of dip direction within the sandstone is attributed to the bedding planes which plunge downcurrent. Dipmeter interpretations should be used with caution because more than one set of geologic conditions can result in a single dip pattern.

INDIRECT PALEOCURRENT DIRECTIONAL INDICATORS

SCALAR PROPERTIES

Several indirect methods can be used to determine paleocurrent flow directions by means of one of the most available tools of the subsurface geologist—bit cuttings.

As previously defined, a scalar property shows a systematic downcurrent change. Table II lists several scalar properties which can be used by the subsurface geologist.

Grain size.—Use of grain size to determine current direction is based on the observation that the diameters of pebbles have a straight-line relation (logarithmic plot) with the distance of trans-

TABLE II. INDIRECT PALEOCURRENT DIRECTIONAL INDICATORS

Scalar Properties
- A. Grain size
- B. Mineral composition
- C. Pollen content
- D. Thickness of individual beds

Miscellaneous
- A. C^{13}/C^{12} ratio of siderite
- B. Absolute age

port—the farther the transport, the smaller the grain size.

Mineral composition.—Stable detrital minerals found in most sandstone can be used to determine a source area. A typical example would be the ratio between tourmaline and zircon. Zircon has a higher specific gravity and therefore should settle nearer the source; thus, the more distant the source, the higher the tourmaline-zircon ratio.

Pollen content.—Pollen content, expressed in the number of grains per gram of sediments, is said to be a function of the distance from the shoreline (Hoffmeister, 1960). As the pollen content decreases, the distance from the shoreline increases.

Thickness of individual beds.—Graded and cross-bedded layers decrease in thickness in a downcurrent direction (Pettijohn, 1961, p. 1481).

MISCELLANEOUS

Ingenious new techniques for determining source areas and thus indirectly paleocurrent flow directions are reported in the recent literature. For example, Weber *et al.* (1964) used the C^{13}/C^{12} ratio of siderite nodules in shale to make a freshwater and marine environmental separation. The location of a shoreline could be made indirectly from this determination. Vistelius (1964) used the potassium/argon method of age determination on the detrital parts of a Cretaceous sandstone, and reconstructed the direction of transport from dispersal centers.

South Pine Hollow Area

INTRODUCTION

The South Pine Hollow area, Pittsburg County, Oklahoma (Fig. 1), was selected as an area where an application of a directional measurement study could answer basic exploration questions about an isolated wildcat discovery.

In 1959, at the time of Carter's lower Hartshorne Sandstone gas discovery, Sec. 24, T5N, R12E (Fig. 2), the South Pine Hollow area was relatively unexplored. Only two other wells had penetrated the Hartshorne Sandstone in this 154-sq mi (398.9 sq km) area. Explorationists needed to determine which of eight possible offset locations should be drilled for the confirmation, and whether or not the trend of the field development could be projected from the discovery well to allow a lease play. The answers could have been found if the explorationist had constructed a paleogeographic map based on environmental and paleocurrent data.

HARTSHORNE SANDSTONE

The Hartshorne Sandstone is of Pennsylvanian age and is the basal formation of the Desmoinesian Stage (Fig. 6). It overlies the Atokan and underlies the McAlester Formation. It may be subdivided into two members, an upper and a lower. The uppermost unit of the lower Hartshorne Member is the lower Hartshorne coal.

The Hartshorne Sandstone in the Arkoma basin is well suited for a directional measurement and environmental study. There are exposures both along the southern basin edge and within the basin on the major structural features—Backbone anticline, Milton anticline, Penitentiary fault, and Savanna anticline (Fig. 7). Figure 7 shows most of the Oklahoma part of the Arkoma basin. The Choctaw fault separates the Arkoma basin from the Ouachita Mountain province. Just north of and parallel with the main Hartshorne exposure is the present basin axis. The South Pine Hollow area is outlined on Figure 7.

DIRECTIONAL MEASUREMENT AND ENVIRONMENTAL STUDY

Directional measurements were made on ripple marks, cross-bedding, and parting lineation. The

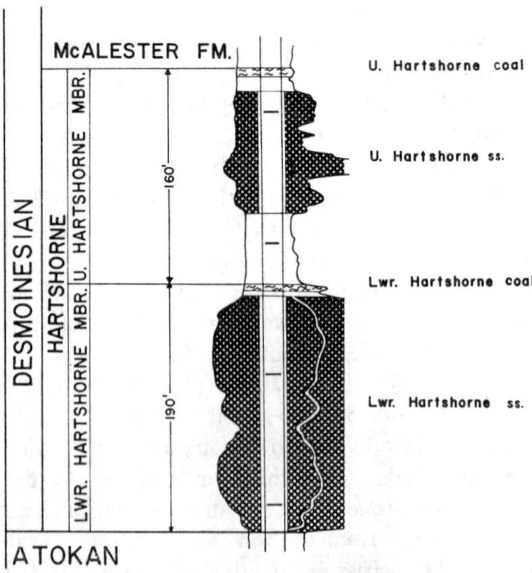

Fig. 6.—Type log, Desmoinesian Hartshorne Formation, Arkoma basin, Oklahoma.

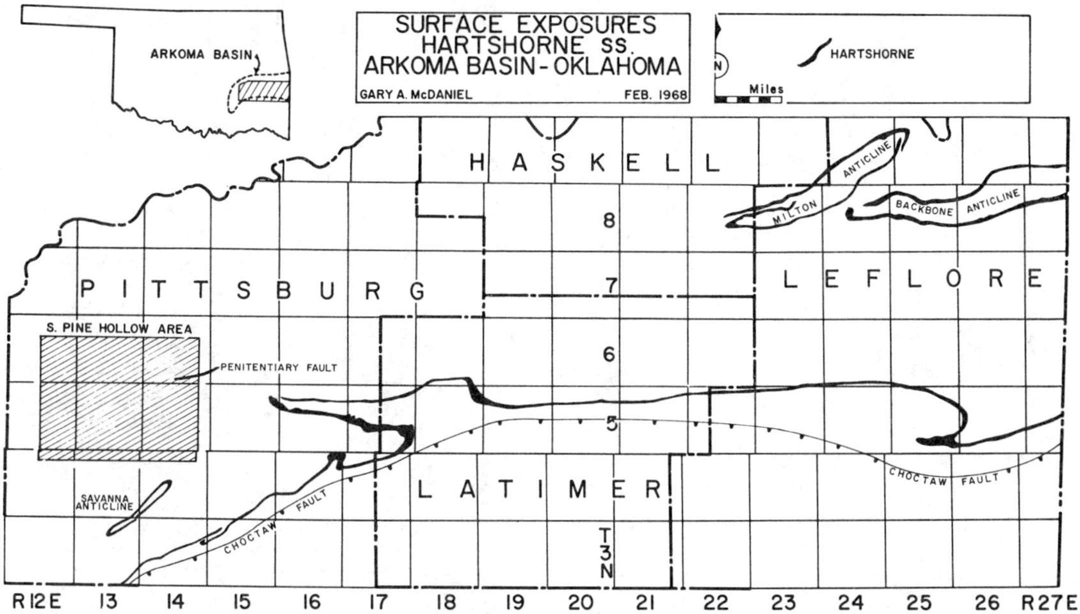

FIG. 7.—Surface exposures of Hartshorne Formation, Arkoma basin, Oklahoma.

arrows on Figure 8 represent a mean of several directional readings. Transport direction of the Hartshorne Sandstone is generally northeast-southwest as determined by these directional measurements.

At the time the directional features were measured, environmental interpretations of the lower Hartshorne member were made (Fig. 8). Factors used in the environmental interpretation were lithologic character, geographical and vertical litho-

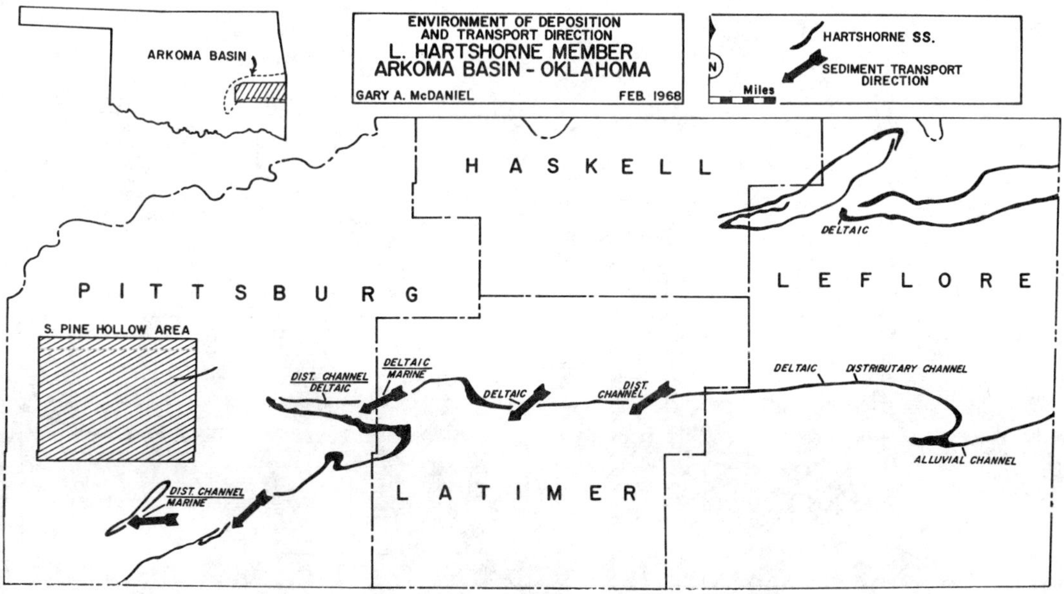

FIG. 8.—Environment of deposition and transport direction of lower Hartshorne member, Arkoma basin, Oklahoma.

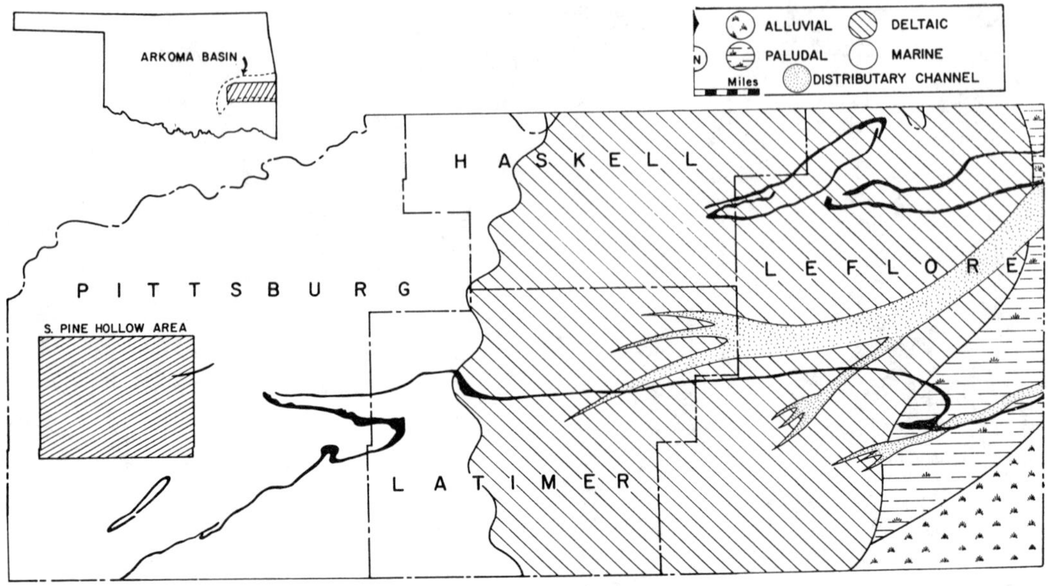

FIG. 9.—Schematic paleogeography, Arkoma basin, Oklahoma, at time of early stage of lower Hartshorne deposition.

logic sequence, sedimentary features, presence or absence of fossils, energy of the environment, *etc.* Four generalized environmental divisions were made: alluvial channel, deltaic, delta distributary channel, and marine.

A summation of the lower Hartshorne member in the Arkoma basin follows.

1. Lithologically, the unit grades from a sandstone to a shale (no carbonates).

2. The unit contains sedimentary features such as

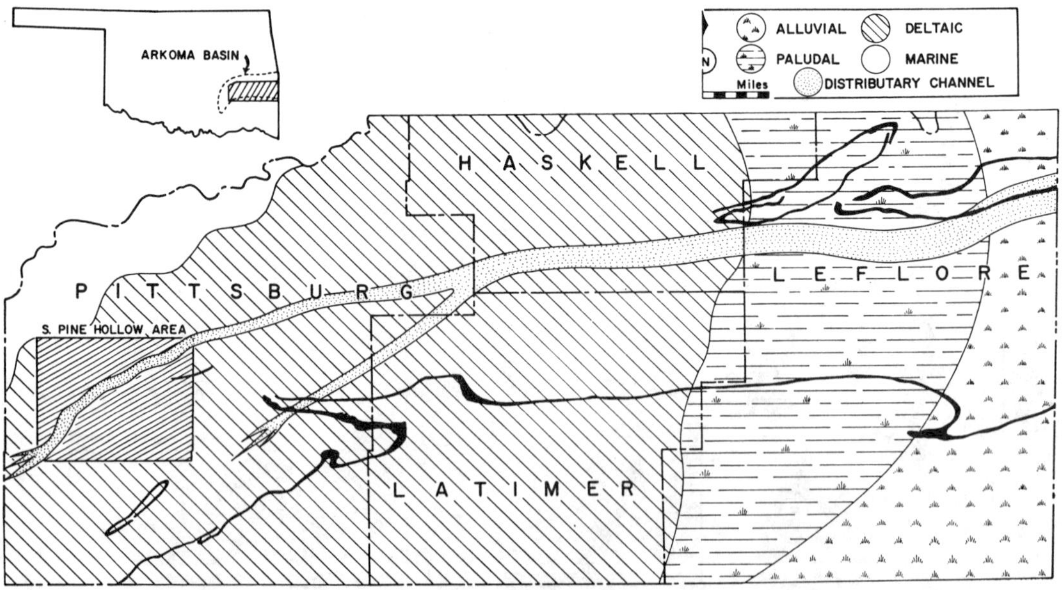

FIG. 10.—Schematic paleogeography, Arkoma basin, Oklahoma, at time of later stage of lower Hartshorne deposition.

boring and churning, cut-outs, cross-bedding, and ripple marks.

3. The unit is unfossiliferous with the exception of abundant plant material.

4. The thicker sandstone bodies are delta-distributary channels.

5. The unit is alluvial on the east and becomes progressively marine westward; it is a regressive unit.

PALEOGEOGRAPHY

A schematic paleogeographic map was prepared from the directional and environmental information shown on Figure 8. Figure 9 illustrates the paleogeography at the time of the beginning of lower Hartshorne deposition. A large delta system is portrayed with an alluvial area on the east. Sediments were transported to the delta by a distributary system aligned NE-SW. In the South Pine Hollow area, marine sediments were being deposited.

Figure 10 illustrates schematically the paleogeography near the end of the time of lower Hartshorne deposition. The delta system continued its westward advance. The South Pine Hollow area was in a marine environment in the early stage of lower Hartshorne deposition, but was receiving deltaic sediments in the later stage. A NE-SW aligned distributary flowed through the South Pine Hollow area.

On the basis of the paleogeographic maps derived from directional measurements and environmental interpretations, Carter's 1959 gas discovery appears to be in a deltaic distributary channel sandstone. The sandstone should trend NE-SW, and the channel should be relatively narrow. An examination of the South Pine Hollow area after the completion of extensive exploration has provided the proof that these assumptions are valid.

EXPLORATION, DECEMBER 1967

Eight years after the discovery in Sec. 24, T5N, R12E, the South Pine Hollow area has been explored extensively (Fig. 11). The study area which before the discovery had only three Hartshorne tests now has 58. There is an apparent NE-SW alignment of the producing wells. The production obviously is not the result of structural entrapment, because it corresponds more closely to the axis of the Talawanda syncline than to the McAlester anticline.

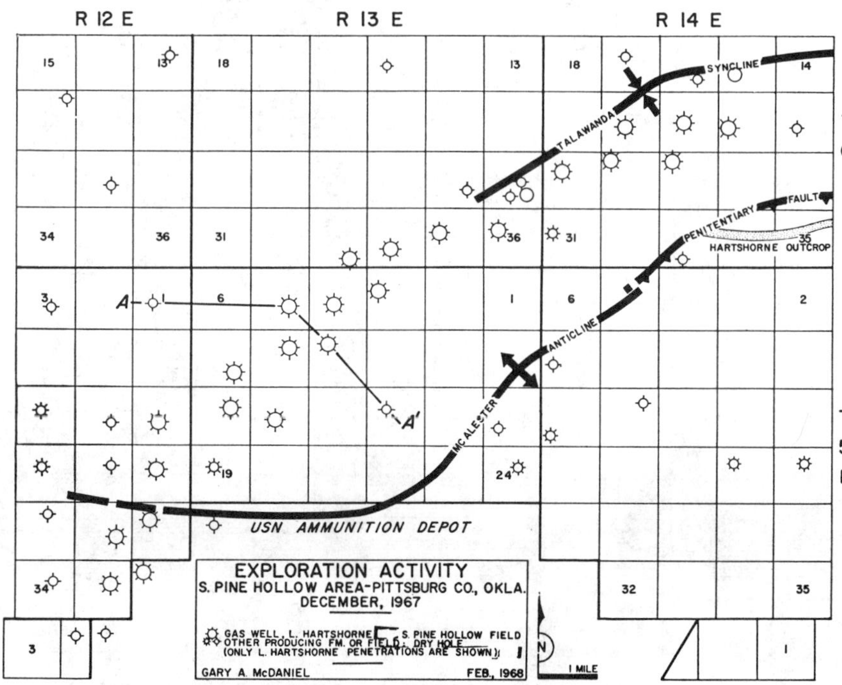

FIG. 11.—Exploration in South Pine Hollow area, December 1967. Major surface structural axes are shown.

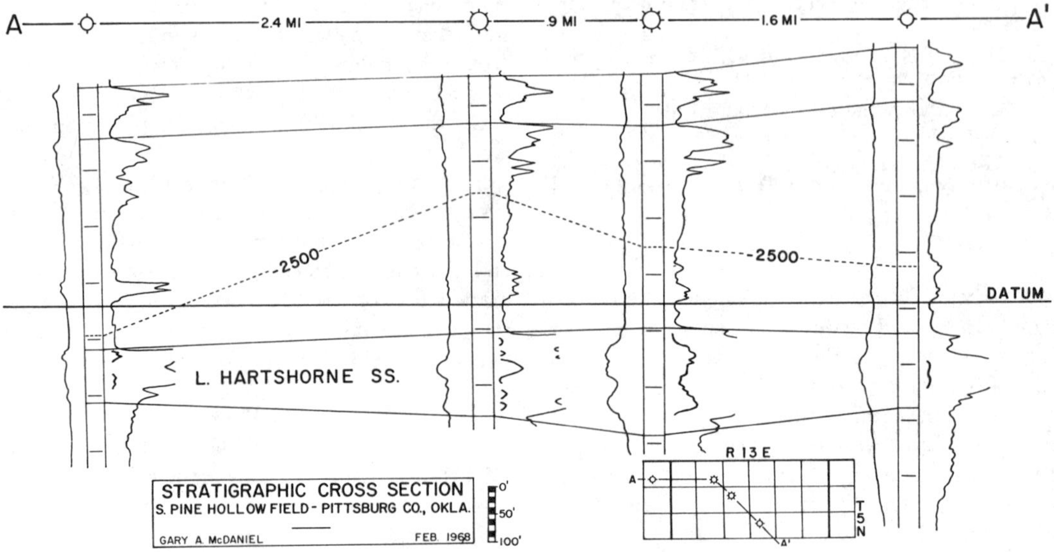

FIG. 12.—NW-SE stratigraphic cross section of lower Hartshorne Sandstone, South Pine Hollow field.

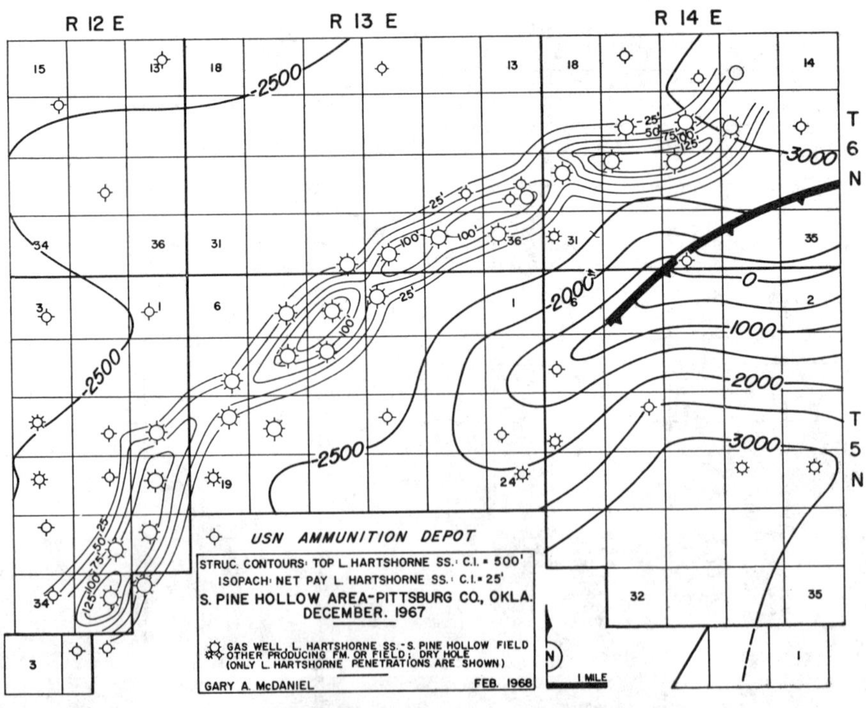

FIG. 13.—Structural contour map. Datum is top of lower Hartshorne Sandstone. CI=500 ft. Net pay isopach of lower Hartshorne Sandstone, CI=25 ft.

Subsurface Geology

A stratigraphic cross section of the lower Hartshorne Sandstone, the productive zone of the South Pine Hollow field, has been constructed (Fig. 12). The line of section is from northwest to southeast, and an easily recognized shale marker is used as datum. The sandstone has a biconvex shape which is the result of channeling at the base and shale compaction in the wells on either side of the producers. The higher energy environment near the channel axis removed the fine sediment fractions. An increase in permeability and effective porosity is reflected by the two gas producers in the thicker or axial part of the lower Hartshorne channel sandstone. The −2,500 ft (−762 m) structural datum line shows that the producing wells are structurally low.

A subsurface structural map, with datum at the top of the lower Hartshorne Sandstone (Fig. 13), contoured with a 500-ft (152.4 m) interval, illustrates the relation of the South Pine Hollow field to the regional structure. The major subsurface structural feature in the area is the McAlester anticline–Penitentiary fault.

The channel sandstone is delineated by an isopach of net lower Hartshorne pay with a contour interval of 25 ft (7.6 m). It is obvious that the productive part of the lower Hartshorne Sandstone is linear, with a NE-SW trend and a width of approximately 1 mi (1.6 km), and is in a structural low which plunges northeastward. A water level has not been established in the field.

Conclusions

The paleogeography of a stratigraphic unit can be reconstructed by the integration of two classes of information—paleocurrent and environmental. Paleocurrent information can be derived directly from sedimentary directional features or indirectly by the use of scalar properties. Although directional features normally are measured on the outcrop, they also can be used in the subsurface by means of the oriented core and the stratigraphic dipmeter.

The careful application of this exploration technique is time consuming; however, in stratigraphic exploration the results warrant the time expenditure. For example, if such a study had been made in the South Pine Hollow area immediately after the discovery, those who conducted the study and acted on the finding would have had a substantial share of the 100+ billion cu ft (3 billion cu m) of gas reserve in the South Pine Hollow field.

Selected References

Hendricks, T. A., 1937, Geology and fuel resources of the southern part of the Oklahoma coal field; pt. 1, McAlester district, Pittsburg, Atoka, and Latimer Counties: U.S. Geol. Survey Bull. 874-A, p. 1–90.

——— 1939, Geology and fuel resources of the southern part of the Oklahoma coal field; pt. 4, Howe-Wilburton district, Latimer and LeFlore Counties: U.S. Geol. Survey Bull. 874-D, p. 255–300.

Hoffmeister, W. C., 1960, Palynology has important role in oil exploration: World Oil, v. 150, p. 101–104.

Knechtel, M. M., 1949, Geology and coal and natural gas resources of northern LeFlore County, Oklahoma: Oklahoma Geol. Survey Bull. 68, 76 p.

McDaniel, G. A., 1961, Surface stratigraphy of the Hartshorne formation, LeFlore, Latimer and Pittsburg Counties, Oklahoma, *in* Arkoma basin and north-central Ouachita Mountains of Oklahoma: Tulsa Geol. Soc.—Fort Smith Geol. Soc. Guidebook, p. 66–71.

Oakes, M. C., and M. M. Knechtel, 1948, Geology and mineral resources of Haskell County, Oklahoma: Oklahoma Geol. Survey Bull. 67, 134 p.

Pettijohn, F. J., 1962, Paleocurrents and paleogeography: Am. Assoc. Petroleum Geologists Bull., v. 46, no. 8, p. 1468–1493.

Schlumberger Technology Corp., Interpretation of continuous dipmeter surveys: South Louisiana Division Training Aid, 9 p.

Vistelius, A. B., 1964, Paleogeographic reconstructions by absolute age determinations of sand particles: Jour. Geol., v. 72, no. 4, p. 483–486.

Weber, J. N., E. G. Williams, and M. L. Keith, 1964, Paleoenvironmental significance of carbon isotopic composition of siderite nodules in some shale of Pennsylvanian age: Jour. Sed. Petrology, v. 34, no. 4, p. 814–818.